HERCULES

Higher European Research Course for Users of Large Experimental Systems

NEUTRON AND SYNCHROTRON RADIATION FOR CONDENSED MATTER STUDIES

Volume II

APPLICATIONS TO SOLID STATE PHYSICS AND CHEMISTRY

Editors

José Baruchel
Jean-Louis Hodeau
Mogens S. Lehmann
Jean-René Regnard
Claire Schlenker

Springer-Verlag Berlin Heidelberg GmbH

The HERCULES Course is organized by :

Institut National Polytechnique de Grenoble (INPG)

Université Joseph Fourier de Grenoble (UJF)

Université Paris-Sud

Centre National de la Recherche Scientifique
Formation Permanente

Institut National des Sciences et Techniques Nucléaires
(INSTN)

It is supported by :

- Ministère de l'Enseignement Supérieur et de la Recherche
- Centre National de la Recherche Scientifique (CNRS)
- Commission of the European Communities
- Commissariat à l'Energie Atomique (CEA)
- European Synchrotron Radiation Facility (ESRF)
- Institut Laue-Langevin (ILL)
- Laboratoire Léon Brillouin (LLB)
- Laboratoire pour l'Utilisation du Rayonnement Electromagnétique (LURE)
- Institut des Etudes Scientifiques Avancées de Grenoble

Secrétariat HERCULES
Marie-Claude Simpson
Maison des Magistères Jean Perrin - CNRS
BP 166 - F-38042 GRENOBLE CEDEX 9

Cover photograph by Balmain

ISBN 978-3-540-57691-4   ISBN 978-3-662-22223-2 (eBook)
DOI 10.1007/978-3-662-22223-2

FOREWORD

This volume belongs to the series of the HERCULES Course on Neutron and Synchrotron Radiation for Condensed Matter studies. This course, coorganized by Universities in Grenoble and Paris, CNRS and INSTN, takes place since 1991 in Grenoble, close to and with the support of the European Synchrotron Radiation Facility (ESRF) and the Institut Laue Langevin (ILL).

The first volume gave general presentation of the theory, instruments and methods used in the fields. This second volume, which corresponds to a two-week course, is devoted to selected applications in Physics and Chemistry of Solids. This domain of applications is extremely wide and no attempt has been made to cover it entirely. It includes fourteen chapters, from general considerations on symmetry in condensed matter to the most recent developments on magnetic excitations and electron spectroscopies in high T_c superconductors. The subjects have been chosen, either for their basic importance or in relation with recent developments. The fifteen authors have been selected on account of their high scientific level and teaching skills. Among them, Jean Rossat-Mignod passed away suddenly in August 1993, and we would like here to honor his memory: he was a very deep physicist and an excellent expert of the applications of neutron techniques to various fields, of condensed matter physics, such as magnetism, correlated metals, superconductivity and phase transitions.

The HERCULES Course and the publication of this volume would not have been possible without the contributions of many scientists, especially the members of the Scientific Advisory Committee and of the Organizing Committee. We would like to acknowledge all of them. Special thanks are due to Eric Geissler for his scientific expertise and advices. We are extremely grateful to Myriam Chakroun for her skill and efficiency in the technical preparation of the typescripts and to Marie-Claude Simpson, Administrative Assistant of the HERCULES Course.

José Baruchel, European Synchrotron Radiation Facility
Jean-Louis Hodeau, Laboratoire de Cristallographie, CNRS
Mogens S. Lehmann, Institut Laue-Langevin
Jean-René Regnard, Université Joseph Fourier and CENG - DRFMC
Claire Schlenker, Institut National Polytechnique de Grenoble and CNRS - LEPES

Contributors

Prof. **Erwin F. BERTAUT**, CNRS, Laboratoire de Cristallographie, BP 166 X, 38042 Grenoble cedex, France

Prof. **Roger A. COWLEY**
Clarendon Laboratory, Parks Road, Oxford OX1 3PU, United Kingdom

Dr. **Salvador FERRER**
ESRF, BP 220, 38043 Grenoble cedex, France

Prof. **Gernot HEGER**
Inst. für Kristallographie, RWTH Aachen, Jägerstraße 17-19, 52056 Aachen, Germany

Prof. **Christian JANOT**
Institut Max von Laue-Paul Langevin, 156 X, 38042 Grenoble cedex, France

Prof. **Diek C. KONINGSBERGER**
Dept. of Inorganic Chemistry, University of Utrecht, Sorbonnelaan 16, PO box 80083, 3508 TB Utrecht, Holland

Dr. **Jose L. MARTINEZ**
Instituto Cienca Materiales, Dept. Fisica, Universidad Autonoma Madrid, Madrid 28049, Spain

Prof. **Yves PETROFF**
ESRF, BP 220, 38043 Grenoble cedex, France

Dr. **Jean-Paul POUGET**
Université Paris-Sud XI, Laboratoire de Physique des Solides, Bâtiment 510, 91405 Orsay cedex, France

Dr. **Jean ROSSAT-MIGNOD**
Laboratoire Léon Brillouin, C.E. Saclay, 91191 Gif sur Yvette cedex, France

Dr. **Michèle SAUVAGE-SIMKIN**

LURE, Centre Universitaire, Bâtiment 209 D, 91405 Orsay cedex, France

Dr. **Jacques SCHWEIZER**

DRF-MDN, CENG, BP 85 X, 38041 Grenoble cedex, France

Dr. **Jean SIVARDIERE**

DRFMC/SPSMS/LIH, CENG, BP 85 X, 38041 Grenoble cedex, France

Prof. **Robert K. THOMAS**

Physical Chemistry Lab., South Park Road, Oxford OX1 3QZ, United Kingdom

Prof. **Adrian C. WRIGHT**

J.J. Thomson Physical Lab., University of Reading, Whiteknights, Reading RG66 2AF, United Kingdom

VOLUME II

CONTENTS

INTRODUCTION

E.F. Bertaut

The logical continuation of volume I of the HERCULES Course on Theory, Instruments and Methods is volume II concerned with Applications to Solid State Physics and Chemistry. There are 14 Chapters written by eminent specialists and in the present introduction it is not my intention to summarize the already highly concentrated content of those chapters.

The existence of Large Facilities for Neutron and Synchrotron Radiation, like ESRF, ILL, LLB, LURE and CENG (Siloé) will certainly favour pluridisciplinary and complementary research. Thus the reader is invited to read with peculiar care the chapter he is mainly interested in, but he should also interconnect chapters illuminating his subject by other aspects and using other techniques.

I shall illustrate these ideas by a few examples. Chapter XII by Ferrer and Martinez deals with artificial, man-made crystals, now called superlattices. They generally consist in alternate layers of two materials which form *one-dimensional periodic* structures, specially treated there. In the case of artificial metallic or semi-conducting constituants, the researcher will also be interested in the structure and quality of the interfaces, which means that he has to apply the principles and techniques of *two-dimensional crystallography* outlined in Chapter III by Sauvage-Simkin.

When one or both materials are magnetic, there are many techniques available for the characterization, such as the surface magneto-optic Kerr effect described in Chapter XII, also neutron diffraction outlined in Chapters V and VI by Schweizer and maybe X-ray resonant scattering (cf. Altarelli, Volume I).

These comments by no means exhaust the aspects of low dimensional crystallography. One dimensional behaviour is found in anisotropic materials, constituted by long chains with large electron overlap along the chain and comparatively small overlap between neighbouring chains. Such systems, considered by Pouget in Chapter XI, illustrate the famous Peierl's instability which is characteristic of one-dimensional metals and here visualized by two techniques. X-ray (or neutron) diffraction detects a lattice distortion incommensurate with the basic lattice. Inelastic neutron scattering detects an anomaly in the acoustic branch, the so-called Kohn anomaly. Both phenomena belong to the same propagation vector $q \sim 2 k_F$. We have here an excellent example of complementarity : the use of elastic *and* inelastic scattering for respectively static and dynamic effects. There are certainly phase transitions of the type commensurate-incommensurate which need studies of static *and* dynamic properties (amplitudons, phasons) to be performed (cf. Chapter VI by Cowley on excitations and vibrations).

The same is true for magnetic structures. They are seen by *elastic* neutron (or magnetic X-ray) scattering and depend mainly on exchange integrals. For instance the antiferromagnetic structures of Fe_2O_3 and Cr_2O_3 are different because their exchange integrals are different. On the other hand, the exchange integrals can be determined with a good precision by the spin wave-spectrum seen by *inelastic neutron scattering* as alluded to in Chapter XIII.

In the context of phase transitions, just mentioned, a subject of irritation for the physicist is the mechanism of the high T_c superconductivity in cuprates, which is still under debate. Superconductivity is due to some attractive pairing effect of electrons (or holes) which overcomes their Coulomb repulsion and gives rise to a Bose condensation. In the BCS (Bardeen-Cooper-Schrieffer) theory the pairing effect is due to an electron-phonon coupling and the energy gap Δ, or the energy difference between the superconducting and normal state, is related to T_c by a critical ratio $2\Delta/(kT_c) = 3.53$. This volume deals with superconductivity in Chapter XIII by Rossat-Mignod who looks for the gap in the spin excitation spectrum of $YBa_2Cu_3O_{6+x}$ by inelastic neutron scattering and in Chapter XIV by Petroff who considers the cuprate $Bi_2Sr_2CaCu_2O_8$ by two methods of electron spectroscopy. These are angle resolved photoemission electron spectroscopy (ARPES) and high resolution electron energy loss spectroscopy (HREELS). The critical ratios in Chapter XIII are below 3.5, whilst in Chapter XIV they are much higher, between 6 and 7. It can be safely predicted that the nature of the gap will be clarified when different methods, applied to the same cuprate, happen to identify the same gap.

Does Volume II invite the reader to make predictions and projections onto the future ? Certainly yes. I indicate some examples.

Apparently there is no correlation between lefthanded and blue-eyed kangaroos on one side, and spin densities on the other side. Schweizer in Chapter VI beautifully illustrates on the kangaroo example what is meant by maximum entropy, and then applies the maximum entropy principle (MEP), to the spin density problem in $YBa_2Cu_3O_7$. The conventional Fourier inversion of the experimental data suffers heavily from truncation effects while MEP gives rise to visible and acceptable spin densities, as shown in figure VI.6.

Such a situation where the crystallographer has a set of good data, but limited in number, at his disposal, is rather frequent and MEP will help him to see details which are lost otherwise. MEP, combined with self-similarity, will perhaps unravel the mysteries of quasi-periodic crystals treated in Chapter IX by Janot ?

Another prediction is already substantiated by the topics of the well known Sagamore conferences (sponsored by the International Union of Crystallography) which are charge, spin and momentum densities. Charge densities of valence electrons are accessible to X-ray studies (cf. Chapter II by Heger) and may be compared to theoretical density maps stemming from molecular orbital functions. Momentum densities of the same valence electrons can be obtained from Compton scattering which is particularly strong in modern synchrotron technology. The important point is that *charge and momentum densities are not independent*. Their respective wave functions, $\psi(q)$ and $\phi(p)$, are related by a Fourier transformation (cf. the "q" and the "p" representation of wave functions in Quantum Mechanics). Thus the momentum densities obtained from Compton scattering are a severe test for the molecular orbital functions proposed for chemical bonds.

A final word on "spin" densities. Orbital and spin contributions cannot be separated in neutron techniques and vary differently with temperature (cf. Figure VI.12). Magnetic X-ray scattering (cf. Altarelli in Volume I) will hopefully separate orbital form factors, spin form factors and their temperature dependence.

I still have to do justice to those authors who do not fit the complementarity scheme outlined above. Chapter I by Sivardière on symmetries and Chapter II by Heger on single crystals are of so general validity that everybody has to read them. They could have been as well in Volume I.

For the study of amorphous systems by X-rays and/or neutrons, described in Chapter VIII by Wright, one needs not only much experimental skill for the sample preparation, but also a lot of imagination for reasonably modelling the three-dimensional "ranges I to IV", defined by Wright, and which have to be inferred from a strictly one-dimensional information, the radial distribution function. (In the case of synchrotron radiation, Compton scattering, so much appraised above, has to be strictly eliminated).

The same is true for EXAFS (Extended X-ray Absorption Fine Structure) which probes the local structure of *one* element near to its absorption edge, as described in Chapter X by Koningsberger, and has the advantage of not depending on crystallinity.

Although neutron scattering is not an obvious choice for surface studies, Thomas in Chapter IV discusses many cases of chemisorbed layers where neutron scattering is even the only choice and proves for instance the mechanism of water decomposition by the Raney-Ni catalyst to be $H_2O \rightarrow O + 2H$ (and not $H_2O \rightarrow OH + H$).

The three chapters VIII, X and IV are rather self-contained, but I have read them with the same pleasure as the other ones. Now it is more up to the reader to make his choice.

CHAPTER I

SYMMETRY OF MATTER FROM MOLECULES TO CRYSTALS
J. SIVARDIERE

I.1.   Introduction

Symmetry is a fundamental concept in physics and chemistry. It provides a concise way of describing the geometrical properties of systems, as well as deriving many qualitative physical properties of these systems using the Curie symmetry principle. In short, symmetry means characterization, classification, prediction and unification.

In this text, we review the basic ideas of group theory, geometrical symmetry of matter and crystallography.

I.2.   Symmetry operations

Symmetry operations are transformations of systems which can be classified in several ways.
-         **Geometrical** operations are given by some correspondence between any point M(x, y, z) and its image M′(x′, y′, z′). Time reversal, permutation of identical particles and charge conjugation (that is reversal of the sign of all electrical charges) are **non-geometrical** symmetry operations.
-         Geometrical operations may be **isometrical** (length-conserving), or **non-isometrical**, such as conformal (angle-conserving), or scale transformations.
-         They are **finite** if the distance between M and M′ is finite, or **infinitesimal** if this distance is infinitesimal.
-         They are **point** operations if they leave at least one point invariant, or **space** operations if they have no **fixed point**.
-         Isometrical operations are **direct** (or **proper**) if they leave the orientation, or chirality, of any chiral object (such as a hand or helix) invariant, and **inverse** (or **improper**) if they reverse it.
-         The **order** of an operation α is the lowest integer n such that α^n = identity. This order may be finite or infinite.

We shall consider here only the geometrical, isometrical and finite symmetry operations which leave molecules and crystals invariant. They can be obtained as products of pure (or direct) rotations by inversions or translations which commute with them, and they are enumerated in the table below.

A roto-inversion (Fig. I.1a) is a rotation around an axis Δ followed by an inversion with respect to a point O of the rotation axis; a reflection is a twofold rotation followed by an inversion; a roto-reflection (Fig. I.1b) is a rotation followed by a reflection in a mirror m perpendicular to the rotation axis. Reflections, roto-inversions and roto-reflections are also called mirrors, inverse rotations and mirror rotations respectively. A helirotation or screw rotation (Fig. I.1c) is the product of a direct rotation by a translation parallel to the rotation axis; a glide reflection (Fig. I.1d) is the product of a mirror reflection by a translation parallel to the mirror.

A direct rotation and a translation transform a right hand (or helix) into a right hand, an inverse rotation and a mirror into a left hand. Notations are given in the table below: the Hermann-Mauguin or international notation is used by solid state physicists, whereas the Schoenflies notation for point operations is used by chemists.

Operation	Point or space	Direct or inverse	International notation	Schoenflies
Rotation	point	direct	n	C_n
Inversion	point	inverse	$\bar{1}$	C_i
Translation	space	direct	t	
Reflection	point	inverse	m	C_s
Roto-inversion	point	inverse	$\bar{n}$	
Roto-reflection	point	inverse	$\tilde{n}$	S_n
Helirotation	space	direct	n_t	
Glide reflection	space	inverse	g	

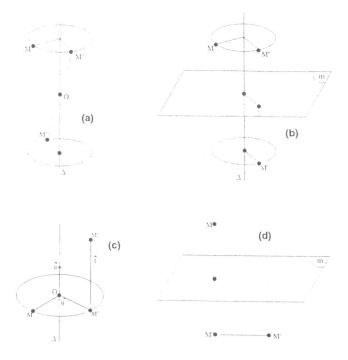

Figure I.1. — Compound symmetry operations: (a) roto-inversion; (b) roto-reflection; (c) screw rotation; (d) glide reflection.

A rotation of angle $\theta = 2\pi/n$ is of order n if n is an integer, and of order infinite if θ is incommensurable with π. An inverse rotation (or a mirror rotation) of angle $\theta = 2\pi/n$ (n integer) is of order n if n is even and of order 2n if n is odd (see Fig. I.2). Note that a mirror m is nothing but an inverse rotation $\bar{2}$. Operations with a translational component are of order infinite.

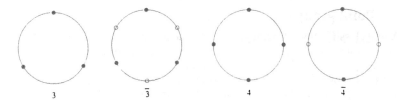

Figure I.2. — The order of an inverse rotation $\bar{n}$ depends on the parity of n.

I.3.   Symmetry groups

The set G of symmetry operations which leave an object invariant (including the identity) is a mathematical **group** since the product of two such operations as well as the inverse of any operation also leave the object invariant. If any two operations commute, the group is said to be **abelian** or **commutative**.

- Operations α, β, γ, ... form a set of **generators** if any operation of the group G can be written as $\alpha^p \beta^q \gamma^r ...$ Such generators are not independent but linked by relations depending on their geometrical nature and relative orientations in space.

- A **finite** group G contains a finite number g of operations, g is the **order** of the group. If h elements of G form a group H, this group is a **subgroup** of G: the ratio g/h is an integer (Lagrange) and is called the **index** of H. An infinite **discrete** group contains an infinite denumerable number of operations, an infinite **continuous** group contains at least one family of operations depending on one or several continuous parameters.

- A **cyclic** group is a finite group with a single generator α. It contains the identity and the operations α, α^2, ... and its order is just the order of α. For instance, a rotation of angle $\theta = 2\pi/n$ generates a cyclic group of order n, denoted by n.

– If a group G has generators α_i which commute with the other generators β_j, it is said to be the **direct product** of the group generated by the α_i and the group generated by the β_j.

- Two groups of order g are said to be **isomorphic** if they are two physical realizations of the same abstract group: there is a one-to-one correspondence between their elements such that, on replacing each element in the multiplication table of the first group by the corresponding element of the second group, the multiplication table of the second group is obtained.

- A **point group** contains only point operations and has at least one fixed point, whereas a **space group** contains space and possibly point operations, and has no fixed point. The symmetry group of a finite object may be only a point group. The symmetry group of an infinite object may be a space or a point group.

- A **chiral** group contains only direct operations, whereas a **non-chiral** group contains inverse as well as direct operations. If an object is chiral (right or left), its symmetry group is chiral. If an object is non-chiral or is a racemic mixture (such as an equimolecular mixture of right and left molecules of the same type), its symmetry group is non-chiral.

I.4.   Point groups

A <u>first classification</u> of point groups may be given.
-       **Proper** groups G_p are chiral ones, they contain only direct rotations.
-       **Improper** groups G_i of order g contain g/2 direct operations (which form a sub-group of index 2) and g/2 inverse operations, the inversion itself does not belong to the group. If the inverse operations are replaced by the corresponding direct ones, we get a proper group G_p isomorphic to G_i.
-       **Centrosymmetric** groups G_c are non-chiral groups similar to the improper ones, except that the inversion belongs to the group. A centrosymmetric group is the direct product of a proper group G_p by the group $(1, \bar{1})$.

Let us discuss first proper groups. We know the cyclic groups n: we may try to combine them and get more complicated groups. In general, we do not get a finite group: the group generated by two rotations n_1 and n_2 around different axes Δ_1 and Δ_2 is the three parameter continuous rotation group, denoted SO3 or ∞/∞, which is the invariance group of the sphere. Let us enumerate the finite proper groups from the cyclic groups of rotations. From a geometrical construction due to Euler (see BUERGER), we know that, in such groups, the product of a rotation of order n_1 and a rotation of order n_2 is a rotation of order n_3 with the same fixed point, and such that:

$$\frac{1}{n_1} + \frac{1}{n_2} + \frac{1}{n_3} > 1$$

The various possibilities are given in the table below (the integer n may take any value).

n_1	n_2	n_3	G_p	g
2	2	n	n2	2n
2	3	3	23	12
2	3	4	432	24
2	3	5	532	60

It can be shown that the axes n and 2 are orthogonal in the **dihedral** group n2: the second symbol of the group designates the n twofold axes, which are equivalent in a rotation n if n is odd. If n is even, the twofold axes are distributed in two non-equivalent classes, and the group is given by the symbol n22.

The groups 23, 432 and 532 are the invariance proper groups of the regular tetrahedron (Fig. I.3a), cube (Fig. I.3b) or octahedron (Fig. I.3c), and dodecahedron (Fig. I.3e) or icosahedron (Fig. I.3f) respectively. The fact that the number of finite multiaxial point groups is finite is related to the existence of a finite number of regular polyhedra.

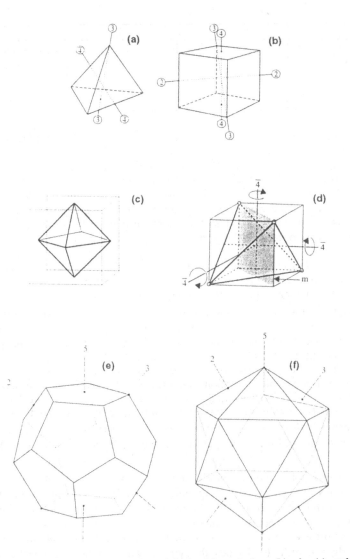

Figure I.3. — Invariance groups of the regular polyhedra: (a) tetrahedron; (b) cube; (c) octahedron; (d) tetrahedron inscribed in a cube; (e) dodecahedron; (f) icosahedron.

If a proper group G_p has a sub-group H of index 2, and if the binary operations of G_p - H are multiplied by the inversion, we get an isomorphic improper group G_i. If G_p has several sub-groups H of index 2, several isomorphic improper group G_i are derived (some of them may be equivalent, for instance $\bar{4}2m$ and $\bar{4}m2$). Any direct product of a proper group by the inversion group is a centrosymmetric group G_c. Some examples are given below.

G_p	H	G_i	G_c
32	3	3m	$\bar{3}$m
422	4	4mm	4/mmm
	222	$\bar{4}$2m	
23			m$\bar{3}$
432	23	$\bar{4}$3m	m$\bar{3}$m

From the above discussion, we get a second classification of point groups:

-       **Binary** or "low-symmetry" groups contain only the identity and possibly binary operations: twofold rotation, inversion, reflection.

-       **Uniaxial** groups contain one and only one rotation of order n > 2. The corresponding rotation axis is called the principal axis. A first class of axial groups includes the cyclic abelian groups n, $\bar{n}$ and the non-cyclic groups n/m. If the group contains also n twofold rotations along axes perpendicular to the principal axis and/or n mirrors containing the principal axis, the non abelian group is a dihedral one.

-       **Multiaxial** groups contain more than one rotation of order n > 2. Whereas the number of axial and dihedral groups is infinite, since n may take any value, the number of multiaxial groups is finite and equal to 7.

The table below sums up the classification of finite point groups. In the binary groups 222, 2mm and mmm (which may be considered also as dihedral groups), the three symbols represent rotations along, or mirrors perpendicular to, three mutually perpendicular directions. In the group n/m, the second symbol represents a mirror perpendicular to the principal axis. In the groups n2 and nm, the second symbol represents a set of n twofold rotations perpendicular to the principal axis, or a set of n mirrors parallel to it. If n is even, this second symbol is generally repeated since the group contains two classes of such twofold rotations or mirrors. The dihedral group $\bar{n}$2m (or $\bar{n}$m2), which exists only for n even, contains n twofold axes perpendicular to the principal axis, and n mirrors parallel to it and perpendicular to a twofold axis (Fig. I.4).

In the cubic group 432, the fourfold rotation axes are along the edges of a cube, the threefold ones along the main diagonals and the twofold ones along the face diagonals. In the cubic group $\bar{4}$3m, the mirrors contain two opposite edges of the cube. In the cubic group m$\bar{3}$m, which is the full symmetry group of the cube, two types of mirrors are found: three are parallel to the faces of the cube, six contain two opposite edges. In order to visualize the symmetry axes of the tetrahedral group $\bar{4}$3m, it is necessary to consider a regular tetrahedron inscribed in a cube (Fig. I.3d): the threefold axes are along the main diagonals of the cube, and the $\bar{4}$ axes are perpendicular to its faces.

Groups	Proper	Improper	Centrosymmetric
binary	1, 2, 222	m, 2mm	$\bar{1}$, 2/m, mmm
cyclic	n	$\bar{n}$ (n even)	$\bar{n}$ (n odd) n/m (n even)
dihedral	n2	nm $\bar{n}$2m (n even)	$\bar{n}$m (n odd) n/m (n even)
multiaxial	23, 432, 532	$\bar{4}$3m	m$\bar{3}$, m$\bar{3}$m, $\overline{53m}$

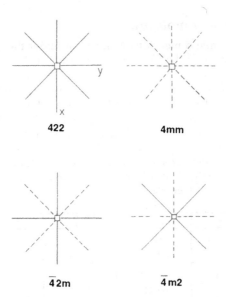

Figure I.4. — Dihedral groups with a fourfold axis of rotation. Full lines are twofold axes, dashed ones are mirrors.

Finally we must consider the 7 continuous point groups that contain one or an infinity of rotations of infinite order, they are called the Curie limit groups (Fig. I.5). The simpler one is denoted by ∞ and is generated by a rotation of infinitesimal angle. ∞/mm is the full symmetry group of the axial ellipsoid, it is a one-parameter continuous group; ∞/∞m is the full symmetry group of the sphere, it is a three-parameter continuous group. A polar vector, such as an electric field, is invariant under ∞m; an axial one, such as a magnetic field, is invariant under ∞/m.

Groups	Proper	Improper	Centrosymmetric
Cyclic	∞		∞/m
Dihedral	∞2	∞m	∞/mm
Multiaxial	∞/∞		∞/∞m

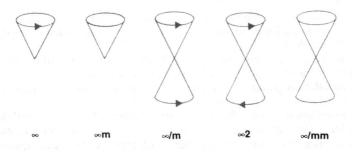

Figure I.5. — The axial and dihedral limit point groups.

I.5.   Symmetry of molecules

A molecule of finite dimensions is invariant under the operations of a point group G, the fixed point is the center of mass of the molecule. G may be of finite order (in general) or continuous (if the molecule is linear); binary, uniaxial (cyclic or dihedral) or multiaxial; chiral or non-chiral. An important point should be emphasized here: the order n of a rotation leaving a finite molecule invariant may take any value. For instance n = 3 for the CH_3Cl molecule; n = 4 for the $[Pt\ Cl_4]^{2-}$ ion; n = 5 for the $C_5H_5^-$ ion; n = 6 for the benzene molecule and n = 8 for the sulphur molecule. A limitation on the value of n will be found for infinite periodic systems. Various examples of molecular symmetry are given in the table below. The CH_4, SF_6 and C_{60} molecules have the shape of a regular tetrahedron, a regular octahedron and a truncated icosahedron respectively.

Type	Group	Molecule	G_p	G_i	G_c
linear	continuous	HCl		∞m	
		H_2			∞/mm
planar	binary	CH_2ClBr		m	
		H_2O			
	dihedral	$PtCl_4^{2-}$			$4/mmm$
		$C_5H_5^-$		$\overline{10}2m$	
		O_3		$\overline{6}2m$	
		C_6H_6			$6/mmm$
non-planar		CHClBrI	1		
	binary	CH_2Cl_2		2mm	
	dihedral	CH_3Cl		3m	
		S_8		$\overline{8}2m$	
	multiaxial	CH_4		$\overline{4}3m$	
		SF_6			m3m
		C_{60}			$\overline{5}3m$

I.6.   Symmetry of infinite systems

Consider an infinite microscopic system such as a linear polymer, a crystal surface or a three-dimensional solid. Three different symmetry groups can be defined for this system.

-       The **space group** takes into account the microscopic structure of the system and describes its translational as well as its orientational invariance properties.

-       The **translation group** is a sub-group of the space group, which describes only the translational invariance properties.

-       The **point group** ignores the microscopic structure of the system and describes only the orientational invariance of macroscopic (mechanical, electrical, magnetic, optical) properties (the point group is not a sub-group of the space group, but a "factor group").

Three types of systems are found in nature as far as the positions of atoms are concerned.

- Some systems, such as liquids, are **disordered**: the atoms are distributed at random, only short-range order may be found. No translation can superpose these systems exactly onto themselves, however they are statistically invariant under any translation since it is impossible to distinguish physically between a system and the translated one. The macroscopic physical properties are homogeneous.

- Other systems are **periodic**. The translations under which they are superposed onto themselves form a lattice.

- Incommensurate (modulated) structures and quasicrystalline solids are ordered **but aperiodic** systems.

Examples are given in the table below: translational symmetry is a very efficient tool for the classification of condensed matter states. In cholesteric liquid crystals, space periodicity is due to orientational ordering, the period is much larger than intermolecular distances. In smectic B and E liquid crystals, the molecules are ordered in the smectic planes, so that both one- and two-dimensional orders are found. A similar situation is found in discotic liquid crystals if the columns are periodic.

Dimensionality	Disorded system	Periodic system	Dimensionality of periodicity
1	discotic column	linear polymer	1
2	liquid film smectic A or C plane	crystal surface smectic B or E plane	2
3	homogeneous liquid	smectic	1
	nematic	cholecteric	1
	amorphous solid	discotic	2
		crystal	3
		plastic crystal	3

I.7.   Symmetry of polymers

Consider an infinite polymer. Suppose it is periodic along x and let **a** be the lattice vector. If it is a one-dimensional system such as an infinite linear molecule, only two space symmetries G_e are possible: p1 and $p\bar{1}$ (Fig. I.6), p designates the one-dimensional lattice (there is only one way of tiling a line with equal intervals) and the second symbol is the point symmetry G.

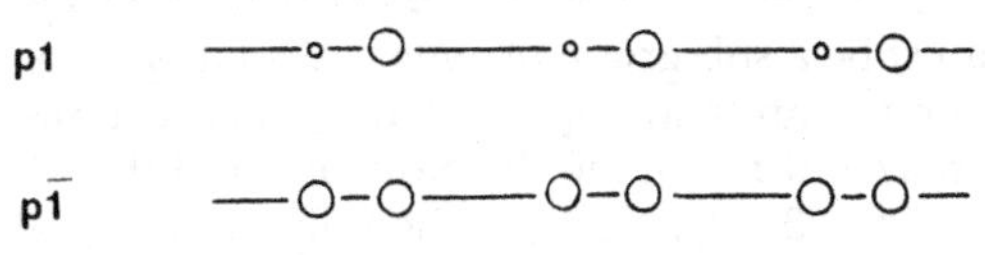

Figure I.6. — The two one-dimensional space groups.

Consider now a planar polymer with lateral bonds along y. The directions x and y must be invariant under the operations of the point group G, so that G must be mmm or one of its 6 sub-groups (1, 2, m, 2/m, 222, mm2). A mirror reflection α containing x may be an ordinary one (m), or a glide one (g) with an associated translation $t = a/2$, since α^2 must be the identity or a lattice translation. Similarly, a binary rotation along x may be an ordinary one (2), or a screw one (2_1) with a translation $t = a/2$.

Three model structures are shown in Fig. I.7. The symbol of a space group G_e gives the rotational axis along (or mirrors perpendicular to) x, y and z. The space groups are $p\ 2_x/m_{yz}\ 2_y/m_{zx}\ 2_z/m_{xy}$ (or $pm_{yz}m_{zx}m_{xy}$ or pmmm in short), $p2_ym_{xy}m_{yz}$ (pm2m), and $p\ 2_{1x}/m_{yz}\ 2_y/g_{xz}\ 2_z/m_{xy}$ (pmg$_{xz}$m or pmam) respectively. The point groups G are mmm, 2mm and mmm respectively. A glide mirror with a translation parallel to x, y or z is noted a, b or c respectively. A space group without helirotations or glide planes is called **symmorphic**. One can find 31 possible space groups for planar periodic polymers, 16 are symmorphic and 15 are non-symmorphic.

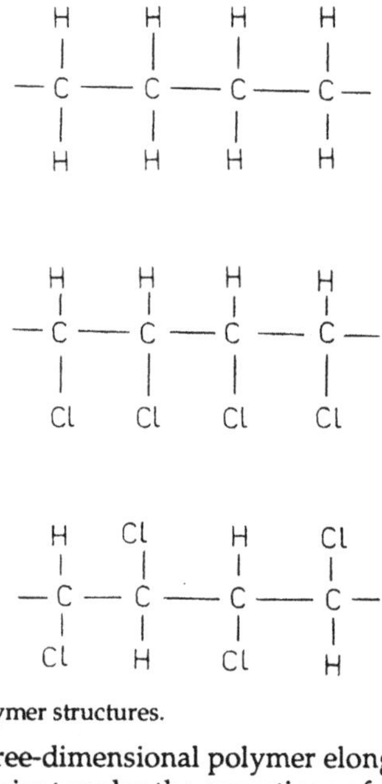

Figure I.7. — Models for planar polymer structures.

Consider now a truly three-dimensional polymer elongated along the direction x. This direction must be invariant under the operations of the point group G of this polymer, so that G must be a sub-group of ∞/mm. A rotation of order n along x may be an ordinary one, or a helirotation n_p with an associated translation t equal to a multiple of a/n: $t = p\ a/n$. If n = 3, the threefold axis is 3 (t = 0), 3_1 (t = a/3) or 3_2 (t = 2a/3). If the rotation is $3_1, 3_2, 4_1, 4_3, ...$ the polymer is chiral (see Fig. I.8). The number of possible space groups is infinite since n may take any value.

A helicoidal polymer may be invariant under a helirotation of angle θ incommensurable with π. Such a chiral system is aperiodic but obviously ordered: it is a special type of one-dimensional incommensurate structure since the carbon skeleton itself is periodic.

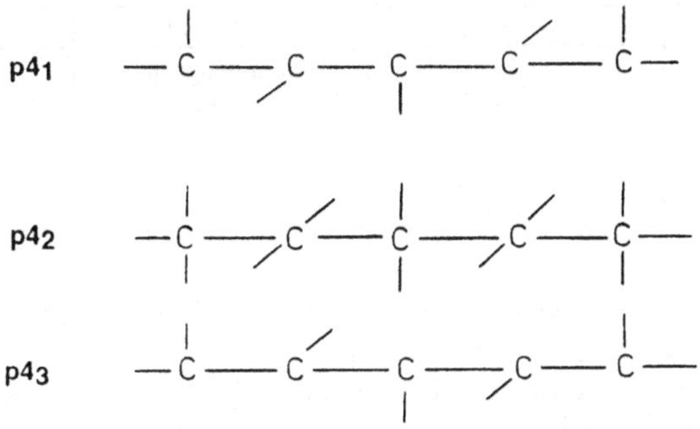

p4₁ ... (figure)

Figure I.8. — Models for three-dimensional polymer structures.

I.8.   Symmetry of periodic surfaces

Consider now a one-sided surface parallel to the xy plane. This plane and the perpendicular direction z must be invariant under the operations of the point group G of the surface, so that G must be a sub-group of ∞m where the infinite order axis is parallel to z.

Suppose that the surface is periodic: its translational symmetry is represented by a mathematical two-dimensional lattice. Let a and b be two independent vectors generating the lattice (Fig. I.9): they define a parallelogram called the basis cell, which contains only one lattice node (each node is shared by four adjacent cells). The xy plane is tiled by doubly periodic repetition of this unit cell. If the angle γ between the two vectors has no special value, the lattice is called oblique. Two lattice vectors $u\mathbf{a} + v\mathbf{b}$ and $u'\mathbf{a} + v'\mathbf{b}$, where u, v, u', v' are integers, define another unit cell, the area of which is equal to $\lambda = \text{abs}(uv' - vu')$ times the area of the basis cell and contains λ lattice nodes. If $\lambda = 1$, the cell is called **primitive** (p), no smaller unit cell can be found; if $\lambda > 1$, it is called nonprimitive or **multiple**. Fig. I.10a shows a centered (c) double unit cell and an edge-centered double cell.

Going from periodic polymers to periodic surfaces introduces a new and fundamental feature of symmetry: if a surface is periodic, it can be invariant only under rotations of order n = 1, 2, 3, 4 and 6. This **compatibility condition** between point and translation symmetry is a consequence of the fact that there exists a minimum distance between the nodes of a periodic lattice. The point group G of the surface is then one of the 10 following groups: the cyclic groups 1, 2, 3, 4, 6; the corresponding dihedral groups m, 2mm, 3m, 4mm, 6mm.

It is interesting to look for lattices invariant under one of these point symmetries. Since any lattice is invariant under a twofold rotation along the direction perpendicular to its plane, only 4 symmetries have to be considered: 2, 2mm, 4mm, 6mm. They are called **holohedral symmetries** (the point group G depends on the atomic pattern inside the unit cell of the lattice, it may be lower than the holohedral symmetry). We have then 4 systems of two-dimensional lattices, called oblique, rectangular, square and hexagonal respectively. However 5 distinct lattices, called **Bravais lattices**, are used for practical reasons and described in the table below.

System	Symmetry	Conventional cell	Type of lattice
oblique	2	parallelogram	p
rectangular	2mm	rectangle	p
			c
square	4mm	square	p
hexagonal	6mm	60° rhombus	p

- Introducing a centered rectangular lattice is not necessary since a primitive lozenge cell with **a** = **b** (arbitrary rhombus) can be used as well. However, although the primitive cell has the holohedral 2mm symmetry, one prefers using a double cell with orthogonal axis (Fig. I.10a).

- In the hexagonal system (Fig. I.10b), it is not possible to find a primitive cell reflecting the holohedral symmetry. Three primitive cells can be associated to form a regular hexagon, but this polygon is not a parallelogram. A hexagonal primitive cell can also be constructed via the Dirichlet-Voronoï or Wigner-Seitz procedure. Draw the lines connecting a given lattice point to all nearby points: the symmetrical WS cell is enclosed by the lines perpendicular to these lines at their midpoint. The WS procedure can be applied to any lattice and gives a primitive cell which is not a parallelogramm but shows the full rotational symmetry of the lattice. This cell comprises all that space around a given lattice point which is closer to that point than to any other lattice point.

The preceding results are related to the possibilities of tiling a plane with regular polygons. It is found that a tiling is obtained only with triangles, squares and hexagons. In particular a tiling with regular pentagons or heptagons is impossible (Fig. I.11).

We discuss finally the space symmetry of one-sided periodic surfaces. From the 10 possible point groups and the 5 possible lattices, 17 possible space groups can be derived, only four of them are non-symmorphic (see table below). In groups p31m and p3m1, the mirrors have different orientations with respect to the lattice.

System	Lattice	Point group	Space group
oblique	p	1	p111
		2	p112
rectangular	p	m	p1m1 p1g1
		2mm	p2mm p2mg p2gg
	c	m	c1m1
		2mm	c2mm
square	p	4	p4
		4mm	p4mm p4mg
hexagonal	p	3	p3
		3m	p31m p3m1
		6	p6
		6mm	p6mm

Point and lattice symmetries cannot be associated at random. The surface point group must be a sub-group of the lattice point group: for instance, a rotation of order 3, 4 or 6 cannot be associated to a rectangular lattice. Moreover pseudo-symmetries, that is accidental ones, must be eliminated: a square lattice is never found if the point group is only binary, except for a special value of external parameters such as temperature; hexagonal symmetry always implies the existence of a threefold or sixfold axis. Once a point group and a compatible lattice are selected, more than one space group may be found.

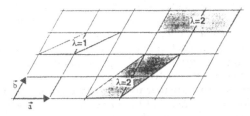

Figure I.9. — Primitive and multiple cells of an oblique lattice.

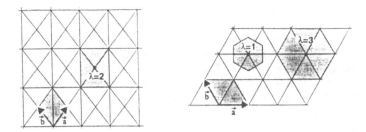

Figure I.10. — Primitive and multiple cells for the centered rectangular and hexagonal lattices.

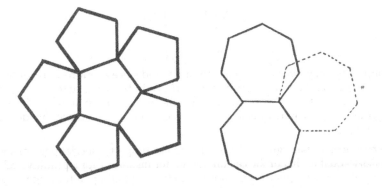

Figure I.11. — Tiling of a plane with regular pentagons and heptagons.

Figure I.12 illustrates the four space groups of the rectangular system with point symmetry 2mm (see International Tables for Crystallography, volume A). Please note that a lattice is a purely mathematical description of the translational symmetry of the surface: its origin may be chosen anywhere in the plane, the lattice nodes have no physical meaning and are not necessarily occupied by atoms. Here the origin of the lattice was chosen on a binary axis.

A point M(x, y) is in a **general position** if it is invariant under no symmetry element: it is not located on a rotation axis or mirror (it may be on a screw axis or glide mirror). If invariant under at least one symmetry element, M is in a **special position**. Equivalent points in a general position (shown on the left part of the figure) are deduced from one of them by application of the symmetries of the space group (black and white points are connected by mirror symmetry). The number of equivalent points - 4 or 8 - in each conventional cell is called the **multiplicity** of the position, it is equal to the order of the point group multiplied by the multiplicity of the cell. Symmetry elements themselves are shown on the right part of the figure: full lines represent mirrors perpendicular to the surface plane, dashed ones represent glide mirrors. It should be noted that the periodicity of some symmetry elements may be different from the periodicity of the surface itself.

Consider for instance the group p2mg. The general position (denoted 4d) is:

$$x, y \qquad \bar{x}, \bar{y} \qquad \bar{x} + \frac{1}{2}, y \qquad x + \frac{1}{2}, \bar{y}$$

The second point is generated from the first one by the twofold axis at the origin, the third one by the mirror m along the y axis at x = 1/4, and the fourth one by the glide mirror g along the x axis at y = 0. The glide translation of g is 1/2, 0. A non zero translation is associated to m, but it is not a glide one since it is perpendicular to m: it has no intrinsic meaning since it depends on the choice of O.

If M is on a mirror, x = 1/4 or 3/4 and we get the special position (2c):

$$1/4, y \qquad 3/4, \bar{y}$$

The points M(0, 0) and M(0, 1/2) are situated on a twofold axis and generate the two other special positions (2b):

$$0, 1/2 \qquad 1/2, 1/2$$

and (2a):

$$0, 0 \qquad 1/2, 0$$

(in group p2gg, only two special positions are found since all mirrors are glide ones). The symbols 2a, 2b, 2c and 4d are the **Wyckoff symbols**: they give the multiplicity of the position, the letter is an arbitrary coding scheme. The **site symmetry** of a position is the group of the symmetry elements on which it lies: it reduces to the identity for the general position.

Symmorphic space groups have a particular property: special positions with a multiplicity equal to that of the cell are found, for instance 1 for a primitive cell. The site symmetry of these points is isomorphic to the point group. If the origin is placed at one of them, a point in a general position with x<<1 and y<<1 has equivalent points close to it and the other nodes of the lattice.

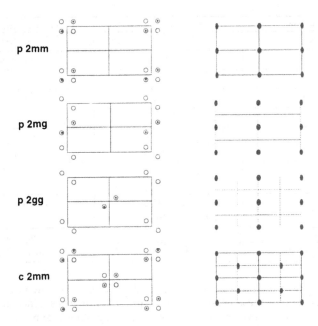

p 2mm

p 2mg

p 2gg

c 2mm

Figure I.12. — Two-dimensional space groups with point symmetry 2mm.

I.9. Symmetry of crystals

As demonstrated by the Von Laue X-ray diffraction experiment in 1912, crystals are triply periodic systems, in agreement with the early hypothesis of Huyghens, Haüy and others.

Let **a, b** and **c** be three independent vectors generating a lattice: they define a parallelepiped called the basis cell, which contains only one lattice node. The whole space is paved by triply periodic repetition of this unit cell. If the lengths and angles of the cell have no special values, the lattice is called **triclinic**. Three lattice vectors $u\mathbf{a} + v\mathbf{b} + w\mathbf{c}$, $u'\mathbf{a} + v'\mathbf{b} + w'\mathbf{c}$ and $u''\mathbf{a} + v''\mathbf{b} + w''\mathbf{c}$, where u, v, ... are integers, define another cell, the volume of which is equal to λ times the volume of the basis cell and contains λ lattice nodes: λ is the determinant of the coefficients u, v, w, u', v', ... If $\lambda \supseteq 1$, the cell is called primitive, no smaller unit cell can be found; if $\lambda > 1$, it is called multiple.

As a periodic surface, a crystal can be invariant only under rotations of order n = 1, 2, 3, 4 and 6. Because of this compatibility relation, a **crystallographic point group** G may not be any sub-group of the limit group $\infty/\infty m$, but only one of the 32 well-known crystallographic point groups enumerated by Hessel as early as 1830. 11 of them are proper, 10 are improper and 11 (Laue groups) are centrosymmetric.

Proper groups	Improper groups	Centrosymmetric groups
$1 = C_1$		$\bar{1} = S_2$ or C_i
$2 = C_2$	m or $\bar{2} = C_h$	$2/m = C_{2h}$
$3 = C_3$		$\bar{3} = S_6$ or C_{3i}
$4 = C_4$	$\bar{4} = S_4$	$4/m = C_{4h}$
$6 = C_6$	$\bar{6}$ or $3/m = S_3$ or C_{3h}	$6/m = C_{6h}$
$222 = D_2$	$2mm = C_{2v}$	$mmm = D_{2h}$
$32 = D_3$	$3mm = C_{3v}$	$\bar{3}m = D_{3d}$
$422 = D_4$	$4mm = C_{4v}$ $\bar{4}2m = D_{2d}$	$4/mmm = D_{4h}$
$622 = D_6$	$6mm = C_{6v}$ $\bar{6}2m$ or $3/mm = D_{3h}$	$6/mmm = D_{6h}$
$23 = T$		$m3 = T_h$
$432 = O$	$\bar{4}3m = T_d$	$m\bar{3}m = O_h$

Following the same procedure as for periodic surfaces, we now look for lattices invariant under one of these crystallographic point symmetries. Since any triclinic lattice is invariant under inversion, only the 11 centrosymmetric symmetries have to be considered. Moreover it can be shown that any lattice with a principal axis is dihedral, so that only 7 **holohedral symmetries** are found:

$$\bar{1}; \ 2/m; \ mmm; \ \bar{3}m; \ 4/mmm; \ 6/mmm; \ m\bar{3}m.$$

We have then 7 **systems** of three-dimensional lattices, called triclinic, monoclinic, orthorhombic, trigonal, tetragonal, hexagonal and cubic respectively. However 14 distinct **Bravais lattices** must be considered (Fig. I.13). Except for the hexagonal symmetry, as already discussed for the two-dimensional case, it is always possible to find a unit cell reflecting the holohedral symmetry. However for some lattices, it happens that this cell is a multiple one: it has a higher symmetry than any primitive one. For instance the primitive cells of centered and face-centered cubic lattices are rhombs which are not cubic. This is the reason why more than one lattice may correspond to a given holohedral symmetry.

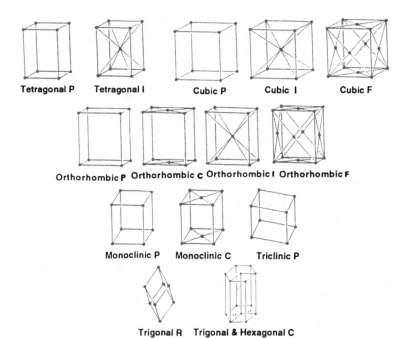

Figure I.13. — The 14 three-dimensional Bravais lattices.

The trigonal lattice deserves a special discussion. The primitive cell is holohedral but, whatever the values of its parameters, it is always possible to find a triple hexagonal cell with a hexagonal symmetry higher than the holohedral one. For this reason, the 7 systems are now classified in 6 **families**, the trigonal and hexagonal lattices belonging to the same family. Some trigonal crystals (no sixfold axis) have a primitive rhombohedral lattice; others, such as quartz, have a hexagonal primitive lattice, no primitive rhombohedral cell can be found.

System	Holohedral symmetry	Type of lattice	Possible point groups
triclinic	$\bar{1}$	P	1 $\bar{1}$
monoclinic	2/m	P C	2 m 2/m
orthoclinic	mmm	P C I F	222 2mm mmm
tetragonal	4/mmm	P I	4 $\bar{4}$ 4/m 422 4mm $\bar{4}$2m 4/mmm
trigonal	$\bar{3}$m	R P$_{hexagonal}$	3 $\bar{3}$ 32 3m $\bar{3}$m
hexagonal	6/mmm	P	6 $\bar{6}$ 6/m 622 6mm $\bar{6}$2m 6/mmm
cubic	m$\bar{3}$m	P I F	23 m$\bar{3}$ 432 $\bar{4}$3m m$\bar{3}$m

From the 32 possible point groups and the 14 possible lattices, 230 possible space groups have been derived by Fedorov and Schoenflies around 1890: 73 are symmorphic and 157 are non symmorphic. The algebraic structure of space groups is not simple: they are <u>not</u> direct products of a point group by a lattice translation group, since any rotation does not commute with any rotation.

Point and lattice symmetries cannot be associated at random. The crystal point group must be a sub-group of the lattice point group: for instance, a rotation of order 3, 4 or 6 cannot be associated to an orthorhombic lattice. Except for the trigonal crystals, which may have a primitive rhombohedral (R) or a hexagonal (P) lattice, lattice pseudo-symmetries must be eliminated. Once a point group G and a compatible lattice are selected, more than one space group may be found: for instance if G is mmm and the lattice is primitive, 16 different space groups are found.

Going from two to three dimensions introduces screw axes : 2_1; 3_1 and 3_2; 4_1, 4_2 and 4_3; 6_1, 6_2, 6_3, 6_4 and 6_5. Pairs of screw axes are enantiomorphic, which means that they can be considered as right- and left-handed respectively, with the same screw translation: 3_1 and 3_2; 4_1 and 4_3; 6_1 and 6_5; 6_2 and 6_4. The other screw axes are neither right- nor left-handed.

For this reason, 11 **enantiomorphic pairs** exist among the 230 space groups, for instance the non symmorphic groups $P6_122$ and $P6_522$, so that only 219 non isomorphic space groups are found. Enantiomorphic crystal structures are described by enantiomorphic space groups.

Here are some practical remarks on point and space groups. If 3 or $\overline{3}$ is found as the second symbol of a point group, this group belongs to the cubic system. If the point group of a crystal is centrosymmetric, the crystal itself is centrosymmetric. Conventions about the assignement of the second, third, ... symbols of a space group depend on the crystal system. Whereas g has been used above as a generic designation for glide reflections in two and three dimensions, the more specific symbols a, b, c, n, d are introduced in the description of crystal symmetry according to the direction of the glide translation. The point group is found easily from the symbol of a space group: suppress the first (lattice) symbol, and replace the following symbols for helicoïdal axes of rotation and glide mirrors by the corresponding symbols for ordinary axes of rotation and ordinary mirrors respectively. The point groups of space groups $P6_122$ and $Fd\overline{3}c$ are 622 and $m\overline{3}m$ respectively (see International Tables for Crystallography, volume A, for further information and discussion and an exhaustive list of 2- and 3- dimensional space groups).

As an example, let us consider the structure of a large variety of dioxides and difluorides RX_2 such as cassiterite SnO_2 and rutile TiO_2. From the Wyckoff's Crystal Structures, we learn that the space group is $P4_2/mnm$ (international notation) or D_{4h}^{14} (Schoenflies notation) with R in position (2a) and X in position (4f) with $x \approx 0.3$ (the exact value is not determined by symmetry and depends on the compound considered). More information is then obtained from the International Tables: a more explicit notation of the space group $P\,4_2/m\,2_1/n\,2/m$ shows that the fourfold axes are parallel to z, the helicoïdal twofold axes parallel to x and y, the non-helicoïdal twofold axes parallel to the directions $[110]$ and $[1\overline{1}0]$. The origin is at a symmetry center, the glide translation associated to the mirror n is $1/2, 1/2, 0$. The coordinates of the R atoms are $0, 0, 0$ and $1/2, 1/2, 1/2$. The coordinates of the X atoms are $x, x, 0$; $-x, -x, 0$; $1/2 + x, 1/2 - x, 1/2$; $1/2 + x, 1/2 - x, 1/2$. Note that the number $Z = 2$ of formula units per unit cell is different from the multiplicity $\lambda = 1$ of the unit cell. The point group $4/mmm$ is centrosymmetric, so that such properties as pyro- and piezoelectricity are forbidden by symmetry.

I.10.   The reciprocal lattice

Finally we introduce the **reciprocal lattice**, which is a fundamental concept in studying diffraction phenomena but was introduced (by Bravais) much earlier than their discovery. The geometrical reason for introducing the reciprocal lattice is the following.

Three non colinear points belonging to a three-dimensional lattice define a lattice plane. If the coordinates are given in terms of the lattice constants, this plane is described by the cartesian equation $hx + ky + lz = m$, where h, k, l and m are integers defined up to a common factor, for instance $6x + 4y + 12z = 12$. Let us always choose for h, k and l the integers with the smallest possible absolute values (irreducible integers or relative primes), the equation becomes $3x + 2y + 6z = 6$. The lattice plane intercepts the coordinate axes at $x = 6/h = 2$, $y = 6/k = 3$, $z = 6/l = 1$ (Fig. I.14).

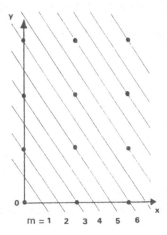

Figure I.14. — The lattice planes $3x + 2y = m$.

All lattice planes parallel to this plane are geometrically equivalent and equidistant. This set of planes contains all lattice points and may be represented by the equation $hx + ky + lz = m$, where the integers h, k, l are relative primes and are called the **Miller indices** of the set, and m is a variable positive or negative integer. The planes which are closest to the origin are given by $m = 1$ and -1. If some indices are negative, the planes are represented by symbols such as $(\bar{h} k l)$, $(h \bar{k} l)$, ...

The (hkl) lattice planes intercept the basis-vectors of the lattice at ma/h, mb/k, mc/l. Varying m from 1 to h, or k, or l, we see that the Miller indices represent the number of planes through which each of the three basis vectors pass.

Let $\mathbf{n}$ be a unit vector perpendicular to the (hkl) planes, d_{hkl} the distance between successive (hkl) planes (this distance is proportional to the density of lattice points in a (hkl) plane), and $\mathbf{h} = \mathbf{n}/d_{hkl}$. From the preceding result, we have: $\mathbf{a}.\mathbf{n} = h\,d_{hkl}$ and two similar relations, or

$$\mathbf{a}.\mathbf{h} = h$$
$$\mathbf{b}.\mathbf{h} = k$$
$$\mathbf{c}.\mathbf{h} = l$$

h, k and 1 cannot be considered as the coordinates of **h** in the "direct" basis **a**, **b**, **c** since this basis is not orthonormal in the general case. Let us however introduce the "reciprocal" basis vectors $\mathbf{a^*} = \mathbf{h}_{100}$, $\mathbf{b^*} = \mathbf{h}_{010}$, $\mathbf{c^*} = \mathbf{h}_{001}$ perpendicular to the planes (100), (010), (001) respectively. They are defined by:

$$\mathbf{a^*.a} = \mathbf{b^*.b} = ... = 1, \mathbf{a^*.b} = \mathbf{a^*.c} = ... = 0$$

We may now write: $\mathbf{h} = h\mathbf{a^*} + k\mathbf{b^*} + l\mathbf{c^*}$, so that the Miller indices are the coordinates of **h** in the reciprocal basis and $d_{hkl} = 1/|\mathbf{h}| = 1/|(h\mathbf{a^*}+k\mathbf{b^*}+l\mathbf{c^*})|$.

Introducing the volume $V = \mathbf{a.(b \times c)}$ of the direct unit cell, we have:

$$\mathbf{a^*} = (\mathbf{b \times c})/V \; ; \mathbf{b^*} = (\mathbf{c \times a})/V, ...$$

(reciprocal basis vectors used by physicists are multiplied by a factor 2π). If L is the dimension of a direct vector, the dimension of a reciprocal vector is $1/L$. The volume $V^* = \mathbf{a^*(b^* \times c^*)}$ of the unit cell of the reciprocal lattice is equal to $1/V$. The reciprocal of the reciprocal lattice is the direct lattice. The direct and reciprocal lattices have of course the same point symmetry. The Wigner-Seitz cell of the reciprocal lattice is called the first Brillouin zone.

In the simplest geometrical situation of a cubic lattice (a=b=c and $\alpha=\beta=\gamma=90°$), the reciprocal lattice basis vectors $\mathbf{a^*},\mathbf{b^*},\mathbf{c^*}$ are parallel to the direct-space basis vectors **a**,**b**,**c**; their lengths are: $|\mathbf{a^*}| = |\mathbf{b^*}| = |\mathbf{c^*}| = 1/a$ and $d^2_{hkl} = a^2/(h^2+k^2+l^2)$; but this is **not** true for other crystal lattices.

We have just seen that the reciprocal lattice points h, k, 1 represent the direct lattice planes having h, k and 1 as Miller indices. What is then the meaning of the other reciprocal lattice points, the coordinates of which are not prime integers, for instance the point nh, nk, nl where n is some integer ? This point represents the set of direct planes containing the lattice planes (hkl) and other planes parallel to them and containing no lattice nodes, the distance between two successive planes being d_{hkl}/n (see Fig. I.15).

An important application of the reciprocal space is the following: the Laue equations for diffraction can be represented by the Bragg construction in direct space, and by the Ewald construction in reciprocal space.

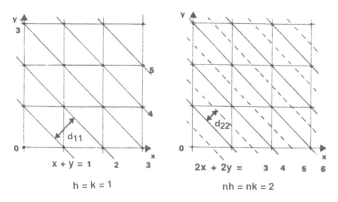

Figure I.15. — Lattice planes h = k = 1 (full lines) and non-lattice planes h = k = 2 (dashed lines).

REFERENCES

BUERGER, M. J., 1956 - Elementary crystallography, John Wiley.

CHEN, N. X., 1986 - "An elementary method for introducing the concept of reciprocal lattice", Am. J. Phys. 54, 1000.

GLASSER, L., 1967 - "Teaching symmetry, the use of decoration", J. Chem. Ed. 44, 502.

International Tables for Crystallography, 1983, volume A, edited by T. Hahn, published for the International Union of Crystallography by Reidel Publishing Company.

KETTLE, S. F. A. and NORRBY, L. J., 1991 - "The Brillouin zone, an interface between spectroscopy and crystallography", J. Chem. Ed. 67, 1022.

KITTEL, C., 1971 - Introduction to Solid State Physics, John Wiley.

MCKIE, D. and MCKIE, C., 1986 - Essentials of Crystallography, Blackwell.

NARAHARI ACHAR, B. N., 1986 - "Reciprocal lattice in two dimensions", Am. J. Phys. 54, 663.

PHILLIPS, F.C., 1971 - An introduction to crystallography, 4th edition, Longman.

RADFORD, L. E., 1975 - "Equal rights for the reciprocal lattice", Am. J. Phys. 43, 697.

SHUBNIKOV, A. V. and KOPTSIK, V. A., 1974 - Symmetry in science and art, Plenum Press.

WYCKOFF R. W. G., 1966 - Crystal Structures, Interscience Publishers.

CHAPTER II

SINGLE CRYSTAL STRUCTURE ANALYSIS
G. HEGER

II.1.   Introduction

Crystal structure determination by diffraction methods on single crystals using X-rays or neutrons is well established. It was obvious from the beginning that the use of highly expensive neutron sources demands special arguments which have to show the need and specific advantages of neutron diffraction experiments in the field of interesting high level research. This type of argumentation is certainly also necessary with respect to the use of synchrotron X-ray radiation. The crucial question remains whether neutron and/or X-ray synchrotron work yields essential information not available by other techniques including X-ray diffraction on small laboratory equipment.

Standard structure analysis on single crystals is the domain of X-ray diffraction using conventional laboratory equipment as far as the determination of space group symmetry, cell parameters and the structure model (of non-hydrogen atoms) are concerned. The aim of this paper is to discuss special cases, for which, in addition to this indispensable conventional part, neutron and/or synchrotron X-rays are required to solve structural problems. Examples of single crystal studies in the field of small molecular systems are presented. A comparison with powder diffraction methods is not considered. The following non-exhaustive list of topics is discussed in detail:

(II.2)   Contrast variation which is important for an experimental distinction of atoms/ions with almost equal scattering amplitudes. Another application is the determination of accurate atomic parameters (positional and thermal parameters, site occupations) of lighter elements in the presence of heavy ones.

(II.3)   The problem of hydrogen localization (proton/deuteron position and corresponding Debye-Waller factors) in X-ray structure analyses which is important in the field of hydrogen bonding. Structural phase transitions may be caused by proton ordering in hydrogen bonds.

(II.4)   Molecular disorder which is of great current interest. The fullerene C_{60} is a prominent example of dynamical disorder of molecules in a crystalline solid.

(II.5)   Electron density analysis which requires complete and accurate X-ray Bragg intensity measurements over a large $\sin\theta/\lambda$ range including weak and very weak reflections. Supplementary information on harmonic and anharmonic contributions to the atomic mean square displacements may be needed.

(II.6)   The absolute configuration of non-centrosymmetric structures which can be determined by using anomalous X-ray scattering.

(II.7)   The phase problem which is the principal difficulty of structure determination by diffraction methods. Special experimental X-ray techniques have been developed to obtain phased structure factors. The "λ-technique" makes use of the variation of the Patterson function in the vicinity of the absorption edge in the case of an anomalous scatterer, whereas the multiple-beam technique is based on dynamical diffraction.

(II.8)   The anisotropy of anomalous dispersion in X-ray diffraction may give new insight into electronic structures.

(II.9)   Experimental constraints may be also strong arguments in favour of the use of X-ray synchrotron or neutron facilities. External parameters (temperature, pressure, electrical field), sample problems (special sensitivities, microsize) or experimental conditions (variable and very short X-ray wavelengths, very high Q resolution) may be easier and better achieved.

II.2.   The contrast variation method

Contrast variation is required in crystal structure analysis to distinguish between atoms/ions with almost equal scattering amplitudes. It might be a problem to attribute special atomic sites or, in the case of mixed systems, to determine the site occupation.

The contrast in conventional X-ray diffraction is directly related to the ratios of the number of electrons Z_j of atoms or ions j. The atomic scattering factor f_j in the structure factor, which represents the Fourier transform of the atomic electron density distribution, is proportional to Z_j ($f_j = Z_j$ for $\sin\theta/\lambda = 0$). Standard X-ray techniques can hardly differentiate between atoms/ions of a similar number of electrons, and only an average structure - including a total occupation probability of mixed occupied sites - may be obtained in such cases.

For neutrons f_j is replaced by the nuclear scattering length b_j, which depends on the isotope and nuclear spin states of the element j. The natural isotope mixture and a statistical spin state distribution, commonly used, lead to $b_j = \alpha \cdot b_{j\alpha} + \beta \cdot b_{j\beta} + \gamma \cdot b_{j\gamma} + ...$ with the sum of the different isotope portions $\alpha + \beta + \gamma + ... = 1$ ($b_{j\alpha}$, $b_{j\beta}$, $b_{j\gamma}$ are the individual scattering lengths of the different isotopes of the element j). Neutron experiments frequently make use of compounds containing single isotope elements, like fully deuterated samples. Incoherent scattering due to a statistical distribution of isotopes and nuclear spin states is not discussed here. It may influence the effective absorption and the background conditions of neutron diffraction studies.

The continuous wavelength distribution of high intensity X-ray synchrotron radiation and a two-wavelength method can be applied to make use of the characteristic wavelength dependence of the anomalous dispersion correction terms $f'(\lambda)$ and $f''(\lambda)$ close to the absorption edges of the elements under consideration. In this way the contrast between atoms/ions of a similar number of electrons can be varied.

The mathematical basis of this difference synthesis (δ-synthesis) is rather simple. The structure factor is in general a function of the reciprocal lattice vector **h** and the wavelength λ

$$F(\mathbf{h},\lambda) = \sum_j f_j(\mathbf{h},\lambda) \cdot \exp[2\pi i(\mathbf{h} \cdot \mathbf{r}_j)] \cdot T_j(\mathbf{h}),$$

where $T_j(\mathbf{h})$ is the Debye-Waller factor and $f_j(\mathbf{h},\lambda)$ the atomic scattering factor of atom j at position $\mathbf{r}_j$

$$f_j(\mathbf{h},\lambda) = f_{oj}(\mathbf{h}) + f_j'(\lambda) + i f_j''(\lambda).$$

The electron density

$$\rho(\mathbf{r},\lambda) = \frac{1}{V} \sum_h F(\mathbf{h}) \cdot \exp[-2\pi i(\mathbf{h} \cdot \mathbf{r})]$$

$$= \frac{1}{V} \sum_h \{ \sum_j f_j(\mathbf{h},\lambda) \cdot \exp[2\pi i(\mathbf{h} \cdot \mathbf{r_j})] \cdot T_j(\mathbf{h}) \} \cdot \exp[-2\pi i(\mathbf{h} \cdot \mathbf{r})].$$

The effect of wavelength variation on the electron density is obtained by differentiation of $\rho(\mathbf{r},\lambda)$:

$$\frac{\delta\rho}{\delta\lambda} = \frac{1}{V} \sum_h \{ \sum_j \delta f_j(\mathbf{h},\lambda)/\delta\lambda \cdot \exp[2\pi i(\mathbf{h} \cdot \mathbf{r_j})] \cdot T_j(\mathbf{h}) \} \cdot \exp[-2\pi i(\mathbf{h} \cdot \mathbf{r})]. \qquad (II.1)$$

The principle of the δ-synthesis is the use of two neighbouring wavelengths, available for diffraction data measurements, which are close to the absorption edge of the interesting element and for which the following conditions hold:

$$\delta f_n/\delta\lambda \cong 0 \quad \text{and} \quad \delta f_e/\delta\lambda \neq 0,$$

where n and e denote the "normal" and the "edge" elements, respectively. The additional condition $\delta f_j''/\delta\lambda \cong 0$ for all elements j simplifies considerably the analysis of the experimental data (no change of the absorption correction). Equation (II.1) then reduces to

$$\frac{\delta\rho}{\delta\lambda} \cong \frac{1}{V} \sum_h \{ \sum_j \delta f_j'(\lambda)/\delta\lambda \cdot \exp[2\pi i(\mathbf{h} \cdot \mathbf{r_j})] \cdot T_j(\mathbf{h}) \} \cdot \exp[-2\pi i(\mathbf{h} \cdot \mathbf{r})]$$

or

$$\delta\rho \cong \frac{1}{V} \sum_h \{ \sum_j \delta f_j'(\lambda) \cdot \exp[2\pi i(\mathbf{h} \cdot \mathbf{r_j})] \cdot T_j(\mathbf{h}) \} \cdot \exp[-2\pi i(\mathbf{h} \cdot \mathbf{r})]. \qquad (II.2)$$

It can be seen from (II.2) that this difference electron density distribution - depending on the derivative of the real part of the anomalous dispersion correction - shows up only at the atomic positions of the edge elements. Quite generally, special atoms can be localized in this way which may help to solve unknown crystal structures.

As an example of this difference synthesis (δ-synthesis) we report here on a recent study by Wulf (1990)[1], who demonstrated the experimental distinction between Pb (Z_{Pb}=82) and Bi (Z_{Bi}=83) in the mineral structure of galenobismutite ($PbBi_2S_4$). Its orthorhombic crystal structure (space group: Pnma; 4 formula units per unit cell) was determined in a previous conventional X-ray diffraction work[2]. All Pb/Bi and S atoms are localized in the mirror planes at z = 1/4 and at z = 3/4. An ordered distribution of the neighbouring elements Pb and Bi was assumed from crystal chemical considerations according to the small differences between the Pb-S (2.85 Å) and the Bi-S (2.70 Å) bond distances. A projection of the crystal structure along [001] is shown in Fig. II.1.

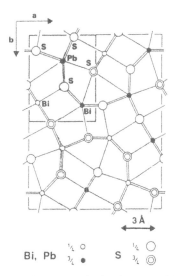

Figure II.1. — Projection of the crystal structure of galenobismutite $PbBi_2S_4$ along [001]. Represented is the asymmetric unit[1].

The anomalous dispersion correction terms f' and f" calculated for Pb and Bi close to the LIII absorption edges are plotted in Fig. II.2. The synchrotron measurements were carried out at λ_F = 0.953 Å and at λ_G = 0.988 Å (indicated by F and G in the plot). For these values the absorption is relatively small ($\lambda_{F,G} > \lambda_{edges}$) and does not change (f"$_{Pb,Bi}(\lambda)$ = const.). Besides the big difference $\Delta f'_{Pb}$ (F-G) $\cong$ 6.4 e⁻ there remains still a $\Delta f'_{Bi}$ (F-G) $\cong$ 1.4 e⁻.

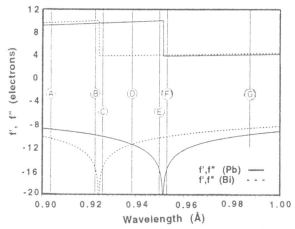

Figure II.2. — Theoretical anomalous dispersion correction terms f'$_{Pb,Bi}$ (lower part) and f"$_{Pb,Bi}$ (upper part) close to the LIII absorption edges of lead and bismuth[3].

Data collection			
Wavelength (Å)	0.7107 (Mo $K\alpha$)	0.953	0.988
X-ray source	Fine focus tube	Synchrotron Hasylab at DESY	Synchrotron Hasylab at DESY
Instrumentation	4-circle diffract. Stoe Stadi 4	5-circle diffractometer Stoe Stadi 4	5-circle diffractometer Stoe Stadi 4
Monochromator	Single crystal Graphite (0001)	Double crystal Germanium (111)	Double crystal Germanium (111)
Degree of polarization (%)	50	~ 93	~ 93
Beam stability control	2 standard reflections	2 standard refl. + real time monitorization of polarization and beam	2 standard refl. + real time monitorization of polarization and beam
Time between standards (min)	120	30	30
Recording technique	ω step scan	ω step scan	ω step scan
Step width ($\Delta\omega$) (°)	0.02	0.003	0.003
Time/step (s)	0.5-2.0	0.2-2.0	0.2-2.0
Max. $(\sin\theta)/\lambda$ (Å^{-1})	0.787	0.803	0.775
Max. $h, k, l, 2\theta$ (°)	18, 22, 6, 68	18, 23, 4, 100	18, 22, 4, 100
Number of F_o's recorded	3501	3186	2473
Unique reflections	1604	1496	1379
Data reduction			
Lp correction	Yes	Yes	Yes
Absorption correction	Ψ scans	Ψ scans	Ψ scans
Absorption coefficient μ (cm^{-1})	717.81	399.30	440.4
Min./max. transmission	0.116/0.801	0.208/0.592	0.208/0.592
Merging R before/after correction	0.372/0.105	0.247/0.140	0.247/0.140
Structure refinement			
Atomic scattering curves			
f_o (neutral Pb, Bi, S)	International Tables (1968)	International Tables (1968)	International Tables (1968)
f', f''	Cromer (1983)	Cromer (1983)	Cromer (1983)
Weighting scheme	$w = 1/\sigma^2$	$w = 1/\sigma^2$	$w = 1/\sigma^2$
Number of F_o's used	1599	1496	1379
$F_o > 1.5\,\sigma\,(F_o)$	1488	1496	1376
$R_{int} = \sum \left\| F^2 - F_{mean}^2 \right\| / \sum F^2 (\%)$	2.81	3.54	3.0
$R = \sum \left\| \left\| F_o \right\| - \left\| F_c \right\| \right\| / \sum \left\| F_o \right\| (\%)$	5.69	5.35	5.65

Table II.1. — Details of the data collection conditions for PbBi$_2$S$_4$[1].

The conditions of these state of the art synchrotron experiments together with those of a conventional data collection using MoKα radiation are summarized in Table II.1. It is obvious that for PbBi$_2$S$_4$ absorption becomes much more important at higher energies, e. g. at $\lambda_{MoK\alpha}$ = 0.7107 Å.

Due to the simple structure of PbBi$_2$S$_4$, only the projection along [001] is necessary for the δ-synthesis. In Fig. II.3, it is seen that the experimental δ-map is in very good agreement with a simulation.

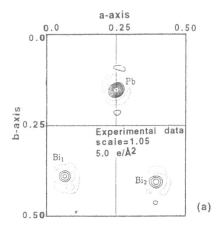

Figure II.3a. — Difference electron density map (δ-map) of $PbBi_2S_4$ based on experimental data at $\lambda_F = 0.953$ Å and $\lambda_G = 0.988$ Å (from Ref.[1]). Projection along [001]; asymmetric unit; contour intervals of 5.0 e⁻/Å².

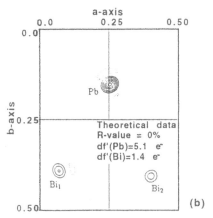

Figure II.3b. — Difference electron density map (δ-map) of $PbBi_2S_4$ calculated from x, y and T parameters of the structure refinement at $\lambda_G = 0.988$ Å[1]. Projection along [001]; asymmetric unit; contour intervals of 5.0 e⁻/Å².

The theoretical calculations are based on the positional and thermal parameters from a standard structure refinement of the data obtained at $\lambda_G = 0.988$ Å and the estimated $\Delta f'(F - G)$ values for lead and bismuth. In this way, the crystal chemical considerations in favour of an ordered distribution of fully occupied Pb and Bi sites are confirmed. The quality of Wulf's synchrotron X-ray data is even sufficient for valuable information concerning the two different Bi sites in spite of a relatively small $\Delta f'_{Bi} (F - G) \cong 1.4$ e⁻.

The contrast in conventional X-ray diffraction is directly related to the ratios of the number of electrons of atoms or ions. In the first example, the galenobismutite ($PbBi_2S_4$), lead and bismuth have almost equal electron numbers ($Z_{Pb}=82$ and $Z_{Bi}=83$) which are about five times larger than those for sulfur ($Z_S=16$). The nuclear scattering lengths b_j of neutron diffraction for $PbBi_2S_4$ are $b_{Pb} = 9.40 \cdot 10^{-15}$ m, $b_{Bi} = 8.53 \cdot 10^{-15}$ m, and $b_S = 2.85 \cdot 10^{-15}$ m. Sulfur has again the smallest scattering power (by a factor of about three). The difference between the scattering lengths of lead and bismuth $\Delta b = b_{Pb} - b_{Bi} = 0.87 \cdot 10^{-15}$ m, about 10% of the individual values, could be sufficient for a precise study of the Pb/Bi distribution by neutron diffraction, if a large single crystal of more than 1 mm^3 would be available. The serious absorption problems in X-ray work on heavy atom compounds do not exist in general for neutrons.

A different possibility of contrast variation, the combination of X-ray and neutron diffraction information, is demonstrated by the example of the intermetallic compounds $(Mn_{1-x}Cr_x)_{1+\delta}Sb$, with $0 \leq x \leq 1$ (Reimers et al., 1982,[4]). This mixed system is of special interest due to its magnetic properties: competing magnetic interactions with isotropic ferromagnetic behaviour for $Mn_{1+\delta}Sb$ and a uniaxial antiferromagnetic structure for $Cr_{1+\delta}Sb$. It crystallizes in the hexagonal NiAs-type structure (space group: $P6_3/mmc$) with some additional partial occupation (≤ 0.14) of the interstitial site 2(d), see Fig. II.4. Conventional X-ray diffraction cannot differentiate between chromium ($Z_{Cr} = 24$) and manganese ($Z_{Mn} = 25$) on sites 2(a) and 2(d):

$$2(a) - 0,0,0; 0,0,1/2 \quad \text{and} \quad 2(d) - 2/3,1/3,1/4; 1/3,2/3,3/4,$$

but yields important information on the overall occupation probabilities $M = (Mn,Cr)$ of these two different sites: M_aM_dSb, where M_a stands for the occupation probability of site 2(a) and M_b for that of site 2(d). The Sb position is assumed to be fully occupied, thus serving as an internal standard.

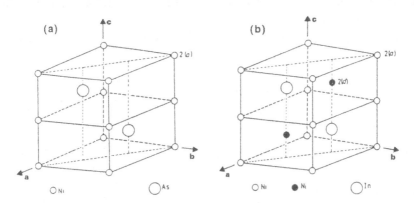

Figure II.4a. — NiAs structure   Figure II.4b. — Ni$_2$In structure (filled NiAs-type)

The corresponding nuclear scattering lengths of neutron diffraction are extremely different with a negative sign for the manganese:

$$b_{Cr} = 3.52 \cdot 10^{-15} \text{ m and } b_{Mn} = -3.73 \cdot 10^{-15} \text{ m}.$$

Remember: A positive value of b_j means that there is a phase shift of 180° between the incident and scattered neutron waves as a consequence of predominant potential scattering. The few negative b_j values - no phase change - result from resonant scattering.

The knowledge of the overall occupation probabilities M_a and M_d - from conventional X-ray studies - allows the evaluation of the Cr: Mn ratios of the different sites 2(a) and 2(d) from the corresponding effective scattering lengths determined by neutron diffraction. In the structure analyses based on the neutron data $b_{eff} = b_{Mn} \cdot PP$ is obtained individually for the two sites (PP_a and PP_d stand for refined pseudo occupation probabilities). According to

$$b_{eff}(2a) = a[(1-y) \cdot b_{Mn} + y \cdot b_{Cr})] \quad \text{and} \quad b_{eff}(2d) = d[(1-z) \cdot b_{Mn} + z \cdot b_{Cr})]$$

we can calculate

$$y = [b_{eff}(2a)/a - b_{Mn}] / [b_{Cr} - b_{Mn}] \quad \text{and} \quad z = [b_{eff}(2d)/d - b_{Mn}] / [b_{Cr} - b_{Mn}].$$

The detailed site occupations lead to the general formula

$$(Mn_{1-y}Cr_y)_a(Mn_{1-z}Cr_z)_dSb$$

$$\text{site 2(a)} \qquad \text{site 2(d)}$$

appropriate to a chemical composition of $Mn_{(1-y)a + (1-z)d} Cr_{y \cdot a + z \cdot d} Sb$. It is evident, that the individual (Cr,Mn) distribution on the two crystallographically different sites 2(a) and 2(d) is not accessible by a chemical analysis. For most of the samples studied, the site 2(a) was found to be fully occupied: $a \approx 1.0$. But the formula used normally $(Mn_{1-x}Cr_x)_{1+\delta}Sb$ is only correct for the special case of equal Cr: Mn ratios on both sites: $x = y = z$ and $1+\delta = a+d$.

The detailed information on the (Cr,Mn) distribution is needed to explain the magnetic properties of these intermetallic compounds, for which only the spins localized on the 2(a) sites are involved in the magnetic ordering (see e.g. the complex magnetic phase diagram in Fig. II.5). An overall Cr: Mn ratio from chemical analysis is not sufficient.

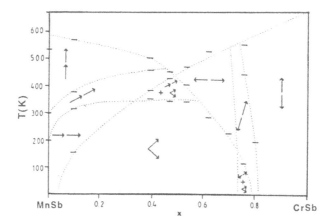

Figure II.5. — Magnetic phase diagram of the system MnSb - CrSb[4]. The vectors indicate the spin orientations in the different magnetic structures.

In general, a mixed occupation of one crystallographic site with <u>three</u> kinds of scatterers - i.e. Mn, Cr, and "vacancies" - requires at least <u>two</u> independent and sufficiently different experimental quantities to determine the fractional occupancies. The two-wavelength method (δ-synthesis), as an application of synchrotron X-ray radiation in crystal structure analysis using anomalous dispersion, would be likewise adequate to solve the (Mn, Cr)-Sb problem. However, the Mn-K absorption edge with its rather high wavelength of $\lambda_K = 1.8964$ Å implies a serious limitation of the data range of accessible Bragg intensities: $(\sin\theta/\lambda)_{max} \approx 0.5$ Å^{-1}, leading to a reduced precision of the analysis.

II.3. The H/D problem

The determination of the structure parameters of hydrogen atoms is a special problem involving different aspects of X-ray and neutron diffraction. It is obvious that H/D atoms with Z = 1 give only a small contribution to the electron density and, therefore, they are hardly visible in X-ray structure analyses. This holds especially when heavy atoms are present. But there is a more general problem: the single electron of H/D is engaged in the chemical bonding and is not localized at the proton/deuteron position. This position, however, is of importance when hydrogen bonds - eventually related to the lattice dynamics or structural phase transitions - are discussed.

The investigation of electron deformation densities on molecular systems has attracted great interest in order to understand better chemical bonding in crystalline solids (this application will be discussed later). Such studies consist normally of an experimental part and model calculations, for which proton/deuteron positions are required when H/D atoms are considered.

One of the most important fields of neutron diffraction is the determination of H/D sites and of their Debye-Waller factors.

Remember: the scattering lengths of the proton and the deuteron are $b_H = -3.74 \cdot 10^{-15}$ m and $b_D = 6.67 \cdot 10^{-15}$ m, respectively. Their magnitudes are comparable to the average of all b_j magnitudes and, therefore, H/D can be considered as "normal" atoms for neutron diffraction. The different signs of b_H and b_D may be of interest in Fourier maps for contrast reasons. Experimental conditions like background and effective absorption are strongly affected by the huge and exceptional incoherent neutron scattering cross section of hydrogen ($\sigma_{inc}(H) = 79.7$ barns for unbound H; for H in a crystal structure the cross section depends on the strength of the bonding and on the wavelength and is typically between 40 and 50 barns). Very often deuterated compounds are preferred in order to profit by the larger b_D value, but mainly to reduce the background from incoherent scattering ($\sigma_{inc}(D) = 2.0$ barns). This volume dependent background becomes crucial for neutron powder diffraction experiments, for which normally sample volumes of more than 1 cm^3 are required.

As an example for a study of a variety of hydrogen bonds, where the structure model was established by conventional X-ray analysis and neutron diffraction served especially to localize the hydrogen atoms, the case of fully deuterated $Na_2S \cdot 9D_2O$ (Preisinger et al., 1982,[5]) was chosen. Its crystal structure (non-centrosymmetric space group: $P4_122$ or $P4_322$) is dominated by discrete $[Na(D_2O)_5]$ and $[Na(D_2O)_4]$ spiral chains which are built up from $Na(D_2O)_6$ octahedra (Fig. II.6).

There are five different water molecules with O-D distances between 0.949 Å and 0.983 Å, and D-O-D angles from 104.6° to 107.5°. These water molecules are involved furthermore in six different O-D...S bridges to the S^{2-} ions, which are completely hydrated and show a slightly disordered icosahedral coordination by 12 D atoms (Fig. II.7).

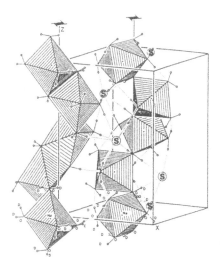

Figure II.6. — Na$_2$S·9D$_2$O: A partial view of the crystal structure[5].

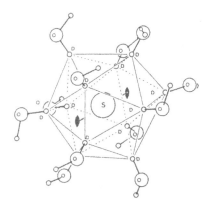

Figure II.7. — Na$_2$S·9D$_2$O: View of the icosahedral coordination of sulfur[5].

Details of the various O-D...O/S hydrogen bridges shown in Fig II.8 are summarized in Table II.2. This information was combined with results of Raman spectroscopy from which the uncoupled O-D(H) stretching frequencies could be reasonably well assigned to the nine different O-D(H) bonds of the crystal structure.

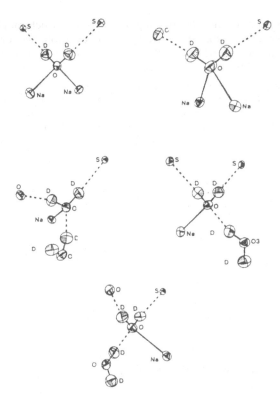

Figure II.8. — Coordination of the D_2O molecules in $Na_2S\cdot 9D_2O^{[5]}$.

A	B	C	A-B	B-C	A-C	∠BAC	∠ABC	∠BAB'	∠CAC'	L	A-L	∠LAL'
O(1)	-D(1)...S		0.961(7)	2.359(5)	3.319(5)	1.4(4)	178.0(6)	106.3(7)	103.4(2)	Na(1)	2.411(4)	116.1(2)
	-D(1')...S		0.961(7)	2.359(5)	3.319(5)	1.4(4)	178.0(6)				Na(1')	2.411(4)	
O(2)	-D(21)...O(5)		0.964(7)	1.793(7)	2.752(7)	4.9(4)	172.4(6)	106.1(7)	111.5(2)	Na(2)	2.588(5)	97.6(2)
	-D(22)...S		0.962(7)	2.550(5)	3.506(5)	5.2(4)	172.8(6)				Na(2')	2.380(5)	
O(3)	-D(31)...S		0.977(7)	2.311(5)	3.284(5)	4.7(4)	173.3(5)	107.5(7)	116.9(2)	Na(1)	2.397(5)	104.8(2)
	-D(32)...O(4)		0.953(7)	1.797(7)	2.730(7)	9.6(4)	165.3(6)				O(5)	2.768(7)	
O(4)	-D(41)...S		0.983(7)	2.294(5)	3.274(4)	3.4(4)	175.1(5)	104.6(6)	104.1(2)	Na(2)	2.418(5)	105.5(2)
	-D(42)...S'		0.973(7)	2.359(5)	3.333(5)	0.3(4)	179.6(5)				O(3)	2.730(7)	
O(5)	-D(51)...O(3)		0.949(7)	1.838(7)	2.768(7)	9.2(4)	166.1(6)	105.5(6)	103.4(2)	Na(1)	2.485(5)	101.7(2)
	-D(52)...S		0.967(7)	2.441(5)	3.401(5)	5.7(4)	172.1(5)				O(2)	2.752(7)	
mean values			0.965					106.0	107.9	<Na-O>	2.447	
mean values O-D...O			0.955	1.809	2.750		167.9					
mean values O-D...S			0.970	2.386	3.353		175.2					

Table II.2. — Interatomic distances (Å) and angles (°) for the hydrogen bonds and the ligands to the water molecules in $Na_2S\cdot 9D_2O^{[5]}$.

The hydrogen problem is of special importance for structural phase transitions driven by proton ordering. As a well known example we present here the ferroelectric transition in KH_2PO_4 (KDP). A characteristic feature of its crystal structure, see Fig. II.9, is the PO_4 groups linked by hydrogen bonds. At room temperature KDP crystallizes in a tetragonal para-electric phase (space group: $I\bar{4}2d$), where the protons in the O⋯H⋯O bonds are dynamically disordered. At T_c = 122 K, KDP transforms to a ferroelectric phase of orthorhombic symmetry (space group: Fdd2) in which the protons order in short asymmetric O-H...O bonds.

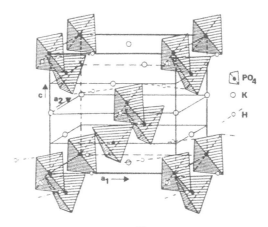

Figure II.9. — Structure of KH_2PO_4 (after West, 1930[6]).

R.J. Nelmes and coworkers have studied this hydrogen ordering very carefully by high resolution neutron diffraction at a series of temperatures near T_c: structure analyses were performed on Bragg intensity data measured with a small neutron wavelength of ≈ 0.55 Å up to a $\sin\theta/\lambda$-limit of 1.6 Å^{-1}. The contour plots of the refined proton distributions are shown in Fig. II.10. Above T_c, the two proton sites in the O⋯H⋯O bond are symmetrically equivalent (related by the two-fold axis of $I\bar{4}2d$).

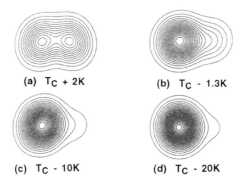

Figure II.10. — Sections through the refined proton distributions in KH_2PO_4 at: (a) T_c + 2 K, (b) T_c - 1.3 K, (c) T_c - 10 K, and (d) T_c - 20 K[7].

Below T_c, one of the sites becomes increasingly occupied at the expense of the other. The importance of the hydrogen ordering for the phase transition of KDP can be also deduced from the large isotope effect on the ordering temperature. For fully deuterated KD_2PO_4 T_c increases to 229 K.

II.4. Molecular disorder

Disordered structures and pseudosymmetries related to dynamical reorientation and/or structural phase transitions are of great current interest. In principal, the dynamical disorder of molecules is due to the fact that the intermolecular bonds are very much stronger than the external ones between the molecular groups and the surrounding crystalline frame. It is obvious that the chemical bonding scheme predicts the symmetry of a crystal structure, and not the other way around. But we can state, however, that in the case of an incompatible point group symmetry of a molecule with respect to its site symmetry in the crystal structure, molecular disorder is the necessary consequence. In order to model the atomic density distributions correctly in a way to obtain physically meaningful potentials, very accurate Bragg intensities over a large $sin\theta/\lambda$ range are required. X-ray experiments are generally more restricted than neutron studies because of the $sin\theta/\lambda$ dependence of the atomic scattering factor f_j. Synchrotron X-rays, however, offer very high intensities and extremely sharp reflexion profiles and permit measurements of very small intensities at large $sin\theta/\lambda$ values.

As an example, related to the H/D problem, the dynamical disorder of the NH_3 group in the cubic high temperature phase of the metal hexamine halide $Ni(NH_3)_6I_2$ (space group: Fm3m) is presented. With the NH_3 tetrahedra (3m symmetry) on crystallographic sites of 4mm symmetry (Fig. II.11) it is obvious that they must be orientationally disordered. At 19.7 K, $Ni(NH_3)_6I_2$ undergoes a first order phase transition to a probably ordered rhombohedral low temperature modification[8].

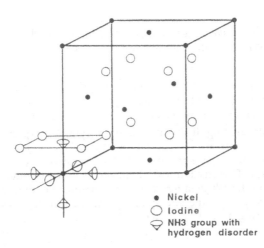

● Nickel
○ Iodine
▽ NH3 group with
   hydrogen disorder

Figure II.11. — High temperature structure of $Ni(NH_3)_6I_2$[8]. The hexamine coordination is shown only for the Ni atom at the origin.

Neutron diffraction studies at 35 K and 295 K by Schiebel et al. (1993)[9] revealed a planar proton density distribution perpendicular to the four fold axes (Fig. II.12). Its four maxima are directed towards the neighbouring iodines according to the influence of N-H...I bonding. This H density can be explained as a consequence of a coupled rotational-translational motion of the amine group. The model calculations, also shown in Fig. II.12, are in very good agreement with the experimental results.

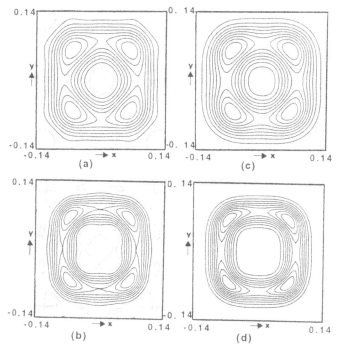

Figure II.12. — Ni(NH$_3$)$_6$I$_2$: Proton density in a [001] section at z = 0.23[9]. (a) and (b) experimental results at 295 K and 35 K (c) and (d) calculated densities at 295 K and 35 K.

Since the discovery of methods to produce and isolate fullerenes in macroscopic quantities in 1990[10], the field of chemical, physical and materials science research on this new molecular systems has exploded. Especially the very stable, almost spherical C$_{60}$ molecule (Fig. II.13) of icosahedral symmetry (point group: $\bar{5}$3m) has attracted a lot of interest. In the high temperature phase, C$_{60}$ crystals have a cubic closed packed structure (space group: Fm3m) characterized by a nearly free rotation of the molecules. This dynamical disorder is a consequence of the small intermolecular van der Waals interactions and the incompatibility of the molecular symmetry with respect to the site symmetry (m3m) in the crystal structure. An orientational ordering phase transition occurs at 261 K.

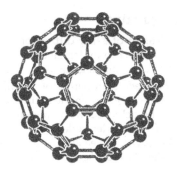

Figure II.13. — View of a C_{60} molecule along its five-fold axis.

The small but significant anisotropic density distribution on the spherical surface of the C_{60} molecule at room temperature was determined independently by different X-ray (conventional and synchrotron experiments) and neutron diffraction studies on sublimation grown single crystals (Chow et al., 1992,[11] and Papoular et al., 1993,[12]). The stereographic projection of the anisotropic part of the electron density in Fig. II.14 shows clearly four lobes nearby the binary {110} axes, indicating preferred carbon orientations, and a section of a density deficit around the [111] direction.

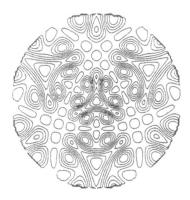

Figure II.14. — Stereographic projection along [111] of the electron probability density distribution of the C_{60} molecule at room temperature evaluated from the spherical harmonic coefficients $C_{l,v}$ refined in the structure analyses of our conventional X-ray data. The spherical term $C_{0,1}$, describing the dominant isotropic distribution ($\approx 95\%$), is omitted (after Ref.[12]).

The partial orientational order of the C_{60} molecules resulting from electron and nuclear densities are in very good agreement. It can be described[12] by a single C_{60} molecule, oriented in a way that a hexagon is slightly tilted around a binary axis away from the [111] direction, and the application of all symmetry operations of the cubic point group m3m.

II.5.   Electron density analysis

A very important application of X-ray diffraction is the investigation of the electron density distribution in order to understand better the chemical bonding in the crystalline state. These studies require high quality experimental structure factors over a large $\sin\theta/\lambda$ range. The specific properties of synchrotron radiation, i.e. high intensity and small divergence, can improve the accuracy of weak and very weak reflections beyond what can be achieved with conventional X-ray equipment (a comparison of the profiles of high order reflections is shown in Fig. II.15). This essential advantage can, however, be easily offset by some instability of the synchrotron source and the related experimental and data processing problems.

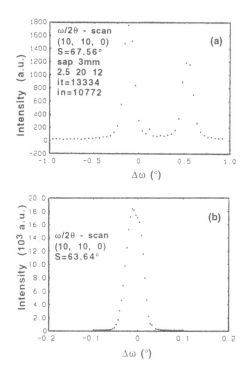

Figure II.15. — $\omega/2\theta$-scan profiles of the high order (10,10,0) reflection on an almost perfect Si crystal (spherical sample of 0.2 mm diameter) recorded using:

(a) a standard CAD4 diffractometer (Enraf-Nonius) and MoKα radiation from a pyrolithic graphite monochromator. The splitting in two well separated peaks at the large scattering angle $\theta = 67.56°$ is due to the presence of the two wavelength components $\lambda_{K\alpha_1} = 0.70926$ Å and $\lambda_{K\alpha_2} = 0.71354$ Å. A total scan width of $\Delta\omega \approx 2°$ is required to obtain integrated Bragg intensities; each peak has a full width at half maximum (f.w.h.m.) of $\Delta\omega = 0.18°$.

(b) the four-circle diffractometer WDIF4C at the LURE synchrotron (Orsay), equipped with a Si (111) double-monochromator, operated at a wavelength of $\lambda = 0.6886$ Å. Even at $\theta = 63.64°$ the peak shape remains well defined and narrow ($\Delta\omega=0.043°$ (f.w.h.m.)), and a total scan width of $\Delta\omega \approx 0.2°$ is sufficient.

Recently, Kirfel and Eichhorn (1990)[13] have performed a feasibility study on two well known reference structures, cuprite (Cu_2O) and corundum (α-Al_2O_3), in order to assess the accuracy of single crystal X-ray diffraction data obtained with synchrotron radiation.

The special cubic structure of Cu_2O with a primitive lattice and two formula units per unit cell (space group: Pn$\bar{3}$m) is built up by a body-centered sublattice of oxygen and a face-centered sublattice of copper (Fig. II.16):

O   (2a) - 0,0,0; 1/2,1/2,1/2

Cu (4b) - 1/4,1/4,1/4; 3/4,3/4,1/4; 3/4,1/4,3/4; 1/4,3/4,3/4.

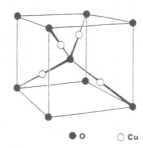

● O   ○ Cu

Figure II.16. — Crystal structure of cuprite (Cu_2O).

For a harmonic structure model of non-deformed electron densities of the atoms, there are only contributions of Cu to the structure factor of (hkl) reflections with all indices h,k, and l of the same parity (consequence of the F sublattice of copper). The oxygens of the I sublattice contribute only to (hkl) reflections, where the sum of h+k+l = 2n (even). The occurence of additional "forbidden" reflections is related to some deformation of the dynamical electron density distribution. (In the case of nuclear densities from neutron diffraction, the observation of such "forbidden" reflections indicates directly anharmonic effects). For Cu_2O, we can distinguish the structure factors and the Bragg intensities of four groups of reflections (e and o stands for even and odd indices):

(eee) reflections - Cu and O contributions of the harmonic model (centered multipolar density distributions); strong intensity

(ooo) reflections - in addition to the dominant centered Cu contributions there are non-centered contributions of O; strong intensity

(ooe) reflections - centered contributions of O and non-centered of Cu; weak intensity

(eeo) reflections - only non-centered contributions of Cu and O; "forbidden" reflections; very weak intensity.

It was shown for Cu_2O[13], that a thorough structure and electron density analysis requires accurate measurements especially of the so-called oxygen (ooe) reflections and the "forbidden" (eeo) reflections. The details of the electron density structure depend mainly on the "forbidden" reflections whose intensities cannot be measured using X-ray tube radiation.

According to Kirfel and Eichhorn[13] we can state the following general rules: synchrotron diffraction yields no gain for strong and medium intensity reflections with respect to standard X-ray measurements, since their signal/noise ratios do not profit from the small beam divergency. This does not hold in the case of selected measurements of strong reflections at various wavelengths in order to improve the quality of the data by experimental assessment of extinction effects. But, an important improvement of data accuracy can be achieved by increasing the reliability of weak reflections. There are almost no "unobserved" reflections in synchrotron X-ray diffraction experiments! More precisely measured weak reflections will not only decrease the statistical noise in the Fourier maps and reduce spurious peaks at special sites of high symmetry, but will also alter the weighting scheme in favour of the high-order reflections.

We can conclude from this work that <u>the combination of conventional and synchrotron X-ray measurements is a powerful possibility to obtain very complete and highly accurate diffraction data</u>. For these combined experiments the same crystal of typically 0.3 mm diameter, which would be normally too large for synchrotron X-ray diffraction, could be used.

Electron density studies on simple molecular crystals, where theoretical calculations for isolated molecules are possible, are of special interest in order to compare experimental and theoretical results. Molecular crystals consist normally of light atoms often including hydrogen. Therefore, only small absorption effects have to be taken into account for the X-ray data. A combination with neutron diffraction experiments is important to determine the structure parameters of the H/D atoms properly. More generally, the structure analysis by neutron diffraction yields separately and independently from the X-ray data the structure parameters of all atoms including the mean square displacements due to static and dynamic (even anharmonic) effects. This complete information can be used in a so-called X-N synthesis to obtain experimental electron deformation densities from the measured X-ray Bragg intensities.

As an example, we report here on a combined X-ray and neutron diffraction study of 4-methylpyridine (C_6H_7N) at 120 K (Ohms et al., 1985,[14]). C_6H_7N shows a sequence of structural phase transitions, which is not yet completely understood, but is probably related to dynamical reorientation of the methyl group:

$$P2_1/c\ (?) \xrightarrow[\longleftarrow]{T_{c1}\approx 100\ K} I4_1/a \xrightarrow[\longleftarrow]{T_{c2}=254\ K} I4_1/amd\ (?) \xrightarrow[\longleftarrow]{T_{mp}=276.8\ K} \text{liquid state}$$

The structure model was determined by direct methods from a conventional X-ray diffraction experiment (space group: $I4_1/a$ with 8 molecules per unit cell). A special challenge for both the X-ray and the neutron analyses is the merohedral twinning of all sample crystals at 120 K. The interchange of the axes $\mathbf{a}_1$ and $\mathbf{a}_2$ of the tetragonal lattice leads to a perfect superposition of (hkl) and (kh$\bar{l}$) reflections, which are not equivalent. The measured intensities are thus $I_{obs}(hkl) = V_1 \cdot I(hkl) + V_2 \cdot I(kh\bar{l})$, where V_1 and V_2 are the volume parts of the two twinning domain orientations with $V_1 + V_2 = 1$.

An ORTEP plot of a 4-methylpyridine molecule is shown in Fig. II.17. The molecular arrangement in the structure is plotted in Fig. II.18. The orientational disorder of the CH_3 group around the two-fold axis of the molecule, demonstrated in Fig. II.19, resulted from the neutron experiment.

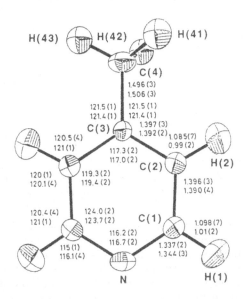

Figure II.17. — ORTEP plot of a C_6H_7N molecule including bond lengths (Å) and angles (°) at 120 K, upper values deduced from neutron data and lower values from X-ray data, e.s.d. in parentheses[14].

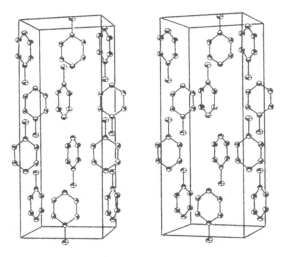

Figure II.18. — C_6H_7N: Stereographic plot of the molecular arrangement in the unit cell. The hydrogen atoms are omitted[14].

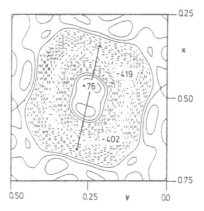

Figure II.19. — C$_6$H$_7$N: proton density map of the CH$_3$ group from neutron data in a plane perpendicular to the two-fold axis at z = -0.05. The projection of the corresponding pyridine ring is marked by a line[14].

The experimental dynamic electron deformation density of C$_6$H$_7$N was determined from the conventional X-ray data and atomic scattering factors f_j for the free atoms using the refined structure parameters of the neutron diffraction analysis (X-N synthesis). A theoretical dynamic electron deformation density map was calculated for the free 4-methylpyridine molecule for which the atomic parameters were also taken from the neutron analysis. A good agreement between experimental and theoretical charge densities was found for all bonding and lone pair regions (Figs. II.20 and II.21). The differences near C(4) are due to the ordered methyl group of the free 4-methylpyridine molecule in the theoretical calculations.

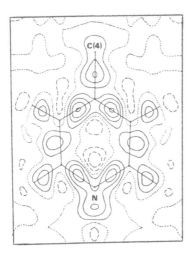

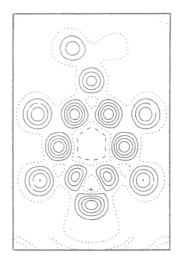

Figure II.20. — X-N deformation density map in the plane of the C$_6$H$_7$N molecule; contour line interval = 0.1 e$^-$/Å^3, negative contours = dashed lines, zero contour = dotted line[14].

Fig. II.21. — Theoretical dynamic deformation density map in the plane of the C$_6$H$_7$N molecule; contours as in Fig. II.20[14].

II.6.   The absolute configuration of non-centrosymmetric structures

It is well known that the structure factor of a molecular configuration (I)

$$F_I(h) = \sum_j f_j(h) \cdot \exp[2\pi i(h \cdot r_j)] \cdot T_j(h)$$

is equal to the conjugate complex $F_{II}^*(h)$ of its enantiomeric configuration (II), where all positions r_j of configuration (I) are transformed to $-r_j$ for (II), if only the real part of the atomic scattering factors $f_j(h)$ is considered:

$$F_{II}(h) = \sum_j f_j(h) \cdot \exp[-2\pi i(h \cdot r_j)] \cdot T_j(h) = F_I^*(h).$$

The measured diffraction intensity $I(h) \propto F(h) \cdot F^*(h) = |F(h)|^2$ and, therefore,

$$I_I(h) = I_{II}(h), \text{ or - according to Friedel's law - } I(h) = I(-h).$$

In case of anomalous diffusion the atomic scattering factor becomes complex

$$f_j(h) = f_{oj}(h) + f_j' + i f_j'',$$

and

$$F_{II}(h) = \sum_j \{f_{oj}(h) + f_j' + i f_j''\} \cdot \exp[-2\pi i(h \cdot r_j)] \cdot T_j(h)$$

$$\neq \sum_j \{f_{oj}(h) + f_j' - i f_j''\} \cdot \exp[-2\pi i(h \cdot r_j)] \cdot T_j(h) = F_I^*(h).$$

The intensities of the two enantiomeric configurations are no longer equivalent, $I_I(h) \neq I_{II}(h)$, and the intensities of Friedel pairs are also systematically different, $I(h) \neq I(-h)$.

The determination of the absolute configuration of a non-centrosymmetric structure from the measurement of so-called Bijvoet pairs, $I(h)$ and $I(-h)$, depends on the quantity of the imaginary correction term $f_j''(\lambda)$ of the anomalous dispersion. The possibility of continuous wavelength variation of X-ray synchrotron radiation can be important to increase $|f_j''(\lambda)|$.

The determination of the absolute configuration of sodium uranyl(VI) triacetate, $Na[UO_2(C_2H_3O_2)_3]$, will be shown as an example of a conventional study afterwards in section (II.8).

II.7.   The phase problem

The measured Bragg intensities $I(h)$ from diffraction experiments yield only the amplitudes of the structure factors, $|F(h)| \propto \sqrt{I(h)}$ with

$$F(h) = \sum_j f_j(h) \cdot \exp[2\pi i(h \cdot r_j)] \cdot T_j(h) = |F(h)| \cdot \exp[i\varphi(h)],$$

and not their phases $\varphi(h)$, which are needed for the Fourier synthesis of a direct structure determination. This phase problem is of special importance for non-centrosymmetric structures where $\varphi(\mathbf{h})$ is not restricted to 0 ($\exp[i \cdot 0] = +1$) and π ($\exp[i \cdot \pi] = -1$), but all values between 0 and 2π are possible.

Fischer (1987)[15] has developed the so-called "λ-technique" as an approach to the phase problem of structure determination. Using the wavelength dependence of the anomalous dispersion correction terms $f'(\lambda)$ and $f''(\lambda)$ close to an absorption edge, we can consider a crystal structure composed of one edge-atom species (anomalous scatterer) and the rest being normal-atoms. In the simplest case 3 (or 4) wavelengths λ_i are selected, for which the following conditions hold:

(1) $f_n' = $ const. and $f_n'' \cong 0$ for all n-atoms and all wavelengths λ_i

(2) $f_e'(\lambda_1) = f_e'(\lambda_2) \neq f_e'(\lambda_3)$ and $f_e''(\lambda_1) \neq f_e''(\lambda_2) = f_e''(\lambda_3)$ for the e-atoms

(3)   the differences $f_e'(\lambda_3) - f_e'(\lambda_2)$ and $f_e''(\lambda_1) - f_e''(\lambda_2)$ should be very large.

For PbBi$_2$S$_4$, treated in section (II.1), and the e-atom Pb, the wavelengths indicated by E, F, and G in Fig. II.2 could be chosen:

$$f_{Pb}'(E) = f_{Pb}'(F) \neq f_{Pb}'(G) \text{ and } f_{Pb}''(E) \neq f_{Pb}''(F) \cong f_{Pb}''(G).$$

But this is not a favorable case because of the important variation of $f_{Bi}'(\lambda_i)$ for the n-atom Bi, Bi being a neighbouring element to Pb.

The Patterson function, which is the Fourier transform of the reflection intensities, $I(\mathbf{h})$ or $|F(\mathbf{h})|^2$, is directly available from the experiment

$$P(\mathbf{u}) = 1/V^2 \cdot \sum_h |F(\mathbf{h})|^2 \cdot \exp[-2\pi i(\mathbf{h} \cdot \mathbf{u})].$$

From the difference of reflection intensities measured at different wavelengths λ_i and λ_j

$$\Delta F^2_{ij}(\mathbf{h}) = |F_i(\mathbf{h})|^2 - |F_j(\mathbf{h})|^2$$

a difference Patterson function can be calculated with $\Delta F^2_{ij}(\mathbf{h})$ as Fourier coefficients

$$\Delta P_{ij}(\mathbf{u}) = 1/V^2 \cdot \sum_h \Delta F^2_{ij}(\mathbf{h}) \cdot \exp[-2\pi i(\mathbf{h} \cdot \mathbf{u})],$$

which becomes complex for a non-centrosymmetric structure

$$\Delta P_{ij}(\mathbf{u}) = \Delta P_{cij}(\mathbf{u}) - i \cdot \Delta P_{sij}(\mathbf{u}).$$

The real part $\Delta P_{c12}(\mathbf{u})$ from data at λ_1 and λ_2 contains sharp peaks representing vectors between e-atoms only (e $\rightarrow$ e). The imaginary part $\Delta P_{s12}(\mathbf{u})$, added to the real part $\Delta P_{c23}(\mathbf{u})$ from data at λ_2 and λ_3, consists of the vectors from the e-atoms to all other n-atoms (e $\rightarrow$ n), the reverse n $\rightarrow$ e vectors being suppressed. With this information a starting model for structure refinement, similar to the case of the "heavy-atom" method, can be obtained.

If standard methods are not sufficient to establish a structure model, the "λ-technique" may be a complicated way (at least 3 complete data sets measured at a synchrotron are required) to solve crystal structures.

An experimental determination of phases using three-beam multiple X-ray diffraction was developed by Hümmer and coworkers (1986/1989)[16] and Chang &

Tang (1988)[17]. The rocking curve of an azimuthal ψ-scan around a reciprocal lattice vector **h** is analysed in the vicinity of a three-beam diffraction case, where in addition to the origin of the reciprocal lattice **O** simultaneously two further nodes corresponding to **h** and another vector **g** lie on the Ewald sphere. The interference between the directly diffracted wave of **h** and the "Renninger Umweg" waves of **g** and **h-g** causes a modification of the ψ-scan profile related to the dynamical theory of X-ray diffraction. Hümmer et al.[16] reported on the determination of the triplet phase sum $\Phi_\Sigma = \varphi(-h) + \varphi(g) + \varphi(h-g)$ of the structure factor product $F(-h) \cdot F(g) \cdot F(h-g)$. Tang & Chang (1988)[18] compared multiple diffraction profiles measured with a conventional X-ray source and synchrotron radiation. They obtained relative phases with errors as small as 15°.

II.8.   The anisotropy of anomalous dispersion in X-ray diffraction

Anomalous X-ray scattering is normally treated in diffraction experiments as an isotropic effect. However, it was pointed out already in 1980 by Templeton and Templeton[19] that this does not hold when dichroism occurs. With an experimental synchrotron study on sodium uranyl(VI) triacetate, $Na[UO_2(C_2H_3O_2)_3]$, they demonstrated in 1982[20] that the anisotropy of anomalous dispersion can be observed in Bragg intensities, and that it agrees with results derived from X-ray absorption measurements using linear polarized synchrotron radiation. They stated that this effect "adds a new dimension of complexity to the theory of X-ray scattering" which may become an opportunity for extracting more information on the electron structure from diffraction experiments.

· The example of sodium uranyl(VI) triacetate, chosen by the Templetons[20], is very interesting because of complementary information, which are available from conventional X-ray diffraction, from polarised anomalous X-ray scattering, and from neutron diffraction. The crystal structure of this cubic compound was obtained from X-ray film data by the "heavy-atom" method as early as 1935. A redetermination was performed more recently by Templeton et al. (1985)[21], who used a standard 4-circle diffractometer and MoKα radiation. At room temperature $Na[UO_2(C_2H_3O_2)_3]$ crystallizes in the non-centrosymmetric space group P2₁3 with 4 formula units per unit cell of a=10.689(2) Å; a view of the structure around the central U atom is shown in Fig. II.22. The absolute configuration of the chiral molecular structure of sodium uranyl(VI) triacetate, which produces negative optical rotation, was obtained by the study of Bijvoet pairs. From these data, the anomalous-scattering term for uranium f"$_U$ at the wavelength λ(MoKα) was refined to 9.7(2) e⁻ per atom, which is in good agreement with 9.654 e⁻ per atom calculated by Cromer and Liberman (1970)[22] for $\lambda(MoK\alpha_1) = 0.70930$ Å.

Near the L₃ absorption edge of uranium, Templeton and Templeton (1982)[20] determined the anomalous X-ray scattering terms f' and f" from diffraction measurements on $Na[UO_2(C_2H_3O_2)_3]$ at five wavelengths as a function of the relative polarization direction of linearely polarized synchrotron radiation with respect to the crystal orientation. As a prerequisite for this delicate investigation an untwinned single crystal is needed, for which the enantiomeric purity is verified.

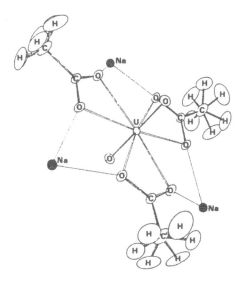

Figure II.22. — Molecular structure of $Na[UO_2(C_2H_3O_2)_3]$ around the central U atom from neutron diffraction data at room temperature indicating the disorder of the CH_3 groups[23].

Using a double-monochromator, consisting of two independent Si (220) crystals, the Templetons achieved an energy spread of just 1 eV (equivalent to $\Delta\lambda = 0.00004$ Å) and a stability of about 1.5 eV (or $\Delta\lambda = 0.00006$ Å) during the measuring time of 12 h at one wavelength. The polarization effects break the symmetry, and each reflection intensity of the data collections had to be treated as an independent observation with its own calculated structure factor. This holds for the symmetry-equivalent reflections from the cubic structure of $Na[UO_2(C_2H_3O_2)_3]$ and even for the same reflection measured at different azimuthal angles.

For an anisotropic atom the scattering factor becomes a tensor $\mathbf{f}$. In sodium uranyl(VI) triacetate each uranyl ion lies on a threefold axis and thus $\mathbf{f}$ must be uniaxial. According to[20]

$$\mathbf{f} = [f_\pi, 0, 0; 0, f_\pi, 0; 0\ 0\ f_\sigma],$$

if z is the molecular axis, with the principal values

$$f_\sigma = f_0 + f'_\sigma + if''_\sigma \text{ and } f_\pi = f_0 + f'_\pi + if''_\pi$$

parallel and perpendicular to the symmetry axis, respectively.

The independent parameters used are the spherical averages

$$f' = \left(f'_\sigma + 2f'_\pi\right)/3 \text{ and } f'' = \left(f''_\sigma + 2f''_\pi\right)/3$$

to compare with ordinary isotropic values, and the anisotropic terms

$$f'_2 = f'_\sigma - f'_\pi \text{ and } f''_2 = f''_\sigma - f''_\pi.$$

The refined anisotropic anomalous scattering terms are listed in Table II.3. Both, f' and f'' change with polarization direction by as much as 2 electrons. The anisotropy parameters f'_2 and f''_2 for the uranyl ion are plotted in Fig. II.23 together with the results derived from X-ray absorption measurements on rubidium uranyl nitrate, $RbUO_2(NO_3)_3$.

λ (Å)	0.72177	0.72170	0.72157	0.72147	0.72124
f'	-19.4 (3)	-19.1 (3)	-16.0 (4)	-14.3 (4)	-13.4 (3)
f'_2	-0.2 (3)	-0.1 (3)	-1.7 (4)	-1.8 (4)	-0.7 (3)
f'_σ	-19.5 (4)	-19.2 (4)	-17.1 (5)	-15.5 (5)	-13.9 (4)
f'_π	-19.3 (3)	-19.1 (3)	-15.4 (4)	-13.7 (4)	-13.2 (3)
f''	10.2 (4)	13.1 (3)	16.1 (5)	13.2 (6)	12.2 (5)
f''_2	-1.7 (3)	-2.1 (3)	-1.1 (3)	0.1 (4)	1.7 (3)
f''_σ	9.1 (4)	11.7 (4)	15.4 (5)	13.3 (7)	13.3 (5)
f''_π	10.8 (4)	13.8 (3)	16.5 (5)	13.2 (6)	11.6 (5)
R (%)	3.4	3.8	3.8	4.4	3.8
n^*	723	1125	643	869	759

* Number of reflections.

Table II.3. — Anisotropic anomalous scattering terms from diffraction experiments on $Na[UO_2(C_2H_3O_2)_3]$[20].

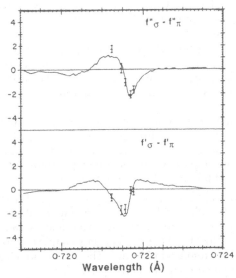

Figure II.23. — Polarization anisotropy of f' and f'' for the uranyl ion near the L_3 edge, determined from diffraction experiments on $Na[UO_2(C_2H_3O_2)_3]$ (points with error bars) and absorption measurements on $RbUO_2(NO_3)_3$ (continuous curves)[20].

A recent neutron diffraction analysis on $Na[UO_2(C_2H_3O_2)_3]$ essentially completes the structure information concerning the H atoms (Navaza et al., 1991[23]).

The refined structure parameters of the non-hydrogen atoms are in reasonable agreement with those of the X-ray study[21]. For the hydrogens of the CH$_3$ group, however, a disordered distribution was found with two equally probable CH$_3$ orientations, which are related by a non-crystallographic local binary axis perpendicular to the uranyl ion passing through C1 and C2 (see Fig. II.22). It is obvious again that even very careful X-ray structure analyses cannot localize hydrogen atoms properly.

II.9.   Experimental constraints

Diffraction experiments for structure analyses on single crystals require very stable conditions (wavelength, polarization, intensity) for relatively long and precise intensity measurements. Conventional X-ray diffractometers and neutron facilities have achieved a high standard. It is still a challenge for the use of synchrotron radiation to guarantee an overall reliability of better than 2% for a whole data set of (relative) Bragg intensities.

An X-ray synchrotron source can offer high intensities at a small beam divergency for a continuous wavelength variation at energies from 5-60 keV (corresponding to $0.2 < \lambda < 2.5$ Å). A (non-exhaustive) list of specific advantages for diffraction experiments, some of which have been discussed already in the frame of this chapter, includes:

- the measurement of very small reflection intensities,
- the use of very small crystals (microcrystals) of μm-size, which may be obtained easier as homogeneous, high quality samples,
- the use of short wavelengths in order to reduce absorption and extinction effects as well as to perform high resolution structure analysis, where data up to large $\sin\theta/\lambda$ are required,
- the combination of small crystals and small wavelengths (low absorption) is very important for experiments with external parameters like temperature, high pressure, electrical and magnetic fields, in-situ irradiation, etc.,
- the use of the anomalous X-ray scattering for contrast variation, phase determination, etc.,
- small and well defined reflection profiles permit a variety of more detailed studies concerning such different subjects as complex systems with very large periodicities (e.g. large molecular compounds and incommensurate structures), systematic extinctions (space group symmetry), twinning, lattice parameter variations, separation of thermal diffuse scattering, etc..

The intensity of neutron beams of present and proposed sources (high flux reactors or spallation facilities) will remain a very important constraint on the size of suitable single crystals for neutron diffraction experiments. Depending on chemical composition, space group, lattice parameters, and sample environment, the required minimum size can vary from 0.5 to 20 mm^3. There is no principal problem to apply external parameters in neutron diffraction experiments.

REFERENCES

[1]   Wulf R., 1990 - Acta Cryst. **A46**, 681.

[2]   Iitaka Y.,Nowacki W., 1962 - Acta Cryst. **15**, 691.

[3]   Cromer D.T., Liberman D., 1981 - Acta Cryst. **A37**, 267.

[4]   Reimers W., Hellner E., Treutmann W., Heger G., 1982 - J. Phys. C: Solid State Phys. **15**, 3597.

[5]   Preisinger A., Mereiter K., Baumgartner O., Heger G., Mikenda W., Steidl H., 1982 - Inorg. Chem. Acta **57**, 237.

[6]   West J., 1930 - Z. Kristallogr., Kristallgeom., Kristallphys., Kristallchem. **74**, 306.

[7]   Nelmes R.J., Kuhs W.F., Howard C.J., Tibballs J.E., Ryan T.W., 1985 - J. Phys. C: Solid State Phys. **18**, L711.

[8]   Eckert J., Press W., 1980 - J. Chem. Phys. **73**, 451.

[9]   Schiebel P., Hoser A., Prandl W., Heger G., Schweiss P., 1993 - J. Phys. I France **3**, 987.

[10]   Krätschmer W., Lamb L.D., Fostiropoulos K., Huffman D.H., 1990 - Nature **347**, 354.

[11]   Chow P.C., Jiang X., Reiter G., Wochner P., Moss S.C., Axe J.D., Hanson J.C., McMullan R.K., Meng R.L., Chu C.W., 1992 - Phys. Rev. Lett. **69**, 2943.

[12]   Papoular R.J., Roth G., Heger G., Schiebel P., Prandl W., Haluska M., Kuzmany H., 1993 - Int. Winterschool on Electr. Prop. of Novel Materials: "Fullerenes and Related Compounds", Kirchberg, Austria , 6. 3. - 13. 3.

[13]   Kirfel A., Eichhorn K., 1990 - Acta Cryst. **A46**, 271.

[14]   Ohms U., Guth H., Treutmann W., Dannöhl H., Schweig A., Heger G., 1985 - J. Chem. Phys. **83**, 273.

[15]   Fischer K.F., 1987 - Z. Kristallogr. **179**, 77.

[16]   Hümmer K., Billy H., 1986 - Acta Cryst. **A42**, 127.

        Hümmer K., Weckert E., Bondza H., 1989 - Acta Cryst. **A45**, 182.

[17]   Chang S.L., Tang M.T., 1988 - Acta Cryst. **A44**, 1065.

[18]   Tang M.T., Chang S.L., 1988 - Acta Cryst. **A44**, 1073.

[19]   Templeton D.H., Templeton L.K., 1980 - Acta Cryst. **A36**, 237.

[20]   Templeton D.H., Templeton L.K., 1982 - Acta Cryst. **A38**, 62.

[21]   Templeton D.T., Zalkin A., Ruben H., Templeton L.K., 1985 - Acta Cryst. **C41**, 1439.

[22]   Cromer D.T., Liberman D., 1970 - J. Chem. Phys. **53**, 1891.

[23]   Navaza A., Charpin P., Vigner D., Heger G., 1991 - Acta Cryst. **C47**, 1842.

CHAPTER III

SURFACES AND INTERFACES: X-RAY STUDIES
M. SAUVAGE-SIMKIN

III.1. Introduction

These lectures are dedicated to the structural aspect of surfaces and interfaces as studied by X-ray diffraction based techniques.

In a first part, the specificity of surfaces from the point of view of crystallography is presented, then a second section describes how X-rays can be made surface sensitive by use of the particular geometry of grazing incidence. However, since the reconstruction in the surface layer is usually accompanied by subsurface atom displacements, both parallel and normal to the surface plane, grazing incidence data have to be complemented by out-of-plane measurements. Accordingly, a short survey of the experimental requirements to be fulfilled by a surface diffraction equipment is given.

The problem of data collection and analysis is then discussed in connection with the quasi two-dimensional (2D) character of the diffracting object and the method for solving surface structures is illustrated with selected examples of clean surfaces and adsorbates in the monolayer range.

In the last sections, the applicability of X-ray diffraction to the study of physical properties in superficial layers, from a few Å to several hundreds of Å is demonstrated for cases as different as surface phase transitions, strained heterostructures or buried interfaces.

The reader may find useful complements to these lecture notes in recently published review articles (Robinson 1991, Feidenhans'l 1989, Robinson and Tweet 1992a).

III.2. Surface structural characteristics

The real surface of a clean material kept in ultra-high vacuum is never the mere truncation of the bulk atomic configuration by a plane parallel to the surface nominal orientation. The surface atoms experience a different bonding scheme which results in a variety of effects: motions normal to the surface plane changing the lattice spacing at the surface which is said to be relaxed or (and) displacements parallel to the plane leading to a novel symmetry for the surface unit cell, the surface is then said to be reconstructed. In materials with highly directional bonding such as all usual semiconductors, the reconstruction process aiming to minimize the dangling bond density is critically dependent on the surface orientation: Fig. III.1 is a [$\bar{1}$10] projection of the diamond or Zinc blende structure showing the dangling bond (DB) configuration at low index surfaces. It is easy to guess from this figure that dimerisation of surface atoms is likely to occur on (001) surfaces whereas adatoms saturating the outward single DB/atom can be expected on (111) surfaces (see sections III-6 and III-7). In addition, reconstruction may often be related to a change of stoichiometry at the surface: missing rows, ordered vacancies or different chemical composition for compound materials. New surface periodicities are also produced by adsorbates in the sub- monolayer and monolayer regime as for example in metal-semiconductor interfaces.

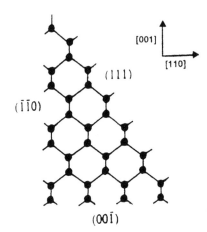

Figure III.1. - [110] Projection of a cubic diamond structure showing the dangling bond configuration on the low index surfaces.

The surface unit cell is depicted by the label "m x n" where m and n are numbers relating the new basis vectors to those of a bulk derived cell:

$$a_1 \text{ surface} = m \times a_1 \text{ volume} \qquad a_2 \text{ surface} = n \times a_2 \text{ volume}$$

To the change in the direct space periodicity corresponds a change in reciprocal space which governs the diffraction properties of surfaces. The detailed analysis of 2D or near 2D diffraction patterns is presented in section III.4.

In addition, it is impossible in practice to prevent the occurence of steps on the surface. When the average orientation is strictly parallel to a dense crystallographic plane, the steps are randomly distributed and their density depends on the quality of the surface preparation. The influence of the step distribution on the diffraction pattern has been discussed by several authors (Andrews et al. 1985, Vlieg et al. 1989). When the surface is cut on purpose at an angle from a simple crystallographic orientation, periodic distributions of steps can be produced.

III.3.   X-Rays and Surfaces

Classically X-rays in the 10 KeV range are considered a highly penetrating probe, contrary to low energy electrons or Helium atoms. However, since most materials have a refraction index n for hard X-rays lower than unity, there exists a critical incidence angle α_c below which total external reflection occurs. Whenever photoelectric absorption can be neglected, the crystal waves become evanescent and their penetration is drastically reduced. Formulae for estimating the relevant parameters are given below:

refraction index $$n = 1 - \delta - i\beta \qquad \text{(III.1)}$$

with $$\delta = \left(e^2 / mc^2\right) \lambda^2 \sum_j \left(Z_j + f'_j\right) / 2\pi V \qquad \text{(III.2)}$$

and $$\beta = \left(e^2 / mc^2\right) \lambda^2 \sum_j f''_j / 2\pi V \qquad \text{(III.3)}$$

where the summation is extended over all the atoms in the unit cell of volume V; Z_j, f_j and f''_j are respectively the atomic number, the real and the imaginary parts of the dispersion correction of atom j.

The critical incidence angle and the penetration depth Λ, defined as the $1/e$ attenuation of the transmitted intensity, are then given by:

$$\alpha_c = \sqrt{2\delta} \qquad \text{(III.4)}$$

and $$\Lambda = \lambda / 4\pi . \, Im\left(\alpha_i^2 - \alpha_c^2 - 2i\beta\right)^{1/2} \qquad \text{(III.5)}$$

where α_i is the incidence angle. Since δ is of the order of 10^{-5}, α_c lies in the mrad range. A typical curve is presented in Fig. III. 2 which shows that Λ can be restricted to a few tens of Å.

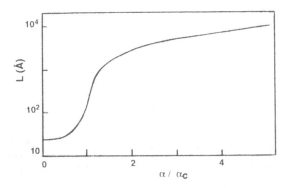

Figure III.2. — X-ray penetration depth as a function of the reduced grazing incidence angle: GaAs, λ=1.488 Å.

In addition, the transmissivity and the reflectivity of the surface are critically dependent on α_i and their variations are given by the well known Fresnel formulae:

$$R^2(\alpha_i) = \left| \frac{\sin\alpha_i - \sqrt{n^2 - \cos^2\alpha_i}}{\sin\alpha_i + \sqrt{n^2 - \cos^2\alpha_i}} \right|^2 \qquad T^2(\alpha_i) = \left| \frac{2\sin\alpha_i}{\sin\alpha_i + \sqrt{n^2 - \cos^2\alpha_i}} \right|^2 \qquad (III.6)$$

Figure III.3 shows the transmission function in reduced coordinates for non-absorbing and absorbing media. Since the transmitted wave acts as the incident wave for the surface diffraction process, it is clear from this figure that when the highest intensity is needed in an experiment for instance when a moderately powerful X-ray source is used (or when very high resolution is required), it will be

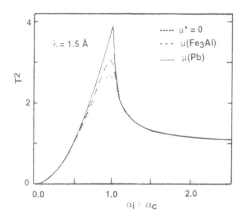

Figure III.3. — Fresnel Transmissivity for transparent and absorbing media (Dosch 1987).

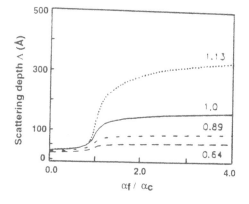

Figure III.4. — Scattering depth variation for several values of the incidence angle calculated for Fe_3Al with $\lambda = 1.5$ Å (Dosch et al. 1988).

profitable to work at $\alpha_i = \alpha_c$, but great care should then be taken to keep the incident angle strictly constant during the whole data collection. When the incident photon intensity is large enough, it is more convenient to record data around $2\alpha_c$ although the bulk related background is much larger. A thorough treatment of X-ray propagation in crystals at grazing incidence has been proposed by Vineyard (1982).

Another way to increase the surface sensitivity and accordingly the depth resolution of the method is to use grazing exit diffraction since refraction effects also take place when the crystal diffracted waves leave the sample. A so-called diffraction depth Λ_d has been introduced by Dosch et al. (1986) and is given as a function of the incident and exit angles α_i and α_f by

$$\Lambda_d = \lambda / 4\pi \ (l_i + l_f) \tag{III.7}$$

with

$$l_{i,f} = \sqrt{2} \left[\left(\alpha_c^2 - \alpha_{i,f}^2 \right) + \left[\left(\alpha_c^2 - \alpha_{i,f}^2 \right)^2 + (2\beta)^2 \right]^{1/2} \right]^{1/2} \tag{III.8}$$

Exit angle resolved diffraction enables depth profiling in crystal with a resolution of about 20 Å (Fig. III.4).

III.4.   Diffraction from Surfaces

III.4.1.   Basic Principles

In the kinematical approximation of single scattering (see Warren 1969), valid for hard X-rays and small objects, the intensity $I(Q)$ elastically scattered in a direction defined by a momentum transfer Q is proportional to the square modulus of the coherent addition of the amplitudes scattered by all the electrons in the diffracting object. In the particular case of a periodic array of atoms built with N_1, N_2, N_3 unit cells it takes on the form:

$$I(Q) \alpha \left| \sum_{n_1=1}^{n_1=N_1-1} e^{in_1q_1a_1} \sum_{n_2=1}^{n_2=N_2-1} e^{in_2q_2a_2} \sum_{n_3=1}^{n_3=N_3-1} e^{in_3q_3a_3} \right|^2 |F_{hkl}|^2 \tag{III.9}$$

where a_1, a_2, a_3 are the basic vectors of the unit cell, q_1, q_2, q_3 are the vector components of Q in the reciprocal space basis a_1^*, a_2^*, a_3^*.

$$Q = q_1 + q_2 + q_3 = ha_1^* + ka_2^* + la_3^* \text{ with } a_i^* = 2\pi \frac{(a_j x a_k)}{(a_i, a_j, a_k)} \tag{III.10}$$

and F_{hkl} is the structure factor of the unit cell defined by:

$$F_{hkl} = \sum_j f_j \ e^{2\pi (hx_j + ky_j + lz_j)} \ e^{-M_j} \tag{III.11}$$

The summations extends over all the atoms in the unit cell; f_j, x_j, y_j, z_j, M_j are respectively the scattering factor, fractional coordinates in the unit cell and Debye-Waller factor of atom j. Formula (III.9) can be written:

$$I(Q) \propto \frac{\sin^2(N_1 a_1 q_1 / 2)}{\sin^2(a_1 q_1 / 2)} \frac{\sin^2(N_2 a_2 q_2 / 2)}{\sin^2(a_2 q_2 / 2)} \frac{\sin^2(N_3 a_3 q_3 / 2)}{\sin^2(a_3 q_3 / 2)} |F_{hkl}|^2 \quad (III.12)$$

It can be shown that constructive interference is only observed around integer values of the hkl coordinates.

However, in the case of a reconstructed surface considered from now on normal to the a_3 bulk axis, to the "m x n" expansion of the direct space corresponds a "1/m x 1/n" contraction of the reciprocal space: diffraction from this 2D periodic object may then take place also at nodes h/m, k/n, l. Due to the lack of periodicity in the normal direction, the summation on n_3 disappears in formula (III.9) and no condition is to be applied on the index l which can be considered as a continuous variable. Diffraction from a perfect 2D reconstructed surface is then rod shaped with a monotonous decrease in intensity as θ increases along the rod (Fig. III.5), the occurence of l modulations in the rod intensity is an evidence for a finite thickness of the reconstructed layer.

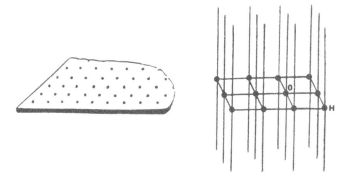

Figure III.5. — Direct (left) and reciprocal (right) lattice of a 2D periodic object.

Formula (III.11), establishing the connection between the diffracted intensity and the atomic arrangement in the unit cell, is the basis for structure analysis by X-ray diffraction.

If one considers first the intensity collected at very grazing incidence (l close to zero) fractional orders can thus be measured which carry the information on the projected atomic coordinates in the surface cell; for integer orders, h multiple of m and k multiple of n, diffraction from the bulk is superimposed with the surface contribution and some analysis is needed to extract the information which relates here to the registry of the surface layer with respect to the underlying bulk.

Modelling the rod intensity variation with l for fractional orders gives information on the z coordinates of the atoms in the reconstructed layer. More generally, rod profiles are needed whenever z related properties are investigated: surface and interface roughness, subsurface atom displacements, stacking sequences.

III.4.2. Integrated Intensity

It is clear from formula (III.12) that the diffracted intensity is distributed around the reciprocal lattice points in a finite ΔQ^3 range which depends on the size of the coherently diffracting object. Moreover, in the case of slightly misoriented diffracting domains (mosaic spread) an additionnal broadening is observed in the intensity distribution. It is of primary importance, for further data analysis, to collect the whole intensity around each node. For the in-plane extension, this is practically achieved by summing the counts produced by rocking the sample in front of a sufficiently wide opened detector. However further corrections need to be applied to these experimental integrated intensities to derive a set of numbers proportional to F_{hkl}:

- normalisation versus a constant incident photon flux
- correction for the Lorenz factor: $1/\sin 2\Theta$
- correction for the size of the irradiated area: $1/\sin 2\Theta$

when the area is no longer determined by the slits but by the sample edges, which may happen at small diffraction angles, a specific correction has to be included.

- absorption correction from the Beryllium windows when required by the geometry of the diffractometer
- corrections due to the anisotropy of the resolution function of the instrument: this point will be discussed in more details when the components of an X-ray surface diffraction set-up are introduced.

A schematic of the vector geometry in a GIXD experiment is shown in Fig. III.6.

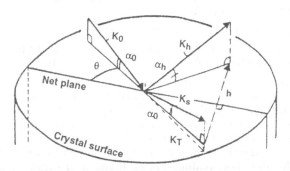

Figure III.6. — Incident K_0, transmitted K_t, specular K_s and diffracted K_h wave vectors.

III.5.   Experimental Set-up

III.5.1.   Beam Line

Since the surface signals are usually small, the beam line should be designed to deliver an incident beam optimized for the geometry, the angle and energy spreads required by the experiment. Standard optical elements could be:
-        a toroidal or a single bend mirror
-        a double crystal monochromator with either a flat or a horizontally focusing second crystal

These elements define the angular divergences $\delta\theta_m$ (in-plane) and $\delta\alpha$ (out of plane) which enter in the instrumental resolution function. The energy spread $\Delta E/E$ brings an additional broadening. Typical values for perfect crystal monochromators and standard mirrors are:

$$\delta\theta m \approx \text{a few } 10^{-4} \text{ rad} \qquad \delta\alpha \approx \text{a few } 10^{-3} \text{ rad} \qquad \text{and } \Delta E/E \approx \text{a few } 10^{-4}$$

These values can be degraded on purpose to increase the photon flux (broad band-pass monochromators). Conversely when a good definition of the angles is required (α resolved experiments) non focused radiation ought to be used.

III.5.2.   The X-ray diffractometer

A surface or interface diffraction experiment adds to the classical requirements of crystallographic data collection several specific constraints:
-        Ultra-high vacuum (UHV) around the sample for clean surfaces and adsorbate structure determination.
-        Control of the incidence angle with respect to the surface or interface plane from grazing (a few mrad) to finite values (several degrees).
-        Optional control of the exit angle.

When UHV is not required, standard 4-circle diffractometers complemented with an adjustment of the α_i angle and a proper displacement of the detector out of the surface plane (or a linear position sensitive detector) can be used.

For UHV studies, the sample chamber must be equipped with vacuum tight beryllium windows which brings a limitation to the reciprocal space exploration particularly in the l direction. Such a chamber can be made portable in order to be transferred from a surface preparation station to a heavy-duty diffractometer (Feidenhans'l 1988). The other choice is an UHV compatible diffractometer. Several solutions have been adopted at the existing synchrotron radiation facilities and their respective merits and shortcomings can be found in dedicated instrumentation papers (Fuoss and Robinson 1984, Vlieg et al. 1987, Claverie et al. 1989).

III.5.3.   The Detector and the Resolution Function

Several types of detectors can be used: scintillation counters have a large linear counting rate but a poor background rejection capability; they are recommended behind a crystal analyzer in the case of a strongly diffracting sample.

Solid state detectors on the contrary are very efficient for separating the diffraction signal from fluorescence or any parasitic background occuring at an energy different from the fundamental wavelength, however their linear range is limited and great care should be taken to account for the dead time of the detection system. They are well adapted for surface structural data collection.

Position sensitive detectors have proved extremely useful for performing rod intensity profiles and exit angle resolved experiments.

The in-plane aperture $\delta(2\theta)_d$ of the detector enters in the longitunal in-plane component of the resolution function. When a good in-plane resolution of the detector is achieved, a co-operative rotation of the sample and detector (θ–2θ) makes it possible to record a scan of the in-plane reciprocal node along the $\bar{Q}$ vector direction (radial scan) whereas a mere sample rotation provides a transverse scan. Once deconvoluted from resolution broadening, the full width at half maximum (FWHM) $\Delta\theta$ of a scan is related to the $1/\Delta Q$ correlation length L_c through the formulae

$$L_c = \frac{\lambda}{2\Delta\Theta_r \cos\theta} \qquad (III.13)$$

and

$$L_c = \frac{\lambda}{2\Delta\Theta_t \sin\theta} \qquad (III.14)$$

for radial and transverse scans respectively.

The out-of-plane aperture $\delta\alpha_f$ determines the normal component $\Delta Q_\perp$ of the resolution function over which the intensity of the diffraction rod is integrated. This length must be equal for all hkl measurements, which is achieved when the rod does not rotate with respect to the detector aperture; whenever the inclination of the rod changes, a special correction has to be included(Robinson 1988).

III.5.4.   Computer control

Collecting surface diffraction data requires bringing both a given reciprocal lattice vector Q in reflection conditions and the normal N to the surface at a selected angle from the incident beam. Each set-up provides a number of degrees of freedom to achieve this goal and adapted computer codes have been developed. For 4-circle instruments, implemented versions of the basic procedure proposed by Busing and Levy (1967) are available (Robinson 1989, Mochrie 1988), whereas the mathematics in use for a 5-circle instrument is given by Vlieg et al. 1987.

III.6.   Structural Data Analysis

Once a set of observed structure factor moduli F_{hkl} has been obtained, together with their standard deviations σ_{hkl} estimated from counting statistics and reproducibility of symmetry equivalent reflections, the structure determination process can start. The first step is to produce a model close enough to the real structure to be least-square refined in a second step. In some cases, such a model is already available from other investigations (Scanning tunnelling microscopy-STM-, Electron diffraction and spectroscopies ...) but the refinement based on X-ray diffraction data has been shown to improve the accuracy of the structure description, particularly concerning the atomic displacements in the subsurface layers.

III.6.1.   Patterson Maps

In the absence of model, the methods classically developed for 3D crystallography can be applied: Patterson and Fourier Difference methods (see Lipson and Cochran 1966). A brief description of the Patterson function method will be given here. Since the phase information is missing in the data, it can be shown that a Fourier transform of the experimental structure factor moduli will not give the electron density map in the unit cell, $\rho(r)$, but the autocorrelation function of $\rho(r)$. Its section in the 2D direct space is referred to as the Patterson map and is expressed as:

$$P(x,y) = \sum_h \sum_k |F_{hk}|^2 e^{2\pi i(hx+ky)} \qquad (III.15)$$

or

$$P(\Delta r) = \int_{unit\ cell} \rho(r) \cdot \rho(r + \Delta r)\ dr \qquad (III.16)$$

with $\qquad\qquad\qquad\qquad \Delta r = xa_1 + ya_2$

A peak in the Patterson map means that two atoms in the unit cell are separated by the vector Δr. Inspecting the various peaks, keeping in mind the assumed atomic content of the unit cell and the expected bonding constraints, one may then try to build a reasonable model. However, the accuracy of the method relies not only on the quality but also on the number of independent structure factor values and since in most cases, surface data set are rather limited, one must be very careful in interpreting the features of the Patterson map. Still this analysis has proved very useful for surface structure determination particularly when heavy atoms are present. For example, in the Sb/Ge (111) 2x1 reconstructed surface, van Silfhout et al. (1992) were able to decouple the study of the Sb atoms arrangement in the very top layer ,which dominates the fractional order Patterson map, from that of the registry and subsurface distortions of the substrate derived from the in-plane and out of plane analysis of integer orders.

The second step of the structure analysis consists in a least-square refinement of the model parameters. The quality of the fit can be assessed either by the reliability factor R:

$$R = \frac{\sum |F_{obs} - |F_{calc}||}{\sum F_{obs}} \qquad (III.17)$$

or by the χ^2 value, where the errors and the number of fitted parameters p are included:

$$\chi^2 = \frac{1}{N-p} \sum \frac{\left(F_{obs} - |F_{calc}|\right)^2}{\left(\sigma_{hk}\right)^2}$$

(III.18)

In both formulae, the summations extend over the N independent measured values.

A satisfactory surface structure determination should yield a precision of a few 10^{-2} Å on the projected atomic coordinates. This accuracy depends on the range in reciprocal space which is accessible and is of the order of .05 x $(2\pi/Q_{max})$. Due to the constraints specific to UHV diffractometers, the accuracy is three to four times better for in-plane than for normal coordinates.

The next sections will be dedicated to a survey of recent applications, representative of the potentialities of X-ray methods, with a due account of the input from other surface investigation techniques.

III.7. Semiconductor Clean surfaces and Interfaces

III.7.1. (111) orientation in elemental semiconductors

As mentioned previously, adatoms saturating the dangling bonds are the basic elements in the reconstruction or interface formation processes on these surfaces. Examples concerned with Ge(111) substrates will be described in some details since they illustrate the major aspects of X-Ray diffraction application and display general characters.

The annealed Ge (111) c(2x8) surface has been studied by LEED (Phaneuf et al. 1985) and STM (Becker et al. 1989), but the interpretation of these data still left open questions: is it a simple adatom structure, and if such, what kind of adatom site is occupied ? The X-ray diffraction data collected in-plane (Feidenhans'l et al. 1988) and out of plane (van Silfhout et al. 1992) were able to bring the answers:

A simple adatom distribution is responsible for the new symmetry (Fig. III.7); moreover, the assignment of a T4 site for the adatom (above an atom of the second bulk layer), inferred from the displacement pattern produced by the in-plane data refinement, was fully confirmed by the rod scan analysis.

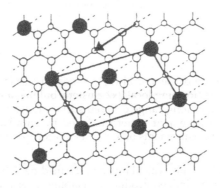

Figure III.7. — Asymmetric unit of the Ge (111) c2x8 reconstructed surface: adatoms (shaded disks) in T4 sites; large circles: first bulk layer; small circles second bulk layer (after Feidenhans'l et al. 1988).

The Si (111) 7x7 reconstructed surface is more complex, Surface X-ray diffraction
(Robinson et al. 1988)) and reflectivity (Robinson and Vlieg 1992)) experiments were
able to confirm the DAS (Dimer-Adatom-Stacking fault) model proposed by
Takayanagi (1983) and also observed with STM (Hamers et al. 1986) adding valuable
quantitative information on the subsurface displacements down to the sixth
subsurface layer. A comparison between the $10l$ rods recorded in the Ge (111) 2x8
surface and in a Si(111) 7x7 interface buried under amorphous silicon (Robinson et
al. 1986) shows the sensitivity of the method to the presence of the stacking fault in
the silicon subsurface layers, responsible for the deep minimum at l = -2.5 (Fig. III.8).

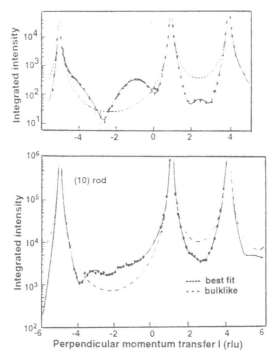

Figure III.8. — $10l$ rod; lower panel: Ge(111) 2x8 (after van Silfhout et al. 1992) upper panel:
Si(111) 7x7 /a-Si (after Robinson et al. 1986).

The common feature of Ge and Si (111) annealed surfaces is thus a top layer
with adatoms on T4 sites, forming local 2x2 clusters, but major differences are found
in the coupling with other long range ordered surface defects. Figure III.9 is a side
view of the general displacement pattern induced by an adatom in a T4 site and
derived from rod scan measurements on both clean reconstructed surfaces and metal
interfaces.

Total energy minimization calculations performed on a 2x2 unit cell by Meade
and Vanderbilt (1989) have indeed produced an atomic arrangement close to the one
displayed in Fig. III.9. However, a more accurate treatment with the c(2x8)
germanium unit cell (Takeuchi et al. 1992) was able to account for the full symmetry
of this surface.

III.7.2.   Metal/ (111) semiconductor interfaces

X-ray diffraction has enabled the solution of a large number of interface structures in these systems which can be divided as follows.

III.7.2.1.   Metal adatoms in T4 sites

The systems concerned are the $\sqrt{3} \times \sqrt{3}$ R30° interfaces formed in the low coverage regime on Si and Ge (111) by non reactive metals such as Pb, Sn (for a review see Feidenhans'l 1989) or In (Finney et al. 1993). A similar local configuration is also found in the Sn/Ge (111) 5x5 and 7x7 interfaces (Pedersen 1988).

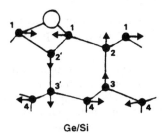

Ge/Si

Figure III.9. — Side view of the atomic displacements induced by an adatom on a T4 site in the cubic diamond lattice (after Pedersen 1988).

As mentioned above, the displacement pattern induced by the adatom is quite close to the one shown in Fig. III.9 and has been deduced from rod scan analysis. In most cases, a strain energy minimization procedure introduced by Keating (1966) has been coupled to the least squares fitting of the experimental data, to compensate for the limited set of measured independent structure factors, in order to derive the atomic positions with a reasonable confidence.

III.7.2.2.   Metal trimers

At 1ML coverage Au and Ag form complex $\sqrt{3} \times \sqrt{3}$ R30° interfaces with Si or Ge (111) which have given rise to a huge amount of controversial results. It is now admitted that the structures can be described by metal trimers arranged on a honeycomb pattern and possibly imbedded in an incomplete Si or Ge layer. The metal-metal distance in the trimer is close to the bulk value for gold interfaces (Dornisch et al. 1992, Howes et al. 1993) but much larger in the case of Ag/Si (111) interfaces (Vlieg et al. 1991) where it is preferable to talk of metal triangles instead of trimers. For these surfaces, the necessity of gathering information from several investigation techniques - STM, photoelectron diffraction, X-Ray standing wave and surface diffraction- was clearly demonstrated.

III.7.2.3.   Reactive Interfaces: Silicide Formation

With more reactive metals such as 3d transition metals or rare earths, epitaxial silicides are formed on annealing. The state of strain in the epilayer and the interface bonding scheme can be derived through rod scan measurements in specular and non-specular conditions. The method has been successfully applied to the interfaces NiSi₂/Si (111) Robinson et al. 1988) and PdSi₂/Si(111) (Robinson et al. 1992b). X-ray diffraction has also been applied to unravel the epitaxial relationship for silicides with complex structures such as β FeSi2/Si (111) (Jedrecy et al. 1993). For heterointerfaces, the coupling of X-ray methods with high resolution transmission electron microscopy (HRTEM) has proved extremely fruitful.

III.8. The (001) surfaces and interfaces of elemental and compound semiconductors

Dimerisation is the basic feature in these surfaces. For Si (001) (Jedrecy et al. 1990a) and Ge (001) (Rossmann et al. 1992) 2x1 reconstructed surfaces it is now admitted that asymmetric dimers are present on the surface.

The variety of phases is much larger in compound semiconductor surfaces since the stoichiometry of the top layers can be adjusted through the growth parameters (substrate temperature, partial pressures, etc ...).

The example of GaAs(001) is particularly well known in connection with the MBE growth of III-V heterostructures. The symmetry and composition of the various surface phases have been characterized by RHEED and electron spectroscopies (Massies et al. 1980). However, the first atomic structure determination was obtained by grazing incidence X-ray diffraction (Sauvage-Simkin et al. 1989) on the As rich c(4x4) surface. A mixture of two ordered structures involving clusters of two or three As dimers, chemisorbed on an As top layer in a c(4x4) square lattice, was proposed (Fig. III.10). The existence of more than one ordered atomic arrangement is quite frequent on semiconductor surfaces. In simple cases where the reciprocal lattices are not completely superimposed, X-ray diffraction is most suitable for extracting the spectra from individual surface phases (Etgens et al. 1991).

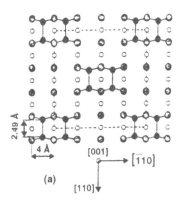
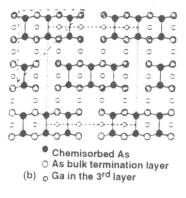

● Chemisorbed As
○ As bulk termination layer
(b) ○ Ga in the 3rd layer

(a)

Figure III.10. — Stable atomic structures on the GaAs (001) c(4x4) arsenic saturated surface (after Sauvage-Simkin et al. 1989).

A transition towards a lower As concentration in the top layers is observed on annealing above 350°C in vacuum or above 450°C under an As pressure, the new symmetry being 2x4. GIXD data recorded during the annealing on diffraction lines pertaining to either phase (Fig. III.11) enable the monitoring of the transformation which appears as a continuous order-disorder transition followed by nucleation and growth of the new ordered phase (Etgens 1991).

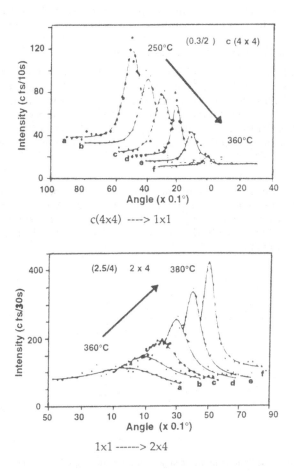

Figure III.11. — Variation with temperature of the strongest diffraction lines in the c(4x4) and 2x4 phases of GaAs(001). Curves are shifted along the axes for the sake of clarity.

This transformation can be made reversible by the application of an As pressure of about 10-6 mBar on the sample. When the atomic supply may come from the bulk by out-diffusion, reversible transitions are observed even under UHV conditions as for instance on CdTe (001) (Etgens et al. 1993a).

Dimers are also observed with interfaces in the monolayer coverage regime as for instance with As/Si(001) 2x1 (Jedrecy et al. 1990) and Sb/Ge(001) 2x1 (Lohmeier et al. 1992). However, in some cases, bridging atoms are found as structural features responsible for the 2x1 symmetry (Te/GaAs (001) 2x1 - Etgens et al. 1991 or CdTe (001) 2x1 - Lu et al. 1991).

III.9.   Surface Roughness

Steps on the surface will give rise to specific features in the diffraction pattern. A random step distribution can be modeled as a diffuse solid- vacuum interface and the corresponding intensity has been calculated both for out of plane (Robinson 1986) and in-plane (Vlieg et al. 1989) measurements.

A maximum sensitivity is achieved in so-called anti-Bragg conditions when adjacent terraces diffract out-of-phase. Such a situation has been selected by van Silfhout et al. (1989) to follow the roughness variation during MBE homoepitaxial growth of Ge (111) and by Fuoss et al. (1992) in the case of organometallic vapor phase epitaxy (OMVPE) of GaAs (001). Figure III.12 shows oscillations of the (1 1 0.05) diffracted beam sampled at 40 ms per data point, demonstrating the layer by layer growth mode of GaAs at the rate of 1ML/sec.

Periodic step distributions on vicinal surfaces are detected by diffraction satellites which can be treated in the same way as superlattice reflections in multilayer structures within the kinematical formalism. When different steps are present on the surface, monoatomic and diatomic steps for instance, an accurate description of the terraces can be derived from the details of the diffraction pattern as has been done by Renaud et al. (1992) for the Si(001) surface.

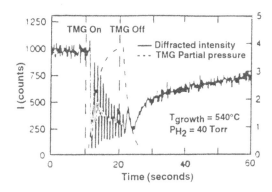

Figure III.12. — X-ray oscillations in anti-Bragg conditions during OMVPE growth of GaAs (after Fuoss et al. 1992).

The roughening transition on metal surfaces has been extensively studied by X-ray diffraction measurements. Evidence for the multiplication of steps at the roughening temperature has been obtained for Pt (110) by Robinson et al. (1989). A modelling of the expected lineshapes for the diffraction peaks in terms of the functional divergence of the height-height correlations on the surfaces at the transition, has been given and compared with experimental data on Ni(113) by Robinson et al. (1990).

III.10. Heteroepitaxy

The tunability of the X-ray penetration depth from a few tens of angströms to several microns finds an optimal use in the assessment of heterostructures: Bragg diffraction can be excited both in the epilayer and substrate, specific features arising from buried reconstructed interfaces can be detected.

III.10.1. Strain evaluation

Nowadays, systems of large technological interest couple materials with lattice mismatch greater than 4%: GaAs-Si (4%), Ge-Si (4%), ZnTe-GaAs (8%), CdTe-GaAs (14.6%). The transition between pseudomorphic (if any) and plastically relaxed growth takes place for epilayer thicknesses around ten Angströms. For such thin overlayers, GIXD performed in-situ on a diffractometer coupled to an MBE growth chamber is a unique technique to measure the in-plane lattice parameter mismatch $\varepsilon_{//}$ as a function of the deposited thickness according to (III.19) and thus to characterize fully the growth mode:

$$\varepsilon_{//} = -\Delta\theta \, \cotan\theta$$

(III.19)

where $\Delta\theta$ is the Bragg angle shift between the epilayer and substrate.

Two different behaviours have been observed: GaAs/Si [Jedrecy et al. 1990b)], does not present a 2D pseudomorphic growth step (Fig. III.13a) and partially relaxed islands are formed from the very beginning whereas a fully strained layer by layer growth takes place up to a specific critical thickness for both ZnTe/GaAs (Etgens et al. 1993b) (Fig. III.13b) and Ge/Si (Williams et al. 1991).

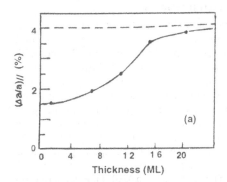
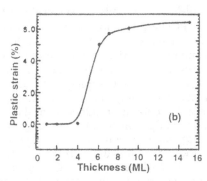

Fig.ure III.13. — Variation of the in-plane strain as a function of the deposited thickness. a) GaAs/Si heteroepitaxial growth; b) ZnTe/GaAs heteroepitaxy, the critical thickness is found at 5 ML.

By adjusting the probing depth with the grazing incidence angle, strain gradients were revealed in the epilayer in the intermediate phase following the on-set of 3D relaxed growth where a full coalescence of the islands is not completed (Fig. III.14).

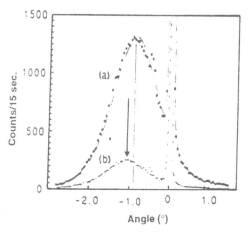

Figure III.14. — In-plane rocking curves of a 15ML (45 Å) ZnTe layer on GaAs recorded in bulk (a) and surface (b) sensitive conditions; reflection 200, λ=1.488 Å.

The range of pseudomorphic growth can be extended by the introduction of foreign atoms at the heterointerface: in the Ge/Si (001) system, the presence of Sb at the growing surface was shown to inhibit island formation leading to a layer by layer growth mode, as demonstrated by the X-ray reflectivity measurements performed by Thormton et al. (1992).

III.10.2.   Periodic structures at buried heterointerfaces

Plastic relaxation of the epitaxial strain is achieved through a network of interface defects. In some systems, the periodicity of the dislocation array is good enough to give rise to a specific diffraction pattern which can be collected with X-rays inclined at an incidence slightly larger than the critical value in order to penetrate through the epilayer. The atomic positions in the interface region can thus be derived through a structural analysis complemented by additional information since the amount of independent data is usually too small compared to the large number of parameters. By coupling the least squares fitting of the data with strain energy minimization conditions, Bourret et al. (1992a) were able to propose a model for the GaSb/GaAs (001) 14x14 reconstructed interface involving square protrusions of 2-3 GaAs layers in the GaSb region. In a second example dealing with the 2x9 supercell formed at the interface between CdTe (111) and GaAs(001), a combination of high resolution transmission electron microscopy (HRTEM) and X-ray diffraction produced a description where three layers build the interface region: a distorted (001)-like As layer, a (111)-like Ga layer and a (111)-like Te layer which is thus identified as the first one of the CdTe film (Bourret et al. 1992b).

A powerful way to enlarge the amount of independent data is to use anomalous scattering. The energy dependent parts f'_j and f''_j of the atomic scattering factors f_j appearing in formula III.11 suffer rapid variations on crossing absorption edges. This effect, already applied in protein crystallography (see Karle 1989 for a review), can be used to sort out the contribution of the various atomic species to the diffraction signal and thus to facilitate the structure determination. The first

successful application to a buried reconstructed interface has been published by
Tweet et al. (1992) and is concerned with the ordering induced by 1/3 ML of Boron
at the $\sqrt{3}\times\sqrt{3}$ R30° interface between a $Ge_xSi_{(1-x)}$ alloy and a 100 Å Si(111) cap. Third
order peaks collected at 31 energies across the Ge Kedge (Fig. III.15) made it possible
to draw partial Patterson maps for the interfacial Si and Ge which gave evidence for
a non random distribution of both species on the lattice sites. Such experiments are
made possible by the continuous spectrum delivered by synchrotron radiation
sources.

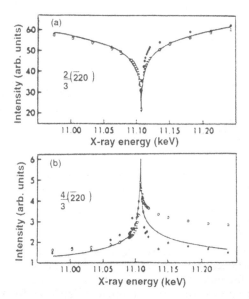

Figure III.15. — Typical energy scans across the Ge Kedge for two fractional reflexions (after Tweet et
al. 1992).

III.11.   Phase transitions at surfaces

As indicated in section III.3 an enhanced depth sensitivity is obtained through
exit angle resolved experiments. Phase transitions in alloys have been studied in
great details by this technique (Mailander et al. 1990). Concerning for example the
order-disorder transition in the Cu_3Au alloy (Dosch et al. 1988), diffraction data
recorded on the superstructure peak (110) show that the behaviour in near surface
layers is different from the bulk: the peak decay is apparently more progressive close
to the surface than in deeper layers. This could be interpreted as due to the
propagation of an order-disorder interface from the surface into the bulk over a
temperature range of about 30K. In particular the sweeping over the first 50 Å is
extended over 25K. Opposite behaviour is reported for the 111 surface of the same
alloy where order is seen to persist longer close to the surface (Zhu et al. 1988).

The method has been equally applied to characterize the onset of surface
melting on the Al (110) surface (Dosch et al. 1991). Three regimes are identified by
modeling the intensity and width of an evanescent Bragg peak as a function of

temperature (Fig. III.16): normal harmonic Debye-Waller behaviour up to 690°K, strong anharmonicity of surface thermal vibrations between 690°K and 770°K, growth of a liquid-like layer of increasing thickness up to the bulk melting temperature of 933°K. By probing the crystalline order in the near surface region, the results are complementary to those obtained by LEED, only sensitive to the very top layer or by ion blocking which measures the amount of disorder.

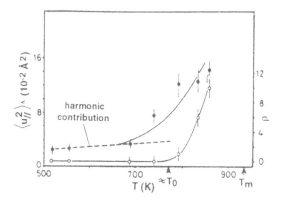

Figure III.16. — Mean-square vibrational amplitudes (full dots) and number of liquid-like layers (empty circles) as a function of temperature at the surface of Al(110) (after Dosch et al. 1991).

III.12. Conclusion

The aim of these lecture notes was to survey both practical and fundamental aspects of X-ray surface diffraction. Through a series of selected examples, it was intended to outline the unique information which can be derived from this method compared to other surface structural approaches such as STM: accurate atomic coordinates, registry between the surface layers and the underlying bulk, estimate of the distortions induced in subsurface layers and possibility of in-situ measurements over a broad temperature range. It was also emphasized that X-ray diffraction is most suitable to probe not only the top surface but also the near surface region, from a few Å to a few thousands of Å. However, the method requires a reasonable long range order and is rather time consuming the advent of the new sources could nevertheless make feasible real time kinetic surface studies.

REFERENCES

ANDREWS S.R., COWLEY R.A., 1985 - J. Phys. C, 18, 6247.

BECKER R.S., SCHWARZENTRÜBER B.S., VICKERS J.S., KLITSNER T., 1989 - Phys. Rev., B39, 1633.

BOURRET A., FUOSS P.H., 1992a - Appl. Phys. Lett., 61, 1034.

BOURRET A., FUOSS P.H., FEUILLET G., TATARENKO S., 1992b - Phys. Rev. Lett., 70, 311.

BUSING W.R., LEVY H.A., 1967 - Acta Crtys. 22, 457.

CLAVERIE P., MASSIES J., PINCHAUX R., SAUVAGE-SIMKIN M., FROUIN J., BONNET J., JEDRECY N. - Rev. Sci. Instrum., 60 (1989) 2369.

DORNISCH D., MORITZ W., SCHULZ H., FEIDENANS'L R., NIELSEN M., GREY F., JOHNSON R.L., LE LAY G., 1992 - Surf. Sci. 274, 215.

DOSCH H. - Phys. Rev.B35 (1987) 2137.

DOSCH H., BATTERMAN B.W., AND WACK P.C. - Phys. Rev. Lett., 56 (1986) 1144.

DOSCH H., MAILÄNDER L., LIED A., PEISL J., GREY F., JOHNSON R.L. AND KRUMMACHER S., 1988 - Phys. Rev. Lett. 60, 2382.

DOSCH H., HÖFER T., PEISL J. AND JOHNSON R.L., 1991 - Europhys. Lett., 15, 527.

ETGENS V.H., 1991 - Thèse de Doctorat, Université P. et M. Curie, Paris (France).

ETGENS V.H., PINCHAUX R., SAUVAGE-SIMKIN M., MASSIES J., JEDRECY N, GREISER N., TATARENKO S., 1991 - Surf. Sci., 251/252, 478.

ETGENS V.H., TATARENKOS S., BASSANI F., KLEIN J.C., SAMINADAYAR K., SAUVAGE-SIMKIN M., PINCHAUX R., - Proceeding of the 4th International Conf. on the Formation of Semiconductor Interfaces, Jülich (1993a).

ETGENS V.H., SAUVAGE-SIMKIN M., PINCHAUX R., MASSIES J., JEDRECY N., WALDHAUER A., TATARENKO S., JOUNEAU P.H., 1993b - Phys. Rev. B47, in press.

FEIDENHANS'L R., 1986 - PhD Thesis, RISO -M Reports 2569.

FEIDENHANS'L R., 1989 - Surf. Sci. Repts, 10, 105.

FEIDENHANS'L R., PEDERSEN J.S., BOHR J., NIELSEN M., GREY F., JOHNSON R.L., 1988 - Phys. Rev. B38, 9715.

FINNEY M.S., NORRIS C., HOWES P.B., VAN SILFHOUT R.G., CLARK G.F., THORNTON J.M.C., 1993 - Surf. Sci., in press.

FUOSS P.H., ROBINSON I.K., 1984 - Nucl. Instrum. Methods, 222, 171.

FUOSS P.H., KISKER D.W., LAMELAS F.J., STEPHENSON G.B., IMPERATORI P., BRENNAN S., 1992 - Phys. Rev. Lett., 69, 2791.

HAMERS R.J., TROMP R.M., DEMUTH J.E., 1986 - Phys. Rev. Lett., 56, 1972.

HOWES P.B., NORRIS C., FINNEY M.S., VLIEG E., VAN SILFHOUT R.G., 1993 - Phys. Rev. B, in press.

JEDRECY N., SAUVAGE-SIMKIN M., PINCHAUX R., MASSIES J., GREISER N. , ETGENS V.H., 1990 - Surf. Sci., 230, 197.

JEDRECY N., SAUVAGE-SIMKIN M., PINCHAUX R., MASSIES J., GREISER N., ETGENS V.H., 1990 - J. Cryst. Growth, 102, 293.

JEDRECY N., ZHENG Y., WALDHAUER A., SAUVAGE-SIMKIN M., PINCHAUX R., 1993 - Phys. Rev., B48, in press.

KARLE J., 1989 - Phys. Today, 42, 22.

KEATING P. N., 1966 - Phys Rev., 145, 637.

LIPSON H., COCHRAN W. - "The determination of crystal structures" (Cornell University Press 1966).

MAILÄNDER L., DOSCH H., PEISL J., JOHNSON R.L., 1990 - Phys. Rev. Lett., 64, 2527.

MASSIES J., ETIENNE P., DEZALY F., N.T. LINH, 1980 - Surf. Sci. 99, 121.

MEADE R.D., VANDERBILT D., 1989 - Phys. Rev., B40, 3905.

MOCHRIE S.G., 1988 - J. Appl. Cryst. 21, 1.

PEDERSEN J.S., 1988 - PhD Thesis, RISO-M Reports 2713.

PHANEUF R.J., WEBB M.B., 1985 - Surf. Sci., 164 167.

RENAUD G., FUOSS P.H., BEVK J., FREER B.S., HAHN P.O., 1992 - Phys. Rev. B45, 9192.

ROBINSON I.K., 1986 - Phys. Rev., B33, 3830.

ROBINSON I.K., 1988 - Austral. J.Phys., 41, 359.

ROBINSON I.K., 1989 - Rev. Sc. Instrum., 60, 1541.

ROBINSON, I.K. - Handbook on Synchrotron Radiation, vol. 3 (ed. D.E. Moncton and G.S. Brown), North Holland Publ. Co., Amsterdam 1991.

ROBINSON I.K., WASKIEWICZ W.K., TUNG R.T., BOHR J., 1986 - Phys. Rev. Lett., 57, 2714.

ROBINSON I.K., WASKIEWICZ W.K., FUOSS P.H., NORTON L.J., 1988a - Phys. Rev. B37, 4325.

ROBINSON I.K., TUNG R.T., FEIDENHANS'L R., 1988b - Phys. Rev., B38, 3632.

ROBINSON I.K., VLIEG E., KERN K., 1989 - Phys. Rev. Lett., 63, 2578.

ROBINSON I.K., CONRAD E.H., REED S.D., 1990 - J. Phys. France, 51, 103.

ROBINSON I.K., TWEET D.J., 1992a - Rep. Prog. Physics, 55, 599.

ROBINSON I.K., VLIEG E., 1992b - Surf. Sci. 261, 123.

ROBINSON I.K., ENG P., BENNETT P.A., DEVRIES B., 1992c - Appl. Surf. Sci., 60, 133.

ROSSMANN R., MEYERHEIM H.L., JAHNS V., WEVER J., MORITZ W., WOLF D., DORNISCH D., SCHULZ H., 1992 - Surf. Sci., 279, 199.

SAUVAGE-SIMKIN M., PINCHAUX R., MASSIES J., CLAVERIE P., BONNET J., JEDRECY N., ROBINSON I.K., 1989 - Phys. Rev. Lett. 62, 563.

TAKAYANAGI K., TANISHIRO Y., TAKAHASHI S., TAKAHASHI M., 1985 - Surf. Sci., 164, 367.

TAKEUCHI N., SELLONI A., TOSATTI E., 1992 - Phys. Rev. Lett., 69, 648.

THORTON J.M.C., WILLIAMS A.A., MACDONALD J.E., VAN SILFHOUT R.G., FINNEY M.S., NORRIS C., 1992 - Surf. Sci., 273, 1.

TWEET D.J., AKIMOTO K., TATSUMI T., HIROSAWA I., MIZUKI J., MATSUI J., 1992 - Phys. Rev. Lett., 69, 2236.

VAN SILFHOUT R.G., FRENKEN J.W.M., VAN DER VEEN J.F., FERRER S., JOHNSON A.D., DERBYSHIRE A.S., NORRIS C., MACDONALD J.E., 1989 - J. Physique, 50, C7-295.

VAN SILFHOUT R., VAN DER VEEN J.F., NORRIS C., MACDONALD J.E., 1990 - J. Chem. Soc. Faraday Discussion N° 89, 169.

VAN SILFHOUT R.G., LOHMEIR M., ZAIMAS S., VAN DER VEEN J.F., HOWES P.B., NORRIS C., THORNTON J.M.C., WILLIAMS A.A., 1992 - Surf. Sci., 271, 32.

VINEYARD G., 1982 - Phys. Rev. B26, 6247.

VLIEG E., VAN'T ENT A., DE JONGH A.P., NEERINGS H., VAN DER VEEN J.F., 1987a - Nucl. Instrum. methods, 262, 522.

VLIEG E., VAN DER VEEN J.F., MACDONALD J.E., MILLER M., 1987b - J. Appl. Cryst. 20, 330.

VLIEG E., VAN DER VEEN J.F., GURMAN S.J., NORRIS C., MACDONALD J.E., 1989 - Surf. Sci., 210, 320.

VLIEG E., FONTES E., PATEL J.R., 1991 - Phys. Rev. B43, 7185.

WARREN B.E. - "X-ray diffraction" (Addison-Wesley, Reading Ma 1969).

WILLIAMS A.A., THORTON J.M.C., MACDONALD J.E., VAN SILFHOUT R.G., VAN DER VEEN J.F., FINNEY M.S., JOHNSON A.D., NORRIS C., 1991 - Phys. Rev. B 43, 5001.

ZHU X-M., FEIDENHANS'L R., ZABEL H., ALS-NIELSEN J., DU R., FLYNN C.P., GREY F., 1988 - Phys. Rev. B37, 7157.

CHAPTER IV

SURFACES AND INTERFACES: NEUTRON STUDIES
R.K. THOMAS

IV.1. Neutrons as a Surface Probe

The many techniques that have been used to great effect to study the properties of gases adsorbed on solids have almost exclusively relied on probes that have been extremely strongly scattered by matter. The most notable example is electron scattering. The rationale is clear; a combination of radiation that is strongly scattered with an exposed surface creates a situation where the scattering is highly surface selective. Neutrons and, to a lesser extent, X-rays, are only weakly scattered by matter and it is not at all obvious that either can be an effective technique for studying interfaces. The purpose of this chapter is to show the many circumstances where neutrons, although not an obvious choice for a surface probe, do give interesting information about surfaces, and in one circumstance, specular reflection, are actually surface specific.

There are three reasons why one might want a probe other than electrons; the energy resolution possible with electrons is not anything like as high as with neutrons; the strong scattering causes difficulties in interpreting diffraction intensities quantitatively because of multiple scattering, and, most important of all, the strong scattering of electrons means that electrons can only reach a surface through an ultrahigh vacuum. Thus the important areas of application of neutrons corresponding to these limitations are inelastic and quasielastic scattering, diffraction from molecular layers, where the structure of the layer is the important feature, and buried interfaces, which includes circumstances where the interface is between solid and vapour at normal vapour pressures (inaccessible to electrons) and between two condensed phases such as solid and liquid.

IV.2. Neutron Diffraction from Vapours Adsorbed on Solids

The difficulty with neutron diffraction from an adsorbed layer is that the signal from the layer is very low and therefore means must be taken to enhance it. This can be done by using high scattering adsorbates (e.g. deuterium containing molecules) and adsorbing materials with a very high ratio of surface area to mass, typically not less than $5 \, m^2 \, g^{-1}$. However, the larger the value of the specific area the less homogeneous the surface, and the less homogeneous the surface the poorer the quality of the long range order and the less well defined the diffraction pattern. It is necessary to achieve a balance. This has so far been done by using only adsorbents (substrates) with relatively well defined surfaces, e.g. graphite and magnesium oxide. An apparent exception to this is the case of molecules adsorbed in zeolites, but here the molecules are isolated species incorporated into a regular lattice and the purpose of the experiment is entirely different.

The nature of materials such as graphite is that they interact relatively weakly with adsorbed species, i.e. the molecules are only held by physical forces (physisorption) as opposed to chemical bonds (chemisorption). A characteristic of physisorption is that adsorption is sufficiently weak that much of the interesting range of adsorption occurs at normal vapour pressures when, for the reasons given above, electron diffraction cannot be used. Both X-ray and neutron diffraction are then suitable techniques and, as will be shown below, the limited number of diffraction peaks observed often means that it is necessary to use both techniques.

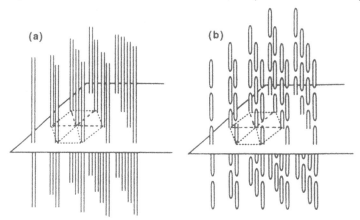

Figure IV.1. — (a) The reciprocal lattice of a 2-D layer of points in the plane of the surface, (b) the same reciprocal lattice when there is structure in the direction normal to the surface. The plane of the surface is drawn and the lattice has hexagonal symmetry.

Because of the requirement of high surface area the adsorbents are either powders or partially oriented powders, e.g. exfoliated graphite. The reciprocal lattice of an adsorbed layer is a set of rods, as shown in Fig. IV.1(a). The lineshape for a 2-D layer adsorbed on a powdered material is obtained by averaging over all the intersections of the Ewald sphere with the reciprocal lattice of rods. The result is the characteristic Warren lineshape

$$I_{hk} = A_{hk} \exp(-2W) \left[\frac{L}{\lambda \sqrt{\pi} \sin^{3/2} \theta} \right] \int_0^\infty \exp\left[-(x-a)^2 \right] dx \tag{IV.1}$$

where

$$a = \left[\frac{2L \sqrt{\pi}}{\lambda} \right] \left(\sin\theta - \sin\theta_{hk} \right) \tag{IV.2}$$

A_{hk} is the two dimensional structure factor, $\exp(-2W)$ the Debye-Waller factor, and L the mean diameter of the 2-D domain on the surface. As for diffraction from 3-D crystals the smaller the domain the greater the peak broadening. However, whereas peak broadening is rare for 3-D crystals, the size of 2-D domains is usually sufficiently small, typically of the order of 100 Å, that there is always some degree of peak broadening. The characteristic peak shape expressed in equation (IV.1) is shown for two domain sizes in Fig. IV.2(a).

The appearance of the line shape is very important in physisorbed systems because a variety of types of structure is possible for the adsorbate. Because the balance of the substrate-molecule and molecule-molecule forces is delicate the adsorbate molecules may be more inclined to form multilayers or even 3-D crystallites rather than cover (wet) the whole surface, and this tendency may depend on temperature or coverage. The shape of a diffraction peak is dramatically different when multilayer adsorption occurs. An example is shown in Fig. IV.2(b) for a multilayer of methane on graphite for 1, 2, and 6 molecular layers. The presence of structure in the direction normal to the surface causes the rods of the reciprocal lattice to break up (Fig. IV.1(b)) and the diffraction peak develops a structure determined by the structure of the layer in the vertical direction. In the case of methane the two embryonic peaks in Fig. IV.2(b) would evolve on further addition of methane into the first two diffraction peaks for the bulk crystal. If methane did not wet the graphite at all two narrow peaks would occur at the positions of the two peaks in Fig. IV.2(b), which would again be quite distinctive.

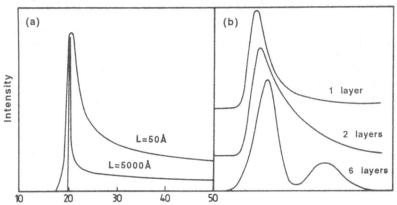

Figure IV.2. — (a) The shape of a diffraction peak from a two dimensional atomic layer on a powdered substrate for two domain sizes of (i) 50 Å and (ii) 5000 Å. The position for a domain of infinite size is given by a vertical line. (b) The effect of additional layers on the calculated peak shape for (i) 1, (ii) 2, and (iii) 6 adsorbed layers of methane on graphite.

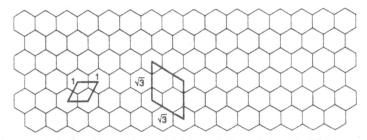

Figure IV.3. — The structure of the surface of graphite. The (1 x 1) structure of the surface is marked on the left and the ($\sqrt{3}$ x $\sqrt{3}$)R30° structure on the right. The nomenclature of the latter results from the size of the unit cell relative to the surface unit cell ($\sqrt{3}$ x the surface lattice parameter) and the relative orientation (Rotated by 30°).

Most neutron diffraction studies of physisorption have used graphite as substrate. The hexagonal network of graphite is shown in Fig. IV.3 and marked on it are the (1 x 1) structure of the surface together with the most commonly observed structure for small molecules, the ($\sqrt{3}$ x $\sqrt{3}$)R30^0 structure. For small molecules, for example krypton, methane, and nitrogen, the interactions which help to determine the structure are the variation of the graphite-molecule interaction across the surface (the corrugation of the potential), the dispersion forces between the molecules, the packing requirements, and any electrostatic interactions, for example quadrupoles in nitrogen and octopoles in methane. Packing requirements almost always favour a hexagonal lattice and electrostatic forces predominately determine the mutual orientation of the molecules. The balance between the corrugation of the surface potential and the dispersion forces determines whether the 2-D layer will form a commensurate or incommensurate layer on the surface.

The first stage in identifying the type of layer structure is from the shape of the observed diffraction peak from the layer. As can be seen from Fig. IV.2(b) this may not necessarily distinguish 1 or 2 layer structures, but there is usually some independent information available which can settle this question. Whether or not the structure is commensurate or incommensurate is then determined by examination of the layer lattice parameter (determined in the usual way from the Bragg condition). For a layer whose structure is commensurate with the substrate the problem may be complicated because of the probable coincidence of commensurate peaks with diffraction peaks of the substrate and since the mass ratio of substrate and adsorbate is typically of the order of 1000 these peaks will be totally obscured. Also, when the period of the commensurability is long there will be little coincidence of the diffraction peaks and this can be equally confusing.

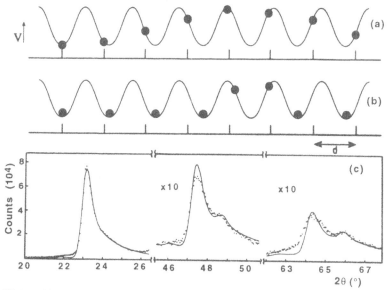

Figure IV.4. — (a) and (b) The competition between the molecule-molecule interactions (favoured intermolecular distance = d) and a corrugated surface potential. In (a) the intermolecule distance is optimised for the molecule-molecule interaction, but (b) is generally the more stable structure. (c) The observed and calculated (continuous line) X-ray diffraction patterns for xenon adsorbed on graphite at a coverage on the incommensurate-commensurate phase boundary.

For many physically interesting systems the analysis finishes at this point because the problem concerns either what makes the layer commensurate or incommensurate or the phase change between the two types of structure. An example of the former is the inert gas series adsorbed on graphite at low temperature (Dash & Ruhwalds, 1980). The diameter of Xe (from the bulk solid) is about 4.4 Å compared with the $\sqrt{3} \times \sqrt{3}$ distance of 4.26 Å. Thus at low coverage Xe forms an incommensurate layer and only becomes commensurate when the surface pressure is increased by loading the surface more heavily. The diameter of Kr is 4.1 Å, slightly less than the favoured commensurate distance. It forms the $\sqrt{3} \times \sqrt{3}$ structure at low coverages but can be squeezed into an incommensurate phase with smaller lattice parameter at high coverage, i.e. the opposite behaviour to Xe. Evidently the slightly unfavourable molecule-molecule interaction in the commensurate phase is more than compensated by the better interaction with the substrate. The diameters of Ar and Ne are 3.8 and 3.2 Å respectively and too much molecule-molecule interaction energy is lost for a commensurate structure to form and both form only an incommensurate phase. However, for the very light He atom the quantum mechanical repulsion becomes sufficiently large that the $\sqrt{3} \times \sqrt{3}$ structure once more becomes favourable. Both ^{3}He and ^{4}He form a commensurate layer at low coverage, and only become incommensurate when the surface pressure is relatively high. Furthermore, the greater repulsion between ^{3}He atoms, because they are fermions, means that the commensurate phase is stable over a wider range of conditions.

The commensurate-incommensurate phase transition may cause the appearance of a diffraction peak from the 2-D layer to be more complex. If the layer is incommensurate the molecules will lie at varying positions in the corrugated potential of the substrate, as shown schematically in Fig. IV.4(a). Such a conformation is not generally the lowest state of free energy of the system. A better structure is that shown schematically in Fig. IV.4(b), where "walls" separate regions of commensurate phase. Generally what is then observed are satellite peaks accompanying the diffraction peak because of the modulation of the lattice by the periodicity with which the walls occur (Freimuth et al., 1990). The separation of the satellite peaks from the main peak in terms of the wavevector is of the order of the fractional misfit of the lattice divided by the unit cell length, but the details of the pattern will depend on the wall structure. An example is shown in Fig. IV.4(c) for xenon on graphite (Mowforth et al., 1986). This has a hexagonal wall structure and the fractional misfit from the figures given in a previous paragraph above is about 4% giving a satellite spacing of the order of 0.04/4.4 Å^{-1}, which corresponds to about two thirds of a degree for the wavelength used for Fig. IV.4(c).

The structure of more complex layers is one of the most effective ways of probing intermolecular forces in condensed phases partly because the surface acts as a template which can hold quite different molecules in a similar structure and partly because the simpler 2-D structures are easier to interpret than their 3-D counterparts. In addition the conditions can be varied over a wider range than for 3-D structures. For example, the surface pressure (approximately equivalent to the 3-D pressure) is varied just by changing the coverage or the temperature. A good example of the contribution of different factors can be found in the halomethanes, CH_3X where X is F, Cl, Br, and I, where the dominant effects are the competition between the dipole/dipole interactions, the dispersion forces between the halogen atoms, and the dispersion forces between halogen and graphite (Knorr, 1992). These contributions change markedly across the series. In order to probe these interactions it is first necessary to interpret the diffraction pattern from the 2-D layer.

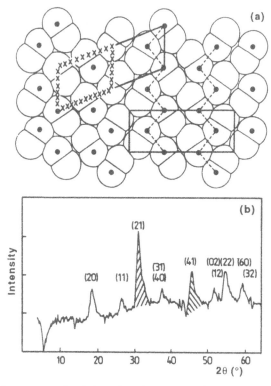

Figure IV.5. — (a) The structure of an iodomethane layer adsorbed on graphite. The unit cell deduced from X-rays is marked with xxx and a larger oblique unit cell, necessary to account for the neutron pattern, is marked as an extension of the X-ray unit cell. The final lattice deduced is the rectangular lattice and it has Pgg symmetry. The lattice is incommensurate with the graphite substrate. (b) Neutron diffraction pattern of iodomethane on graphite. The shaded peaks are completely missing from the X-ray pattern.

The interpretation of diffraction patterns of adsorbed layers differs in two respects from those of crystals; there are not usually many diffraction peaks and the number of molecules in the unit cell is not generally known (for 3-D crystals it is deduced from the bulk density). There is therefore more guesswork and more scope for ambiguous assignments of the peaks. It is essential to make use of as much data as possible and this means using both neutron and X-ray diffraction patterns. We consider here the case of iodomethane, which illustrates the typical problems that arise in such a study (Bucknall et al., 1989). The 2-D monolayer structure is shown in Fig. IV.5(a) and one of the patterns on which it was based (the neutron pattern) is shown in Fig. IV.5(b). For a molecule containing a heavy atom such as iodine X-ray diffraction is superior to neutrons and, although the X-ray pattern is not given here, more X-ray peaks are observed at higher momentum transfer than in the neutron pattern of Fig. IV.5(b). The X-ray pattern indexes perfectly on the cell marked with crosses in Fig. IV.5(a) and gives the correct peak intensities for such a structure. However, two of the most intense peaks in the neutron pattern, the (2,1) and (4,1), as indexed on the rectangular cell in Fig. IV.5(a), do not occur at all in the X-ray pattern. Even though the neutron pattern is of poorer quality than the X-ray pattern, this result shows that the unit cell deduced from the X-rays is wrong and has to be doubled. It is shown as the tilted unit cell in Fig. IV.5(a). A more careful analysis shows that the higher symmetry rectangular cell of Fig. IV.5(a) with Pgg symmetry accounts for both patterns. The reason for the extra sensitivity of the neutrons to the larger unit cell is that it is only in terms of the methyl groups that there are two different types of iodomethane molecules. This is most clearly seen in the oblique lattice of Fig. IV.5(a). In the rectangular unit cell of Fig. IV.5(a) the two glide planes have the effect that only the coordinates of one molecule are needed to describe the structure. This reduces the number of independent parameters sufficiently that it was possible to deduce the exact angle of tilt of the molecule with respect to the surface, even though only a relatively small number of diffraction peaks had been observed. The iodomethane molecule is almost flat on the surface, the angle of tilt being only 10°. The effect of the dipolar interactions on the orientation of the molecules within the layer is quite complex. There are two zig-zag chains of iodine atoms arranged so that the iodine atoms in each chain interact closely but the dipole moments in each chain are opposed. The iodine-iodine distance in the chains is actually shorter than twice the van der Waals radius, a feature which occurs in one or two other molecular iodine compounds. The different balance between dipolar and dispersion forces is illustrated by the fact that the four halomethanes form a total of seven monolayer structures, no two of them the same. The determination of the details of the in-plane structure of an adsorbed layer cannot generally be done with electron diffraction.

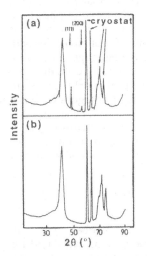
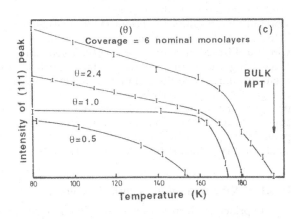

Figure IV.6. — (a) The diffraction pattern of a nominal 0.5 monolayers of ammonia adsorbed on graphite including background scattering from graphite and cryostat. (b) Graphite and cryostat only. (c) The intensity of the ammonia (111) peak as a function of temperature on graphite at different coverages of (i) 6, (ii) 2.4, (iii) 1.0, and (iv) 0.5 nominal monolayers.

At the beginning it was stated that an adsorbate may not cover the adsorbent surface before it forms multilayers or even bulk crystals. This is partial or non-wetting. There is considerable argument about the experimental evidence for partial wetting and wetting transitions and also evidence that such phenomena in physisorbed systems may be sensitive to the larger scale structure of the material, e.g. the presence of pores. However, a clear case of non-wetting, which illustrates some of the features accessible with neutron or X-ray diffraction is ammonia on graphite (Gamlen et al., 1979). The NH_3-NH_3 interactions are stronger than the NH_3-graphite interactions and, as a result ammonia hardly wets graphite at all. Figure IV.6(a) shows the sharp peaks characteristic of 3-D solid ammonia, appearing even at the low nominal coverage of 0.5 monolayers. The peaks are very narrow and therefore characteristic of 3-D crystals. They also show the difficulty of extracting an adequate signal in such experiments when the scattering from substrate and container is very large. The melting of the ammonia crystallites may be followed easily with neutron diffraction (Fig. IV.6(b)) and is highly anomalous with a wetting temperature that does not appear to coincide with the triple point, which is the normally expected behaviour. In this case the value of neutron or X-ray diffraction is not that it is being used to obtain structural information but simply that it is able to follow a simple physical phenomenon under conditions when no other technique could give such unambiguous information.

IV.3.   Neutron Inelastic Scattering

The range of accessible energies and the high resolution of neutron scattering make it suitable for investigating intramolecular vibrations, lattice vibrations, and rotational or tunnelling transitions. This may be used in two ways for the study of surfaces. First, it offers a more direct probe of surface intermolecular forces than a structural determination and this is the way neutron inelastic scattering has been used to investigate physisorbed layers. Second, the study of intramolecular vibrations has traditionally been used to investigate the symmetry of molecules and, in the case where an adsorbate does not form a structure with long range order, from which a diffraction pattern can be obtained, this offers an effective method for elucidating the nature of the surface species. This is particularly useful in chemisorption where the common problem is that a molecule undergoes chemical change when it is chemisorbed and the identification of the chemisorbed species and its symmetry is the information most required. Given also that neutrons easily penetrate catalytic environments and most catalysts, neutrons have a considerable potential, as yet not fully exploited, for the study of catalysis under real industrial conditions. We consider first physisorbed systems.

IV.3.1.   Vibrations of Physisorbed Layers

In principle, the best way to study intermolecular forces in adsorbed layers is to measure the phonon dispersion curves in the layer. These may be calculated for given models of the intermolecular forces and then tested by direct measurement. However, it has proved more or less impossible to obtain the necessary detail because the high surface area required to give an adequate signal means that randomly oriented domains are present which scramble the in-plane vectors and make it impossible to obtain anything useful. The situation is improved when the density of vibrational states of the layer is considered. First, the averaging over the different orientation of the wavevectors is now of no consequence. Second, the density of states is obtained from the incoherent scattering and, since protons have a high incoherent scattering cross section, the signal from a proton containing layer is two orders of magnitude larger than from a deuterated layer. The signal to noise for any incoherent scattering experiment from a proton containing adsorbate is therefore generally 100 times better than for diffraction. What is measured in such an experiment is not exactly a density of states. The differential scattering cross section from a vibrational excitation is approximately proportional to the number of protons involved in the motion and to the mean square amplitude, b, of motion of their motion

$$\frac{d\sigma}{d\Omega} \propto \frac{k}{k_0} \frac{\kappa^2 b^2}{2} \exp\left(-\frac{\kappa^2 b^2}{2}\right)$$

(IV.3)

where k_0 and k are the wave vectors of the incident and scattered neutrons respectively and κ is the momentum transfer.

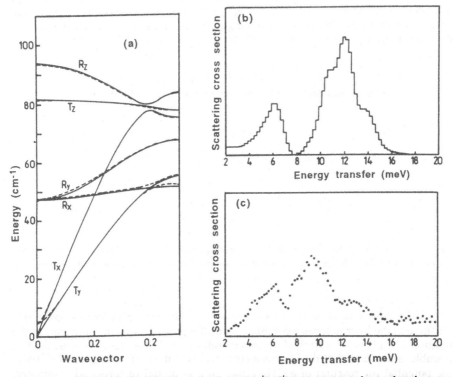

Figure IV.7. — (a) The calculated dispersion curves of a $\sqrt{3} \times \sqrt{3}$ commensurate layer of methane on graphite. R and T denote librational and translational modes respectively, z is the direction normal to the surface. (b) The amplitude weighted density of states calculated from the curves in (a) when the momentum transfer is perpendicular to the surface. (c) The observed neutron incoherent scattering spectrum from adsorbed CH_4 in its commensurate phase.

The density of states for methane on graphite has been studied by incoherent inelastic scattering. Methane behaves like krypton on graphite, forming a commensurate $\sqrt{3} \times \sqrt{3}$ structure at low coverage and an incommensurate structure at high coverage. The dispersion curves calculated for a widely used model of the atom-atom interaction potentials are shown in Fig. IV.7(a). The individual phonon dispersion curves cannot be measured but the calculated weighted density of states is shown in Fig. IV.7(b) and is to be compared with the observed neutron spectrum in Fig. IV.7(c). In this case the spectrum is from a partially oriented graphite sample with the momentum transfer perpendicular to the surface and then only excitations with displacements normal to the surface are observed. Some differences are immediately obvious, which reveal the shortcomings of the calculation. The first is that there is a sharp peak in the calculated spectrum at 12 meV which corresponds to the motion of a molecule against the surface ((T_2) in Fig. IV.7(a)) and which is not observed in the actual spectrum. It is thought that in the real system this motion is damped by coupling with modes of the graphite, but no quantitative explanation has yet been given. A more careful comparison also shows that the rotational (librational) modes do not compare well between theory and experiment. This may be because of failings in the potential used, but as the experiment in the next section will show, the more likely problem is that these rotations are highly anharmonic and the calculation had to use the harmonic approximation. What is usefully observed in this experiment is the dramatic change in the density of states across the commensurate-incommensurate phase transition, which we will return to in the next section.

IV.3.2.   Hindered Rotation in Methane and Hydrogen on Graphite

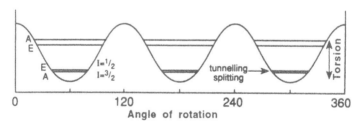

Figure IV.8. — (a) Schematic diagram of the energy levels for the hindered rotation of a CH_3 group in a threefold potential. The ground A and E levels are marked, as are the first excited torsional state levels.

The hindered rotation of species such as H_2, CH_4, and CH_3 groups is extremely sensitive to the intermolecular potential controlling orientation. The splittings of the rotational levels of an adsorbed species therefore constitute the most sensitive test possible of any model of the intermolecular forces at the surface. Figure IV.8 shows the potential for rotation of a CH_3 group in a potential of threefold symmetry. Overlap of the wavefunctions of the three equivalent ground states leads to tunnelling between the different orientations and causes the levels to be split. In a threefold potential, group theoretical arguments show that there will be one non-degenerate level (A symmetry) and one degenerate one (E symmetry). The A symmetry wavefunction is the in phase combination of the three ground state wavefunctions and therefore has no nodes. As always the case in quantum mechanics the state with no nodes has the lower energy. The total nuclear spin of the three protons in the CH_3 group can be 3/2 or 1/2. Symmetry considerations show that the maximum spin is associated with the totally symmetric A level.

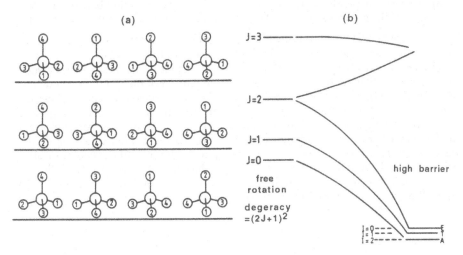

Figure IV.9. — (a) The twelve equivalent configurations of a methane molecule on the surface of graphite. (b) The correlation of the free rotation levels of CH_4 on the left with the levels when the barrier to rotation is infinitely high.

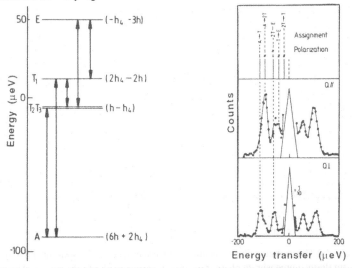

Figure IV.10. — (a) The energy level splitting in the ground state of methane adsorbed on graphite. (b) Neutron scattering spectra of methane adsorbed on a partially oriented graphite showing the separate spectra with the momentum transfer perpendicular and parallel to the surface. The splitting is characterised by two parameters h and h_4, which are related to the barriers to rotation about the unique and tripod axes respectively.

For neutrons there are two important features. First, the A-E splitting, or
tunnelling splitting, is in the range 0 - 5 cm^{-1} (0 - 0.6 meV), which is easily accessible
in a high resolution experiment. Second, the selection rule for incoherent scattering is
$\Delta I = \pm 1$ or $\Delta I = 0$ if $I \neq 0$ where I is the spin of the protons. Note that for infrared,
Raman, and microwave spectroscopy $\Delta I = 0$ (the existence of ortho- and para- H_2 as
separate species depends on this selection rule). Thus the transition between the two
split levels of the ground state of a CH_3 group is allowed only in neutron scattering.

No case of a CH_3 group tunnelling on a surface has yet been observed, but the
more complicated case of CH_4 on graphite has been studied in detail (Smalley et al.,
1981). There are twelve possible configurations of methane on graphite (Fig. IV.9(a)).
To obtain a qualitative idea of what the splitting pattern of the ground state will be
when there is a barrier to rotation a diagram can be constructed which correlates the
levels when there is no barrier, i.e. free rotation, and the twelve equivalent ground
states when the barrier is infinitely high. Figure IV.9(b) shows this correlation for
methane in a tetrahedral field and it can be seen that the ground state is split into
three levels. The lowest A state has nuclear spin 2, the ninefold degenerate T state
has spin 1 and the twofold degenerate E state has spin 0. There are two allowed
transitions, between A and T and between T and E. On graphite the potential field
controlling the rotation of CH_4 is of trigonal symmetry and the central T level is

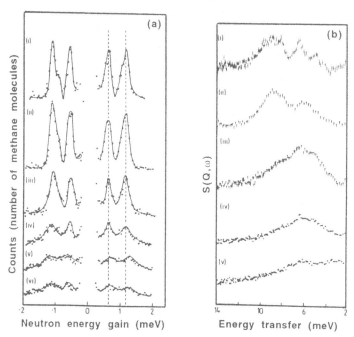

Figure IV.11. — (a) The change in the tunnelling spectrum of methane on graphite on going through
the commensurate-incommensurate phase transition, κ perpendicular to the surface. The coverage is
0.2 monolayers in (i) changing through (iii) 1.0 monolayers to (vi) 1.5 monolayers. The intensities are
normalised to the number of methane molecules in the system. (b) The corresponding changes in the
vibrational density of states of adsorbed methane on a powdered graphite on going from (i) 0.4, (ii)
0.7, through the phase change to (iii) 1.06 and (v) 1.51 monolayers.

further split into two levels. There are then five transitions; each transition involving the T levels is split into two, and there is an allowed transition between the two T levels.

Figure IV.10(a) shows the levels for CH_4 on graphite and Fig. IV.10(b) shows the neutron spectrum with the momentum transfer parallel and perpendicular to the surface (an oriented graphite was used). The splitting of the T levels is determined by the ratio of the barrier heights for rotation about the axis normal to the surface and about one of the three equivalent axes of the tripod of CH bonds that point towards the surface. A quantitative analysis of the splittings gave accurate values of the two barrier heights to rotation of 200 and 175 cm^{-1} respectively.

An interesting feature of this tunnelling spectrum is what happens on going through the commensurate-incommensurate phase transition. First considerations would suggest that, since the tunnelling splitting is sensitive to the potential created by the surrounding molecules, the spacings in the spectrum should change dramatically. However, as shown in Fig. IV.11, no shift in the frequencies is observed at all (Humes et al., 1988). There is only a sharp diminution in the intensities of the transitions. This is in complete contrast to the behaviour of the density of states (Fig. IV.11(b)) which shifts markedly to lower frequencies. The explanation of this behaviour is that the commensurate-incommensurate phase transition proceeds through domain structures, just as suggested schematically in Fig. IV.4(b). This is the only evidence for such a mechanism in the case of methane, although it has been demonstrated for a number of other systems.

IV.3.3.   Vibrations in Chemisorbed Layers

In chemisorption the main interest is in identifying the species adsorbed and this is often quite different from the original species in the gas phase. For example, C_2H_4 and C_2H_2 both give an identical species, the ethylidyne radical CH_3CC-, when adsorbed on platinum. Also, since the surfaces of real catalysts are not of a quality to give single surface domains with sufficient long range order to generate diffraction patterns, the most effective means of deducing the nature of the adsorbed species is to determine its vibrational spectrum. There are several ways of doing this, by electron energy loss for a flat surface, or by infrared and Raman spectroscopy for powdered materials, but each technique has limitations which prevent it being of general application. The features of neutrons that make them useful in this area are (i) most catalysts, often metals, are transparent to neutrons, (ii) most adsorbed species of chemical interest contain protons and therefore give a large incoherent scattered signal, (iii) the transparency of most materials to neutrons means that the adsorption may be studied under the extreme conditions of real catalysis, and (iv) the dependence of the intensity of an excitation in the neutron spectrum on the amplitude of the motion (eq. IV.3) gives a powerful additional method for assigning the vibrational modes.

A simple example is water adsorbed on Raney nickel. Raney nickel is a widely used catalyst in the hydrogenation of many organic compounds. It is a spongy form of nickel, which contains a small percentage of aluminium, and its surface cannot be studied by the conventional spectroscopies. When water is adsorbed on the surface it decomposes, but there are two possible routes:

$$H_2O = OH + H$$
$$H_2O = O + 2H$$

In the second case the neutron incoherent spectrum will just be that of adsorbed H atoms but in the first case there will also be a contribution from the OH group. In fact the spectra of H_2 and H_2O on Raney nickel are identical, showing that the second dissociation path is the correct one. Further separate magnetic experiments indicate that the role of the aluminium in the catalyst is to mop up the O atoms released in the dissociation.

A second example illustrates property (iv) above. In general, there are more unknown force constants determining the vibration frequencies of a molecule than there are vibration frequencies, even when isotopic substitution is used to generate more frequencies, i.e.; there are more unknowns than knowns. This has the consequence that the model used to assign the vibrational spectrum is not necessarily unique and the problem becomes more acute as the molecule becomes more complicated. However, the force field determines both the frequencies and the amplitudes of vibration and the neutron intensity is directly dependent on the amplitude (eq. IV.3). Thus there is additional information in the neutron spectrum, not available for other spectroscopies, which can often resolve ambiguities. In practice this information is not used directly. What is done is to compare the frequencies and intensities of calculated and observed spectra and adjust the force field parameters until the two match. The results for such an experiment are shown for benzene on Raney nickel in Fig. IV.12 (Jobic et al., 1980). Points to note are that the spectrum of adsorbed benzene is totally different from solid benzene, showing how difficult it is to try and solve the force field problem by transferring parameters from the solid phase, and the remarkably good agreement between the calculated and observed spectra. This gives considerable confidence to the conclusion that the benzene molecule lies flat on the surface of the nickel and is centred over a single nickel atom.

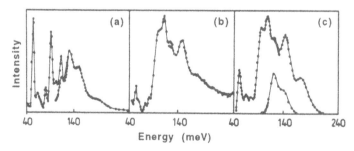

Figure IV.12. — (a) The inelastic spectrum of solid benzene, (b) benzene on Raney nickel after subtracting the nickel background, and (c) the calculated spectrum for benzene on Raney nickel including a 15% contribution from adsorbed hydrogen.

IV.4.   Neutron Quasielastic Scattering

The timescale of diffusive processes on a surface depends on the nature of the binding to the surface. Physisorbed layers generally melt from 2-D solids to 2-D liquids in a fashion parallel to 3-D materials and the rate of the diffusion in the 2-D liquid or vapour is comparable with that in a 3-D liquid. The diffusion is therefore accessible to study by quasielastic scattering and a limited number of such experiments have been done (see, for example, Thomas, 1982). Chemisorbed species diffuse much more slowly and, except for molecules in zeolites, have been little studied.

IV.5. Neutron Reflection

In the previous four sections the difficulty in using neutron scattering has been the weak scattering by the surface species and none of the cases examined could be considered to provide a general method of examining a particular area of surface physics or chemistry. Neutron reflection is completely different; it is a general method for investigating a wide range of interfaces, especially when one or other of the bulk phases bordering the interface is disordered, e.g. a polymer or a liquid. The key to the experiment is that the neutrons are incident on a flat surface at sufficiently grazing incidence that a significant proportion of the beam is reflected at the surface. Under these circumstances the experiment only probes a small depth of material and the signal contains information dominated by the interfacial inhomogeneities.

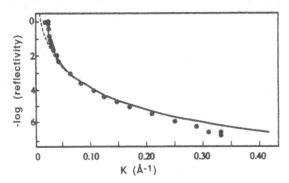

Figure IV.13. — The reflectivity profile for a smooth surface (continuous line) and actual data for water (points. The water has a roughness of 2.8 Å.

The reflection experiment may be divided into specular and off-specular reflection. In specular reflection the angle of incidence equals the angle of reflection and, since the change in momentum transfer is exactly perpendicular to the surface, only information about structure in the direction normal to the interface is obtained. In off-specular scattering there is also a component of momentum transfer in the plane of the surface and it is possible to obtain information about the in-plane structure while using grazing incidence to confine the region probed to the interface. The low level of signal to background in the neutron experiment is such that this second mode of operation has not yet been much explored, although it has proved very powerful with X-rays (Als-Nielsen & Mohwald, 1992). Here, we therefore just consider the specular experiment. The neutron refractive index for most condensed materials is less than for air and this leads to the situation where total reflection of the beam may occur at angles below a critical grazing angle of incidence. Above these critical angle the reflectivity falls off rapidly. In terms of the momentum transfer at the critical angle, κ_c, the reflectivity of a perfectly smooth surface is given by

$$R = \left[\frac{\kappa - \left(\kappa^2 - \kappa_c^2 \right)^{1/2}}{\kappa + \left(\kappa^2 - \kappa_c^2 \right)^{1/2}} \right]^2$$

(IV.4)

The value of κ at the critical angle is given by

$$\kappa_c = 4\pi^{1/2} \left(\rho_2 - \rho_1 \right)^{1/2}$$

(IV.5)

and ρ_1 and ρ_2 are the scattering length densities of the two bulk media either side of the interface, given by

$$\rho = \sum_i n_i \, b_i$$

(IV.6)

where n_i and b_i are the number density and empirical scattering length of atom i respectively. For D_2O κ_c has the value 0.0179 Å^{-1} but there is no critical angle for H_2O because its scattering length density is less than 0. The shape of the reflectivity profile from D_2O is shown in Fig. IV.13. The observed reflectivity is lower than predicted by eq. (IV.4) because it has a finite roughness from thermal motion of the surface, which reduces the reflected intensity by a factor resembling the Debye-Waller factor in diffraction, $\exp(-\kappa^2 \sigma^2)$.

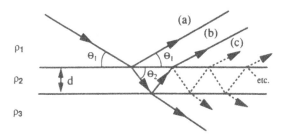

Figure IV.14. — The reflection of a neutron beam when there is a thin film on a substrate. The dashed lines represent continuing multiple reflection and transmission of the beam.

When there is a uniform layer between the two bulk phases the neutron beam may be reflected at the first and second interfaces and the two beams will interfere with one another to give fringes in the reflectivity profile. The calculation of the exact reflectivity for this situation uses standard optical methods, which involve the Fresnel coefficients for the reflected and transmitted amplitude at each interface (Born & Wolf, 1975) and the summation of all the multiply reflected beams shown in Fig. IV.14. If the coefficients used are those for light polarised at rightangles to the reflection plane the expression derived for the reflectivity is exact, but is somewhat cumbersome because of the need to include multiple scattering and a more convenient, approximate, expression which brings out the essential physics is

$$R = \frac{16\pi^2}{\kappa^4}\left[\left(\rho_2 - \rho_1\right)^2 + \left(\rho_3 - \rho_2\right)^2 + 2\left(\rho_2 - \rho_1\right)\left(\rho_3 - \rho_2\right)\cos\left(\kappa\tau\right)\right]$$

(IV.7)

where τ is the thickness of the layer. This formula shows that the reflectivity of such a layer consists of a series of fringes with spacing $(2\pi/\kappa)$ and amplitude depending on the differences in scattering length density between the different layers, superimposed on a reflectivity level decaying as $(1/\kappa^4)$. Equation (IV.7) is only accurate when κ is at least double κ_c. An example of the effect of a layer on the reflectivity is shown for an organic film 740 Å on glass in Fig. IV.15.

It is evident from eq. (IV.7) that the specular reflectivity profile leads to information about both the dimensions and composition of heterogeneities across an interface in the normal direction. However, the resolution of the experiment is limited by background scattering which principally arises from scattering by the second medium. The background level is of the order of 10^{-6} and the maximum momentum transfer at which the reflectivity can be measured varies between about 0.1 and 0.4 Å^{-1}. Thus it is impossible to probe interatomic distances and the method is only appropriate for larger scale inhomogeneities. However, a large range of problems fall into this category, mainly involving systems containing larger molecules such as polymers and surfactants.

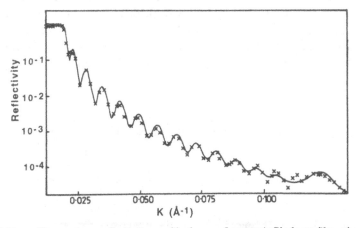

Figure IV.15. — The neutron reflectivity profile from a Langmuir-Blodgett film of deuterated cadmium arachidate on a glass substrate. The film is 740 Å thick. The film has internal structure because it is built up from molecular layers and this internal structure gives rise to a first order Bragg diffraction peak at about 0.125 Å^{-1}. The continuous line is a calculated profile.

The most powerful feature of neutron reflection is its ability to distinguish the contributions of different parts of an interface. Thus, it was noted above that D_2O and H_2O have positive and negative scattering length densities respectively, owing to the different phase of the scattering by deuterons and protons. It is therefore nearly always possible to mix them in a ratio such that $(\rho_2-\rho_1)$ in eq. (IV.5) is zero, when the reflectivity (eq. (IV.4)) becomes zero, i.e. the interface vanishes. This is shown for reflection at the air/water interface in Fig. IV.16. On the left is the scattering from D_2O as observed on a square multidetector. The specular peak, at the lower centre, contains a total of about 10^7 counts. On the right is the signal from a 9:1 molar mixture of H_2O and D_2O and there is no reflection at all. In both cases the background scattering can be observed and it is higher from the H_2O containing sample because of the greater incoherent scattering of the protons.

Once the surface has been made to vanish, isotopic labelling may be used to highlight any material adsorbed. The simplest example of this is the adsorption of deuterated surface active material at the air/water interface. If the solution is sufficiently dilute there will be no reflection unless there is adsorption. Any reflected signal is then entirely from the surface layer. Under these circumstances neutron reflection becomes truly surface specific. It is worth considering the sensitivity of the experiment. Using eq. (IV.4) with null reflecting water ($\rho_1 = \rho_3 = 0$) gives

$$R = \frac{16\pi^2}{\kappa^4} \rho_2^2 \left(4 \sin^2(\kappa\tau / 2)\right)$$

(IV.8)

Substitution of typical values of $\tau = 20$ Å and $\kappa = 0.05$ Å^{-1} shows that the reflectivity is 10^{-5} when ρ_2 is 0.3×10^{-6} Å^{-2}. This is approximately the minimum scattering length density for the layer to be observable. For the sake of illustration we take the layer to be a fully deuterated hydrocarbon such as CD_4 which has a scattering length of about 3.3×10^{-4} Å. Since the layer is 20 Å thick there must be 1 molecule every $(b/\tau\rho_1)$ Å^2, i.e. every 50 Å^2. Since the area illuminated by the experiment is typically 100 cm^2 the total mass of CD_4 in the beam is about 0.6 µg! This is to be contrasted with the methane on graphite tunnelling experiment where the lowest observable

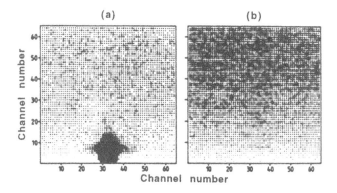

Figure IV.16. — Neutron scattering patterns registered on a multidetector for (a) D_2O and (b) null reflecting water (0.088 mole fraction of D_2O in H_2O). The dots in increasing order of size represent intensities in 100 count intervals from 0 to 800. The specular peak from D_2O reaches a maximum of 244000 counts. The specular reflection is at a momentum transfer of 0.036 Å^{-1} ($\kappa_c = 0.0179$ Å^{-1} for D_2O).

amount of methane in the beam is of the order of tens of milligrams. An example of the observation of an adsorbed layer at the surface of null reflecting water is shown in Fig. IV.17 for a mixture of two surface active solutes, one of which is deuterated (dodecanol) and one of which is also contrast matched to air (SDS). The background has not been subtracted.

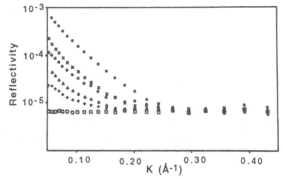

Figure IV.17. — Reflectivity profiles of deuterated dodecanol and protonated SDS mixtures in null reflecting solution in water. The dodecanol/SDS ratio is fixed for all the profiles except the ones with open circles (spread dodecanol on water) and open squares (SDS solution). The only contribution to the reflectivity is from dodecanol adsorbed at the surface. It varies as the overall concentration, which is 0.012(+), 0.009(Δ), 0.00675(.), and 0.0055 M (x).

The use of contrast variation to highlight a surface layer is not limited to the air/water interface. Quite the contrary, it becomes easier to match scattering length densities of the two bulk phases and adsorbed species at interfaces between solid and liquid. This is because the two limiting scattering length densities of proton and deuteron containing species are approximately -0.5 and 6×10^{-6} Å^{-2} respectively and nearly all solids, e.g. silicon and quartz, lie in the middle of this range, whereas air at 0 is too close to the pure proton containing materials. Thus, if the proportion of protons in one of the materials is not sufficiently high it may not be possible to match that material to air. A simple example is CH_2Cl_2. The effectiveness of contrast variation at a solid/liquid interface is illustrated in Fig. IV.18 for four different contrasts of water at the silicon/water interface. Silicon usually has a layer of oxide on the surface. This can be observed directly by matching the water to the underlying silicon, when only the reflectivity from the oxide layer is observed. On the other hand the oxide layer may be made to disappear by matching the water to the oxide layer. Since the composition of the oxide layer is not known beforehand it is a matter of guesswork to match the oxide layer exactly and the match in Fig. IV.18 was just a matter of chance. When the solvent is pure D_2O total reflection is observed but there is no total reflection in H_2O. The four profiles taken together allow one to determine the composition and structure of the oxide layer with some certainty.

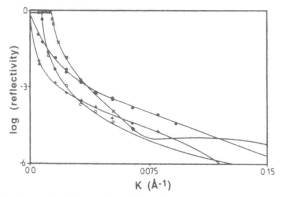

Figure IV.18. — Neutron reflectivity profiles of the silicon/water interface at four different water contrasts, (x) D_2O, (o) water matched to silica, (•) H_2O, and (+) water matched to silicon. The continuous lines are calculated for a layer of oxide of stoichiometry SiO_2, density the same as amorphous silica, and thickness of about 35 Å.

An important area where neutron reflection is uniquely sensitive to inhomogeneities at a surface is in certain magnetic systems. For atoms with a magnetic moment the neutron may be scattered by the nucleus, which has a fixed scattering length, and the magnetic moment of the atom. This leads to a spin dependent scattering length density for magnetised materials. For materials magnetised in the plane of the surface the spin dependent scattering length density for neutrons polarised parallel (+) and antiparallel to the applied field is

$$\rho_\pm = \sum n_i \left(b_i \pm C\mu \right)$$
(IV.9)

where μ is the average magnetic moment per atom and C is a constant. In reflection the effects are most dramatic in between the critical angles for the two states of polarisation and the ratio of the (+) and (-) neutrons reflected from the surface, referred to as the spin-flip ratio, may be very different from unity in this region. This not only provides an effective way of polarising neutrons but is also a sensitive way of probing the magnetic profile across an interface.

GENERAL READING

THOMAS R.K., 1982 - Prog. Sol. State Chem. 12, 1.

PENFOLD J., THOMAS R.K., 1990 - J. Phys. Cond. Matter 2, 1369.

REFERENCES

ALS-NIELSEN J.A., MOHWALD H., 1992 - Handbook of Synchrotron Radiation, Volume 5 - North Holland (Amsterdam).

BORN M., WOLF E., 1980 - Principles of Optics - Pergamon Press (Oxford).

BUCKNALL R.E., CLARKE S.M., SHAPTON R.A., THOMAS R.K., 1989 - Mol. Phys. 67, 439.

DASH J.G., RUHWALDS J., 1980 - Phase Transitions in Surface Films - Plenum (New York).

FREIMUTH H., WIECHERT H., SCHILDBERG H.P., LAUTER H.J., 1990 - Phys. Rev. B42, 587.

GAMLEN P.H., THOMAS R.K., TREWERN T., BOMCHIL G., HARRIS N.M., LESLIE M., TABONY J., WHITE J.W., 1979 - J. Chem. Soc. Far. Trans I 75, 1535.

HUMES R.P., SMALLEY M.V., RAYMENT T., THOMAS R.K., 1988 - Can. J. Chem. 66, 557.

JOBIC H., TOMKINSON J., CANDY J.P., FOUILLOUX P., RENOUPREZ A.J., 1980 - Surf. Sci. 95, 496.

KNORR K., 1992 - Physics Reports 214, 115.

MOWFORTH C.W., RAYMENT T., THOMAS R.K., 1986 - J. Chem. Soc. Far. Trans. 82, 1621.

SMALLEY M.V., HULLER A., THOMAS R.K., WHITE J.W., 1981 - Mol. Phys. 44, 533.

CHAPTER V

NEUTRON SCATTERING AND MAGNETIC STRUCTURES
J. SCHWEIZER

V.1.   Magnetic structures and propagation vectors

V.1.1.   Examples of magnetic structures

In a crystal, the magnetic moments due to the unpaired electrons of the atoms interact, and the exchange energy H depends on the mutual orientation of the moments. For the whole crystal this energy is written in the following way:

$$H = - \sum_{jj'} J_{jj'} \, m_j.m_{j'} \qquad (V.1)$$

where $J_{jj'}$ is the exchange integral between the atoms j and j', which depends on the distance between the atoms and usually falls off very rapidly with increasing distances. At low temperatures this interaction leads to an order of the magnetic moments, the period of which being possibly different from that of the crystal.

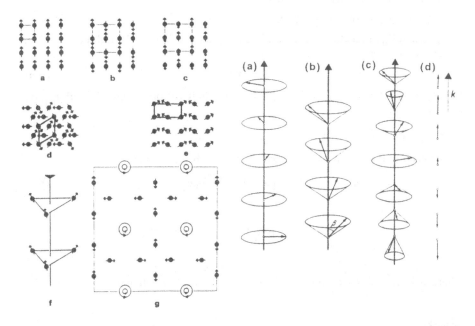

Figure V.1. — Different types of magnetic structures: (a) ferromagnet, (b) antiferromagnet, (c) ferrimagnet, (d) triangular, (e) weak ferromagnet, (f) umbrella, (g) multiaxial[1].

Figure V.2. — Different types of incommensurate magnetic structures: (a) simple spiral, (b) conical spiral, (c) complex spiral, (d) sinusoidal[1].

There are 2 languages to represent the magnetic structures. The first one is mainly illustrative and describes the mutual orientations of the magnetic moments as in Fig. V.1 for magnetic structures which correspond to integral multiples of the chemical cell or in Fig. V.2 for those for which the periodicity is not integral.

In most of the examples, the symmetry of the magnetic structure is lower than the symmetry of the crystal.

The second language is more rigorous and better adapted to neutron scattering. It considers the difference between the translational and the rotational symmetries in the crystal, and therefore it describes the structures in terms of propagation vectors, characteristic of the eigenfunctions of the translation group.

V.1.2.   Propagation vectors

As the magnetic structures are periodic, with a period which may be different from that of the chemical cell, the moment distribution m_{lj} on position j of cell l, can be Fourier expanded:

$$m_{lj} = \sum_k m_j^k \, e^{-ikl} \tag{V.2}$$

where **k**'s are the propagation vectors.

We use the following notations:

The position R_{lj} of an atom corresponding the j-th atom in the cell, and the l-th cell of the crystal is given by:

$$R_{lj} = l + r_j \tag{V.3}$$

where $l = n_1 \, a + n_2 \, b + n_3 \, c$, when **a**, **b**, **c** are the unit vectors of the cell, with n_1, n_2, n_3 integers;

where $r_j = x_1 \, a + x_2 \, b + x_3 \, c$, with x_1, x_2, x_3 fractional.

k is a vector of the reciprocal space, inside the first Brillouin zone:

$$k = k_1 \, a^* + k_2 \, b^* + k_3 \, c^* \tag{V.4}$$

with

$$a^* = \frac{2\pi}{v_0} b \wedge c \; ; \qquad b^* = \frac{2\pi}{v_0} c \wedge a \; ; \qquad c^* = \frac{2\pi}{v_0} a \wedge b$$

where v_0 is the volume of the unit cell and with k_1, k_2, k_3 fractional.

Remarks

1.   The magnetic moment m_{lj} is a real vector. As e^{-ikl} is a complex quantity except for certain particular values of **k** (**k** = 0, **k** = (1/2, 0, 0) for instance), to any vector **k** is associated a vector -**k** with $m_j^{-k} = \left(m_j^k \right)^*$ such as:

$$m_{1j} = m_j^k \, e^{-ikl} + m_j^{-k} \, e^{+ikl} = 2 \left| m_j^k \right| u \cos (kl + \phi) \qquad (V.5)$$

where u is a unit vector.

2.   Note the analogy between the Fourier expansion (V.2) and the well known Bloch functions:

$$\phi_k \, (R) = u_k \, e^{-ikR} \qquad (V.6)$$

V.2.   Magnetic structures and neutron scattering cross section

Neutron are scattered both by the nuclei and by the magnetic moments resulting from the unpaired electrons around the nuclei.

V.2.1.   Nuclear scattering cross section

The scattering amplitude of an individual nucleus j, scattering a neutron wave at a scattering vector Q is:

$$a_{Nj} \, (Q) = b_j + A_j \, \sigma . I_j \qquad (V.7)$$

where $1/2 \ \sigma$ is the neutron spin operator and I_j the spin of the nucleus. This scattering amplitude does not depend on the Q vector, which implies that it does not depend on the scattering angle.

In most of the cases the nuclear spins I_j are not aligned (they order at temperatures well below 10^{-3} K) and the second term does not contribute at all to the interference effects or coherent scattering, and we are left with:

$$a_{Nj} \, (Q) = b_j$$

For a crystal, that is for all the nuclei three-dimensionally ordered and located at positions $R_{1j} = 1 + r_j$, the neutron cross section which corresponds to the number of neutrons scattered, is proportional to the square of the modulus of the scattered amplitude.

$$\left(\frac{d\sigma}{d\Omega} \right)_N = \left| \sum_{1j} b_j \, e^{i \, Q \, R1j} \right|^2$$

$$= \frac{(2\pi)^3}{v_0} \sum_\tau |F_N|^2 \, \delta (Q - \tau) \qquad (V.8)$$

where τ are the positions of the nodes of the reciprocal lattice:

$$\tau = N_1 \, a^* + N_2 \, b^* + N_3 \, c^* \, (N_1, N_2, N_3 \text{ integer}) \qquad (V.9)$$

and where

$$F_N(Q) = \sum_j b_j \, e^{i \, Q.r_j} \, e^{-W_j} \qquad (V.10)$$

F_N is the nuclear structure factor obtained by a sum $\sum_j$ in the unit cell only where

the exponentional term is the Debye Waller factor.

The above formulae state that for a perfect crystal, there is no nuclear scattered intensity, except delta functions at the nodes of the reciprocal lattice, which correspond to the Bragg reflections.

V.2.2.   Magnetic scattering cross sections

V.2.2.1.   The magnetic scattering amplitude for an isolated moment

The interaction between the neutron spin and an individual magnetic moment m_j is a pure dipole-dipole interaction. The scattering amplitude which results of this interaction is

$$a_{Mj}(Q) = p\,\sigma \cdot M_{\perp j}(Q) \tag{V.11}$$

where $1/2\,\sigma$ is the spin operator of the neutron

and

$$p = \frac{\gamma e^2}{2m_e c^2} = \frac{\gamma r_0}{2} = 0.2696\ 10^{-12}\ cm \tag{V.12}$$

where $M_j(Q)$ is the Fourier transform of the magnetization $M_j(r)$ around the nucleus j. and where $M_\perp(Q)$ is the projection of vector $M(Q)$ onto the plane perpendicular to the scattering vector Q (Fig. V. 3).

$$M_\perp = \hat{Q}_\wedge(M_\wedge\hat{Q}) = M - \hat{Q}(M.\hat{Q}) \tag{V.13}$$

with the unitary vector

$$\hat{Q} = \frac{Q}{|Q|} \tag{V.14}$$

Because of the term $M_\perp$, the magnetic scattering amplitude is highly anisotropic, as expected from the dipole-dipole interaction. This anisotropy depends on the respective orientations of the magnetic moment m and the scattering vector Q.

V.2.2.2.   The magnetic form factor

$M(r)$ represents the spatial distribution of the magnetic density around the nucleus. It comes from the moments associated to the spin of the unpaired electrons $M_s(r)$ and also to the moments $M_L(r)$ resulting from the current loops (the orbits) of these unpaired electrons:

$$M(r) = M_s(r) + M_L(r) \tag{V.15}$$

and

$$a_{Mj}(Q) = p\,f_j(Q)\,\sigma \cdot m_{\perp j}$$

where $m_{\perp j}$ is the projection of the magnetic moment m_j onto the scattering plane and $f_j(Q)$ is the magnetic form factor of atom j.

Magnetic form factors may be regarded in two ways:

. the first regards them as characteristic functions of each magnetic ion, Cu^{2+} or Tb^{3+} for instance. It is necessary to know these functions to calculate the magnetic cross sections in order to solve the magnetic structure. Such functions are tabulated in Boucherle [2], Lisher et al [3] and in the International Tables for Crystallography[4].

. the second considers the accurate measurement of the form factors in order to reveal the fine details of the magnetic distribution on the different atoms in a particular crystal. This approach will be developed in Chapter VI.

V.2.2.3. The magnetic cross section

The magnetic cross section for a crystal is obtained by squaring the modulus of the amplitude scattered by all the magnetic moments over all the crystal

$$\left(\frac{d\sigma}{d\Omega}\right)_M = \left| \sum_{lj} a_{Mlj}(Q) e^{iQR_{lj}} \right|^2 \tag{V.17}$$

If these moments are ordered in a magnetic structure one can use in the expression of the magnetic amplitude the Fourier expansion (V.2)

$$m_{lj} = \sum_k m_j^k e^{-ikl} \tag{V.2}$$

and one ends up with the expression of the magnetic cross section

$$\left(\frac{d\sigma}{d\Omega}\right)_M = \frac{(2\pi)^3}{v_0} \sum_k \sum_\tau |F_{M\perp}(Q)|^2 \delta(Q-k-\tau) \tag{V.18}$$

with the magnetic structure factor given by

$$F_M(Q) = p \sum_j f_j(Q) m_j^k e^{iQr_j} e^{-W_j} \tag{V.19}$$

where the sum $\sum_j$ is over all the moments within the unit cell.

Note that the magnetic structure factor is a vector (a complex vector) while the nuclear structure factor is a scalar (a complex scalar).

$$F_{M\perp} = \hat{Q} \wedge \left(F_M \wedge \hat{Q} \right) = F_M - \hat{Q}\left(F_M \cdot \hat{Q} \right) \tag{V.20}$$

$$|F_{M\perp}|^2 = F_M \cdot F_M^* - \left(\hat{Q} \cdot F_M \right)\left(\hat{Q} \cdot F_M^* \right) \tag{V.21}$$

While for a perfect crystal the intensity due to nuclear scattering corresponds to delta functions at the nodes of the reciprocal lattice, for a perfect magnetic structures, the intensity due to magnetic scattering concentrates on delta functions located at:

$$Q = \tau + k \qquad\qquad (V.22)$$

for the different τ nodes of the reciprocal lattice and for the different k'_s which enter in the Fourier expansion (V.2).

V.2.2.4. Examples of magnetic intensities

k = 0 ferromagnetism (ferrimagnetism,antiferromagnetism)

$$\left(\frac{d\sigma}{d\Omega}\right)_M = \sum_\tau |F_{M\perp}(Q)|^2 \, \delta(Q\text{-}\tau) \qquad\qquad (V.23)$$

The magnetic reflection are on the nodes of the reciprocal lattice, and their intensities add to the intensities of the nuclear reflections (Fig. V.3a).

k = k$_x$, k$_y$, k$_z$: modulated or helical structure

To each vector **k** with a component m_j^k si associated a vector -k with a component

$$m_j^k = \left(m_j^k\right)^*$$

For **k** $Q = \tau + k$: reflection τ^+ or (hkl)$^+$

For **-k** $Q = \tau - k$: reflection τ^- or (hkl)$^-$ (Fig. V. 3b)

k on symmetry points of the surface of the Brillouin zone:

ex: **k** = (1/2, 0, 0)

In such a case, it is not necessary to associate a vector -k to the vector k in order to make m_{1j} real.

One may note that a magnetic reflection could be considered either as a (hkl)$^+$ or as a (h'k'l')$^-$ (Fig. V.3c)

multiaxis structure: Several vectors **k** ($k_1 = k_x^1,\ k_y^1,\ k_z^1;\ k_2 = k_x^2,\ k_y^2,\ k_z^2\ ...$) are present in the expansion (V.2), as a result of a coupling between them.

There are more than one pair of magnetic reflections around each node of the reciprocal lattice. (Fig. V.3d)

distorted modulated structures where the modulation is no more sinusoidal (squaring of the modulation for instance) and requires more than one pair of harmonics to represent it.

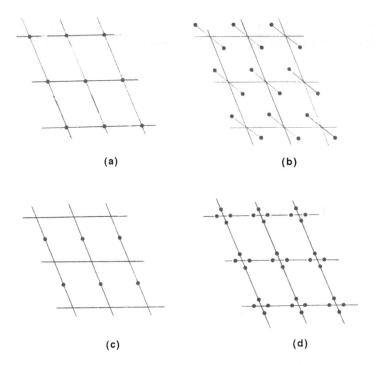

(a)   (b)

(c)   (d)

Figure V.3. — Location of the magnetic reflections in the reciprocal lattice (a) ferromagnetism (b) modulated or helical structure, (c) antiferromagnetism, (d) multiaxis structure.

V.3.   Instrumentation

V.3.1.   Single crystal diffractometer: two types exist

V.3.1.1.   The 4 circle diffractometer

The crystal is mounted in an eulerian cradle (three angles: ω, γ, ϕ, as shown in Fig. V.4) which is able to bring any vector of the reciprocal lattice in a reflection position in the horizontal plane. The counter remains in the horizontal plane and collects the intensity, while the crystal is rotated around the vertical axis (ω scan or $\omega/2\theta$ scan) in order that the reflection nodes cross the Ewald sphere.

In such a case the integrated intensity of a Bragg reflection corresponds to an integration of the differential cross section through the δ function and is given by

$$I_{int} = \Delta(\omega) \sum_i J(\omega_i) = \int \left(\frac{d\sigma}{d\Omega}\right) d\Omega = \frac{N}{v_0} \frac{\lambda^3}{\sin 2\theta} |F|^2 A \qquad (V.24)$$

where N is the number of unit cells in the crystal, v_0 the volume of one of the unit cells, $\Delta(\omega)$ the scanning step, $J(\omega_i)$ the number of counts (background removed) for each position ω_i of the crystal, and A the absorption factor. The Lorentz factor

$1/\sin 2\theta$ expresses how the integration of the δ function in the reciprocal lattice is transformed in an integration in the (ω, χ, 2θ) experimental angles.

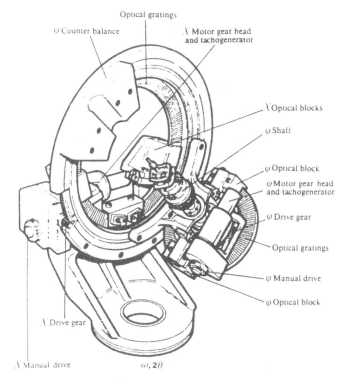

Figure V.4. — 4 - circle diffractometer[5]

V.3.1.2. The lifting counter diffractometer allows a more elaborate sample environment around the crystal, such as cryostat, furnace or magnetic field.

In such an arrangement the only motion of the crystal is a rotation around the vertical axis (ω), but the counter which moves around 2 axes (γ in the horizontal plane and ν in the vertical plane) can be lifted above the horizontal plane to collect the intensities.

Formula (V.24) is then replaced by

$$I_{int} = \Delta(\omega) \sum_i J(\omega_i) = \int \left(\frac{d\sigma}{d\Omega}\right) d\Omega = \frac{N}{v_0} \frac{\lambda^3}{\sin\gamma \sin\nu} |F|^2 A \qquad (V.25)$$

where the new Lorentz factor reflects the change of diffraction geometry (normal beam mode), and where one has to keep in mind, in the evaluation of $F_{M\perp}$, that the scattering vector Q is no longer in the horizontal plane.

V.3.2. Powder diffractometer

The powder sample is contained in a vertical cylindrical holder bathed fully in the monochromatic neutron beam. The diffraction diagram is recorded by a detector which can rotate around the sample, and at each position intercepts part of the Debye Scherrer cones (see Fig. V.5).

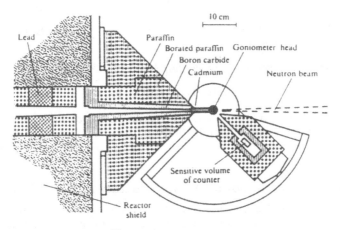

Figure V.5. — powder diffractometer[5]

In these conditions, the integrated intensity of a Bragg reflection, corresponds also to an integration of the differential cross section around the δ function in the reciprocal space.

$$I_{int} = \Delta(2\theta) \sum_i J(2\theta_i) = \int \left(\frac{d\sigma}{d\Omega}\right) d\Omega = \frac{N}{v_0} \frac{\lambda^3}{8\pi \sin\theta \sin2\theta} \frac{l}{r} \sum |F|^2 A \qquad (V.26)$$

where $\Delta(2\theta_i)$ is the scanning step of the counter, $J(2\theta_i)$ the number of counts (background removed), l the height of the counter, r the distance from specimen to counter, F being the structure factor of the Bragg reflection which is measured and where the former Lorentz factor $1/\sin 2\theta$ has become $1/\sin\theta \sin 2\theta$ as it includes also the probability of finding a cristallite in a reflection position, and the portion of the Debye Scherrer cone which is intercepted by the detector.

The neutron intensity in formula (V.26) may arise from nuclear, magnetic or both nuclear and magnetic scattering.

Different auxiliary equipment may be mounted around the sample to impose particular conditions of temperature, pressure or magnetic field.

More and more the unique detector is being replaced by a rotating block of detectors or by a non rotating position sensitive detector, enormously improving the efficiency.

Differential cross section

In the absence of long range order, the number of neutrons which enter the counter (non integrated intensity) at position $2\theta_i$ is proportional to the differential cross section of the sample:

$$J\left(2\theta_i\right) = N' \frac{1}{r} \frac{d\sigma}{d\Omega} A \qquad\qquad (V.27)$$

where N' is the number of diffusing centers in the sample and $d\sigma/d\Omega$ the differential cross section of each of these centers.

For paramagnetic scattering where $d\sigma/d\Omega = 2/3\ p^2 f^2(Q)\ m^2$

$$J\left(2\theta_i\right) = N' \frac{1}{r} \frac{2}{3} p^2\, f^2\,(Q)\ m^2 A \qquad\qquad (V.28)$$

Formulae (V.26) and (V.28) allow a direct comparison of diffuse and integrated intensities which represent differential (in barn/steradian) and integrated (in barn) cross section.

V.4.   Determination of magnetic structures

The knowledge of a magnetic structures implies the determination of the propagation vector, the direction of the magnetic moments and, if necessary, the mutual orientations of the moments of the atoms belonging to different Bravais lattices.

V.4.1.   Identification of the propagation vector

The identification of magnetic reflections is usually performed by a careful comparison of powder diagram below and above T_c. These magnetic reflections have to be indexed with scattering vectors:

$$Q = \tau \pm k$$

and the difficulty in the identification of the vector **k** results from the vectorial nature of this relation.

-       There is an intuitive identification for very simple vectors **k** as **k** = (0, 0, 0), **k** = (1/2, 0, 0) or **k** = (1/2, 1/2, 1/2)

-       There is a graphic method where one plots in some appropriate sections of the reciprocal lattice the circles of radii H_i corresponding to the observed reflections $H_i = (4\pi \sin \theta_i)/\lambda$, and one looks for all the possible pairs, symmetrical, around the nodes τ of the lattice (Fig. V.6).

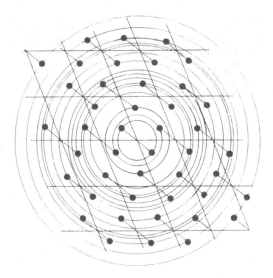

Figure V.6. — Graphic method to find out the propagation vector.

-       There are systematic programmes[6] which slice the first Brillouin Zone in small subcells labelled k_j (for instance $1/100$ of a^*, b^*, c^*) and select all the possible k_j which can fit all the observed magnetic reflection H_i, within their error bars.

V.4.2.   Direction of the magnetic moments

Neutron scattering is sensitive to the direction of the magnetic moments. Formula (V.11) states that the effective part of the magnetic structure factor is the projection of this structure factor on a plane perpendicular to the scattering vector:

$$\left(\frac{d\sigma}{d\Omega}\right)_M \sim |F_{M\perp}(Q)|^2 \qquad (V.29)$$

$$F_{M\perp}(Q) = \sum_j p_j\, f_j(Q)\, m_{j\perp}^k\, e^{iQ\cdot r_j} \qquad (V.30)$$

For a collinear structure, all the $m_{j\perp}^k$ are parallel and

$$|F_{M\perp}(Q)| = \sin\alpha \left| \sum_j \pm p_j\, f_j(Q)\, m_j^k\, e^{iQ\cdot r_j} \right| \qquad (V.31)$$

where α is the angle between the m_j^k and Q and where $\pm$ represents the sign of m_j^k in this structure (parallel or antiparallel).

The integrated intensity, measured on a powder diffractometer implies not a unique vector Q, but all the equivalent vectors Q' deduced from Q by the crystal symmetries.

As the magnetic structure is often less symmetrical than the crystal structure, those vectors Q' may be equivalent or non equivalent as regard to the configurational symmetry of the magnetic structure. If they are equivalent they correspond to a same value of

$$\left| \sum_j \pm p_j \, f_j(Q) \, m_j^k \, e^{iQ.r_j} \right| \qquad (V.32)$$

but to different values of $\sin \alpha$.

The powder integrated intensity, which is the sum of all the possible magnetic reflection at the same angle, can be written

$$I_{int} \sim \sum_{\substack{Q \\ inequiv.}} z_Q <\sin^2 \alpha> \left| \sum_j \pm p_j \, f_j(Q) \, m_j^k \, e^{iQ.r_j} \right|^2 \qquad (V.33)$$

where z_Q is the multiplicity of the "magnetic equivalent" vector Q and $<\sin^2 \alpha>$ the average of $\sin^2 \alpha$ on these equivalent reflections.

Shirane[7] has shown that for a cubic magnetic symmetry $<\sin^2 \alpha> =2/3$, and that for an uniaxial magnetic symmetry, $<\sin^2 \alpha>$ depends only on the angle between this axis and the direction of the collinear structure.

Is it then possible to determine completely the direction of the magnetic moment by scattering on a single crystal ? One has at this point to examine the problem of magnetic domains.

V.4.3.   Magnetic domains

A single crystal, magnetically ordered, consists usually of several magnetic domains. These domains are of 2 types : K and S.

1)   K domains correspond to the different possibilities for the propagation vector k to select one among the equivalent crystal directions.

Example : In MnO, $k = (1/2, 1/2, 1/2)$, with 4 equivalent directions [1,1,1] (Fig. V.7).

2)   S domains which may exist inside a single K domain when there are several equivalent directions for the alignment of m^k, equivalent regarding the crystal axis, but equivalent also regarding the k direction.

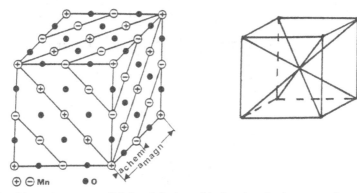

Figure V.7. — Magnetic structure of MnO and the 4 possible directions for the propagation vector.

In a single crystal experiment, different K domains will provide different sets of magnetic reflections, but each magnetic reflection will imply an average on the S domains, with the same conclusions as for the average on powders: unless an external perturbation is applied which unbalance the domain distribution, if the magnetic structure has a unique axis of symmetry, only the angle between the magnetic moments, and this unique axis can be determined.

V.4.4.   Mutual orientations of the moments

For crystals with only one magnetic atom per unit cell, the knowledge of the propagation vector(s) and the direction of the moments determine completely the magnetic structure. For several magnetic atoms per unit cell, the mutual orientation of their moments has also to be found, from the experiment, by the use of formulae (V.19), (V.20) and (V.21), but also with some help given by the theory of magnetic interactions.

V.5.   Magnetic structures and magnetic interactions

V.5.1.   The magnetic interaction operator

If in one magnetic cell (labelled l) there are several magnetic atoms j belonging to the same cristallographic site (several Bravais lattices) the magnetic interaction (V.1) may be generalized at the second order by:

$$H = - \sum_{ll'} \sum_{jj'} \sum_{\alpha\beta} J_{ll'jj'\alpha\beta} \, m_{lj\alpha} \, m_{l'j'\beta} \qquad (V.34)$$

where, α, β are the components x y z

Introducing the Fourier expansions

$$m_{lj} = \sum_{k} m_j^k \, e^{-ik \, l}$$

$$m_{l'j'} = \sum_{k'} m_{j'}^{k'} \; e^{-ik' \, l'}$$

$$H = - \sum_{ll'} \sum_{jj'} \sum_{\alpha\beta} J_{ll' \, jj' \alpha \beta} \sum_{kk'} m_{j\alpha}^k e^{-ikl} \; m_{j'\beta}^{k'} e^{-ik' \, l'} \tag{V.35}$$

we introduce: $e^{-ikl'} \, e^{+ikl'} = 1$

$$\tag{V.36}$$

$$H = - \sum_{ll'} \sum_{jj'} \sum_{\alpha\beta} \sum_{kk'} m_{j\alpha}^k \; m_{j'\beta}^{k'} \; e^{-il'(k + k')} J_{ll' jj' \alpha \beta} \; e^{-i \, k(l - l')}$$

defining $$J_{jj'\alpha\beta}(k) = \sum_l J_{ll' jj' \alpha\beta} \, e^{-ik(l - l')} \tag{V.37}$$

as the Fourier transform of J, it is independant of l' and

$$H = - \sum_{jj'} \sum_{\alpha\beta} \sum_{kk'} m_{j\alpha}^k \; m_{j'\beta}^{k'} \; J_{jj'\alpha\beta}(k) \sum_{l'} e^{-il'(k + k')} \tag{V.38}$$

with

$$\sum_{l'} e^{-il'(k + k')} = \delta(k + k' - \tau)$$

In the first Brillouin zone this implies $k + k' = 0 \Rightarrow k = -k'$ which finally gives

$$H = - \sum_{jj'} \sum_{\alpha\beta} \sum_k J_{jj'\alpha\beta}(k) \, m_{j\alpha}^k \, m_{j'\beta}^{-k} \tag{V.39}$$

which represent, in the Fourier space, the general interaction at the order 2. The propagation vector **k** and the Fourier components of the magnetic structure will be those which minimize this interaction.

These consideration will give a method to determine the possible configurations of the magnetic moments within a chemical cell.

V.5.2.   The Landau Theory applied to magnetic structures

Expression (V.39) may be also written, using any linear combination of the components $m_j^k \, \alpha$.

$$V_\lambda^k = \sum_{j\alpha} a_{\lambda j\alpha}^k \; m_{j\alpha}^k \tag{V.40}$$

which gives

$$H = - \sum_k \sum_{\lambda\lambda'} A_{\lambda\lambda'}^k \; V_\lambda^k \, V_\lambda^{-k} \tag{V.41}$$

The Landau theory of the second order phase transitions, applied to magnetic structures, classifies the linear combination V_λ^k which allow to build a magnetic hamiltonian (V.41) which, in the paramagnetic state, is invariant under all the symmetry operations of the group. These vectors V_λ^k are the basis vectors of the irreducible representation Γ_λ^k of the group.

In the paramagnetic state, magnetic fluctuation exists, represented by vectors V_λ^k in order to keep the symmetry of the hamiltonian. The number of vectors V_λ^k belonging to one irreducible representation Γ_λ^k is the same as the order of the representation. At the critical temperature, one of these fluctuations develops to give an ordered magnetic structure, while the other fluctuations die out.

The magnetic fluctuation which develops is that corresponding to the lowest energy: lowest as regard to the propagation vector k and lowest as regard to the linear combination (V.40). The propagation vector k being determined by the neutron experiment, one has to list the basis vector of all the irreducible representations of the symmetry group and check from the measured intensities which magnetic structure is compatible with the data.

A detailed description of the method is given by Rossat-Mignod[7].

V.6.   Classification of the magnetic structure

V.6.1.   Incommensurate structures

In the general case, the k vector may be any point of the first Brillouin zone, and will define either an incommensurate structure or a long period commensurate one (e.g. $k=1/7$). The latter behaves like the former from the neutron diffraction point of view. Two types of incommensurate structures exist, depending on the dimension of the corresponding irreducible representation.

V.6.1.1.   Sine wave modulated structures

The magnetic moments order according to

$$m_{lj} = u_j \cos \left(kl + \varphi_j\right) \qquad (V.42)$$

comparing to the general Fourier expansion (V.2), the Fourier coefficients m_j^k, corresponding to an irreducible representation of dimension 1 are:

$$m_j^k = \frac{u_j}{2} e^{i\varphi_j} \qquad (V.43)$$

Expression (V.42) is reconstructed by associating the Fourier coefficients m_j^k and m_j^{-k}, which makes m_{lj} real.

These Fourier coefficient m_j^k (one has to be careful with the factor $\frac{1}{2}$ which is present when using amplitude $\bar{u}_j$ in the expression of $\bar{F}_M$) enter the expression of structure factor expressed in (V.19):

$$F_M(Q) = p \sum_j f_j(Q) m_j^k \, e^{iQr_j} \, e^{-W_j} = p \sum_j f_j(Q) \frac{u_j}{2} \, e^{i(Q.r_j + \varphi_j)} e^{-W_j} \qquad (V.44)$$

A sine wave modulated structure implies that all the moments have different lengths. Such structures are often observed at higher temperature, in the vicinity of T_N, where the differences in the length of ordered moments is due to thermal disorder. At lower temperature, unless the ground state is a singlet, the lengths of the ordered moments tend to become equal by a squaring of the modulation, and the occurrence of new harmonics of k.

V.6.1.2.   Helical structures

The magnetic moments are ordered as an helix according to:

$$m_{lj} = u_j \cos\left(kl + \varphi_j\right) + v_j \sin\left(kl + \varphi_j\right) \tag{V.45}$$

where u_j and v_j are 2 vectors orthogonal of the same length.

The Fourier coefficients m_j^k correspond here to an irreducible representation of dimension 2:

$$m_j^k = \left(\frac{u_j}{2} - i\,\frac{v_j}{2}\right) e^{i\varphi_j} \tag{V.46}$$

The 2 basis vectors being $\frac{u_j}{2}\, e^{i\,\varphi_j}$ and $-i\,\frac{v_j}{2}\, e^{i\,\varphi_j}$

Here also m_j^k is associated to m_j^{-k} to make m_{lj} real and the structure factor is expressed in terms of the Fourier coefficients:

$$F_M = p \sum_j f_j\,(Q) \left[\frac{u_j}{2} - i\,\frac{v_j}{2}\right] e^{i\,\varphi_j}\ e^{iQr_j}\ e^{-W_j} \tag{V.47}$$

Such a two dimension representation may be found for propagation vectors parallel to a 3, 4 or 6 fold symmetry axis with moments rotating in an easy plane perpendicular to this axis.

In contrast to a sine wave modulation, a helical structure can remain stable down to the lowest temperature.

V.6.2.   Commensurate structures

Although the general case for the propagation vector is any point of the first Brillouin zone, certain points of this zone are symmetrical for all the properties of the crystal, including the magnetic Hamiltonian. Therefore the Hamiltonian presents an extremum for these k values, possibly a minimum which defines a commensurate magnetic structure.

V.6.2.1. The center of the zone

For $k = (0, 0, 0)$, it is the case of a ferromagnetic structure if there is only one magnetic atom per unit cell, but it may be antiferromagnetic or triangular when there are several magnetic atoms.

The Fourier expansion (V.2) reduces to:

$$m_{lj} = m_j^k = m_0$$

The Fourier coefficient is equal to the moment.

V.6.2.2. Symmetry points at the surface of the Brillouin zone

For each of the 14 Bravais lattice there exist special symmetry points. For the cubic lattices they are:

	Type I	Type II	Type III
Primitive cubic P	0 0 1/2	1/2 1/2 0	1/2 1/2 1/2
Body centered cubic I	0 0 1	1/2 1/2 0	
Faces centered F	0 0 1	1/2 1/2 1/2	

For such structures the Fourier expansion reduces to

$$m_{lj} = e^{-ikl} m_j^k = (-)^n m_j^k$$

There is an alternance, along the propagation vector, of parallel and antiparallel moments.

An important point to notice, in the case of these commensurate structures, is that the Fourier coefficients are real, and imply the moduli of the magnetic moments. The structures factor are written

$$F_M (Q) = p \sum_j f_j (Q) m_j e^{iQr_j} e^{-W_j} \qquad (V.48)$$

Comparing for instance with expression V.44 one can see that there is no continuity in the expression for the magnetic structure factor from incommensurate to commensurate structure, and this is the result of the difference in the average moment which scatters the neutron.

V.7. Single k and multi k structures

Let us assume for simplicity that there is only one magnetic atom per cell. The Fourier expansion for the magnetic moments given by (V.2) reduces here to:

$$m_l = \sum_k m^k e^{-ik.l} \qquad (V.49)$$

We have already mentioned that for a general vector k of the first Brillouin zone, the vector $-k$ should be associated to k with $m^{-k} = (m^k)^*$ to have m_l real.

In the case of several vectors k corresponding each other by the symmetry of the crystal : k_1, k_2, k_3 ... we have to distinguish between:

-       a single k structure where only one of the k vector (eventually associated with -k) enters the sum (V.49).

-       a multi k structure where at least 2 of the equivalent vectors: k_1, k_2... (eventually associated with $-k_1, -k_2$ ...) enter the sum (V.49).

V.7.1.   Single k structure

When the crystal in cooled down, a magnetic structure develops for any of the equivalent k vectors of the crystal. There will be different domains (K domains).

In a 1st domain:

$$m_l = m^{k_1} e^{-ik_1 l} + cc \qquad (V.50)$$

with an energy (at the order 2)

$$H_1 = - J(k) m^{k_1} m^{-k_1} \qquad (V.51)$$

In a 2nd domain

$$m_2 = m^{k_2} e^{-ik_2 l} + cc \qquad (V.53)$$

with an energy

$$H_2 = - J(k) m^{k_2} m^{-k_2} \qquad (V.54)$$

equal to H1.

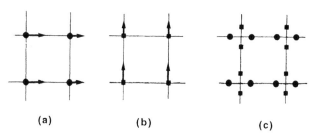

(a)   (b)   (c)

Figure V.8. — Illustration of the scattering due to 2 different single k domains (a) first domain, (b) second domain, (c) magnetic peaks in the reciprocal space.

Each domain will give rise to its own set of magnetic reflections, as illustrated in Figure V.8. The intensity of each set may allow to get the relative proportion of each domain (in volume).

V.7.2.   Multi k structure

For a double k structure equation (V.49) becomes

$$M_l = M^{k_1} e^{-ik_1 l} + M^{k_2} e^{-ik_2 l} + cc \qquad (V.55)$$

with an energy (at the order 2)

$$H_m = -J(k)(M^{k_1}M^{-k_1} + M^{k_2}M^{-k_2}) \tag{V.56}$$

This energy is exactly the same as that of a collinear single k structure as expressed by H_1 or H_2, at the condition that component M^k obeys relation:

$$M^k = \frac{1}{\sqrt{2}}m^{k_1}$$

The condition should have been $\frac{1}{\sqrt{3}}m^{k_1}$ for a triple k structure.

With an hamiltonian limited to order 2, there is no difference of energy between a single k and a multi k structure. It is the higher order terms which will make the difference in the following expression:

$$H = -J\sum_{k_i} m^{k_i}m^{-k_i} + A_4\sum_{k_i}\left(m^{k_i}\right)^4 + A'_4\sum_{k_i k_j}\left(m^{k_i}\right)^2\left(m^{k_j}\right)^2$$

$$+ A_6\sum_{k_i}\left(m^{k_i}\right)^6 + A'_6\sum_{k_i k_j}\left(m^{k_i}\right)^4\left(m^{k_j}\right)^2 + A''_6\sum\ldots \tag{V.57}$$

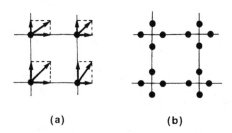

(a)   (b)

Figure V.9. — Illustration of a double k structure (a) the magnetic structure in the real space (b) the magnetic peaks in reciprocal space.

As an illustration, a double k structure and its magnetic pattern in the reciprocal space is represented in Fig. V.9. It is not collinear and there is only one magnetic domain which corresponds to k_1 and k_2. The magnetic pattern is the same as that obtained in the case of a single k structure with 2 magnetic domains equally populated. To distinguish between the two cases it is necessary to apply an external perturbation as a magnetic field or a uni axial stress. Such a perturbation would modify the balance between the domains for single k structure, and would not change, at least for small perturbations, the multi k structure.

V.8. Nota bene: the phase convention on the Fourier component m^k

In this course the Fourier components have been defined by

$$m_{lj} = \sum_k m_j^k e^{-ikl} \tag{V.2}$$

where the phase in the exponential is obtained from the coordinates of the origin of the cell of the magnetic atom (lj).

The corresponding cross section is

$$\left(\frac{d\sigma}{d\Omega}\right)_M = \frac{(2\pi)^3}{v_0} \sum_k \sum_\tau |F_{M\perp}|^2 \, \delta(Q-k-\tau) \tag{V.18}$$

with

$$F_M(Q) = p \sum_j f_j(Q) \, m_j^k \, e^{-Qr_j} \, e^{-Wj} \tag{V.19}$$

where the phase in the exponential is obtained from the vector Q at the position of the magnetic reflection

$$Q = \tau + k \tag{V.22}$$

There exists in the literature another convention for the Fourier components

$$m_{lj} = \sum_k \left(m_j^k\right)' e^{-ik(l+r_j)} \tag{V.2'}$$

where the phase in the exponential corresponds to the coordinates $R_{lj} = l + r_j$ of the magnetic atom.

Comparison of (2) and (2') yields

$$\left(m_j^k\right)' = m_j^k \, e^{ik\,r_j} \tag{V.58}$$

The magnetic reflections are obviously located at the same position of the reciprocal space and equation (18) still stands, but the expression (V.19) of the magnetic structure factor is replaced by

$$F_M(Q) = p \sum_j f_j(Q) \left(m_j^k\right)' e^{i\tau.r_j} \, e^{-Wj} \tag{V.59}$$

where the phase in the exponential corresponds to the position of the node τ of the reciprocal lattice and not to the position of the magnetic reflection.

REFERENCES

(1)   PRANDL W., 1978 - "Neutron Diffraction" edited by H. Dachs Springer (Berlin, Heidelberg New York).

(2)   BOUCHERLE J.X., 1988 - "A compilation of Magnetic Form Factors and Magnetization Densities", Neutron Diffraction Commission of the International Union of Cristallography (Grenoble).

(3)   LISHER E., FORSYTH J.B., 1971 - Acta Cryst. A27, 545.

(4)   International Tables for Crystallography.

(5)   BACON G.E., 1975 - "Neutron Diffraction" Clarendon Press (Oxford).

(6)   KNAPP. 1974 - J. Appl. Cryst 7,370.

(7)   SHIRANE G., 1959 - Acta Cryst 12, 282.

(8)   ROSSAT MIGNOD J., 1987 - "Neutron Physics", edited by Skold and Price, Academic Press.

CHAPTER VI

MAGNETIC FORM FACTORS AND MAGNETIZATION DENSITIES
J. SCHWEIZER

VI.1.   Introduction

In Chapter V we have seen how neutrons can determine magnetic structures, that is the arrangement of the moments of the different magnetic atoms in the crystal. On a smaller scale the neutrons are also a unique tool to reveal the details of the distribution of the magnetization density around the nuclei of the magnetic atoms.

Such investigations require very accurate measurements of the magnetic cross sections, and therefore imply the use of polarized neutrons, as we shall see in the next paragraph. We shall then examine how it is possible to get informations on the magnetic electron distributions, first without the help of any physical model and then using such a model.

VI.2.   The accurate measurement of the magnetic amplitudes with polarized neutrons

The magnetization density $m(r)$ being periodic the measurements are performed at the Bragg reflections. For unpolarised neutrons the intensity of these reflections is given by:

$$I = |F_N|^2 + |F_M|^2 \qquad \text{(VI.1)}$$

where F_N and F_M are the nuclear and magnetic structure factors. For small F_N the sensitivity is very much improved by the use of polarized neutrons as the intensity of Bragg reflections depends on the polarization of the beam in the following way:

$$I^{\pm} = |F_M \pm F_N|^2 \qquad \text{(VI.2)}$$

To illustrate the sensitivity of polarized neutrons for small magnetic amplitudes, let us take an example where $F_M = 0.1\ F_N$. The unpolarized neutrons would be scattered with an intensity $I = F_N^2 + (0.1\ F_N)^2 = 1.01\ F_N^2$ while the polarized neutrons would provide $I^+ = (F_N + 0.1\ F_N)^2 = 1.21\ F_N^2$ or $I^- = (F_N - 0.1\ F_N)^2 = 0.81\ F_N^2$ for the two possible directions of the polarization. The improvement is tremendous.

An experiment with polarized neutrons, in order to measure the spin (or magnetization) density, consists in a collection of flipping ratios at the Bragg reflections (see Fig. VI.1), the flipping ratio being the ratio of the intensities for the 2 polarizations up and down (+ and -) of the incident beam:

$$R(Q) = \frac{I^+(Q)}{I^-(Q)} = \frac{|F_N(Q) + F_M(Q)|^2}{|F_N(Q) - F_M(Q)|^2} \qquad \text{(VI.3)}$$

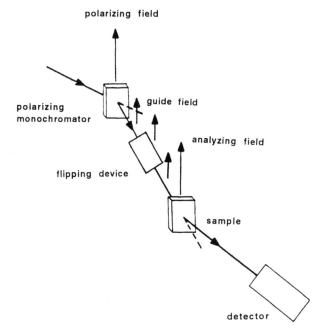

Figure VI.1. - Experimental set up for a polarized neutron diffraction experiment.

As the crystal structure, and hence the set of F_N, are supposed to be known, equation (VI.3) provides, at least for the centric structures where F_N and F_M are real quantities, the values of the magnetic structure factors F_M.

The question which arises now is how to retrieve the spin (or magnetization) density from the experimental $F_M(Q)$.

VI.3.   The Fourier Method: a model free reconstruction

VI.3.1.   The Fourier inversion

The magnetic structure factors F_M are the Fourier coefficients of the periodic function $m(r)$:

$$F_M(Q) = \iiint m(r) e^{iQr} d^3r \tag{VI.4}$$

It is then quite natural to use the Fourier inversion to obtain the magnetization density

$$m(r) = \frac{1}{V} \sum_Q F_M(Q) e^{-iQr} \tag{VI.5}$$

where V is the volume of the unit cell and where the sum $\sum_Q$ implies all the reciprocal vectors. In practice it is not possible to measure all the intensities and the series (VI.5) is performed over a number of reflections large enough to see the details of interest in the magnetization map. This method, first applied by Shull et Yamada[1] for the magnetization of iron in 1962 (see Fig. VI.2) has been used since, almost systematically, for all the studies with polarized neutrons. A more recent example is presented in Fig. VI.3 which shows the coexistence of non magnetic and magnetic cerium atoms in the compounds Ce_2Sn_5 and Ce_3 Sn_7, two superstructures of the intermediate valence system $CeSn_3$ (Boucherle et al[2]).

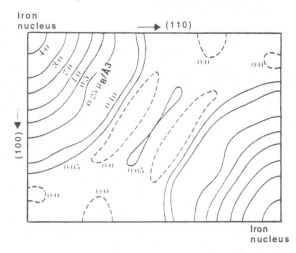

Figure VI.2. — Magnetization density distribution in the (110) plane of Fe obtained by Fourier inversion[1].

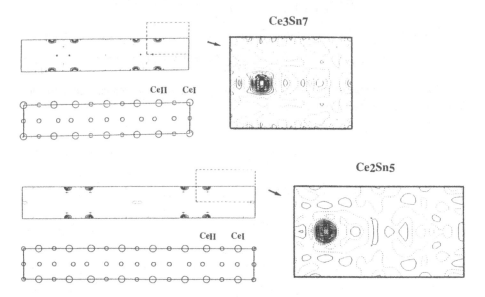

Figure VI.3. — Magnetization density projected along the $\bar{a}$ axis of Ce_2Sn_5 and Ce_3Sn_7, obtained by Fourier inversion. Ce_I is non magnetic (as in $CeSn_3$) while Ce_{II} is magnetic[2].

VI.3.2.   Criticism of the Fourier inversion

The Fourier inversion, so simple in practice, suffers from several fundamental drawbacks as the lack of completeness of the data and the ignorance of the experimental uncertainties in the treatment.

VI.3.2.1. Lack of completeness of the data

The Fourier inversion expressed by relation (VI.5) is exact only if the sum $\sum\limits_{Q}$ includes all the reciprocal lattice points. As this number is infinite, the magnetic structure factors are evaluated up to a limit values of Q: Q_{max}. On the other hand, some reflections inside the sphere $Q = Q_{max}$, cannot be measured simply because their nuclear amplitude is too weak to yield a reliable value of the flipping ratio. For these reasons the sum (VI.5) is replaced by a partial sum where all the missing terms may not be negligible. This implies both biased values for the calculated densities but also series terminaison errors.

To reduce the troubles due to the limitation $Q < Q_{max}$, one has to keep in mind that such measurements cannot yield details of the densities with dimensions d smaller than d_{min}[3] ($d_{min} = 3.8 \, / Q_{max}$). Then, instead of reconstructing the point density m(xyz), one considers the average density $\overline{m}$(xyz)[4], average of the density m(xyz) over a box of volume $(2\delta)^3 \dfrac{V}{abc}$ and centered at xyz. V, a, b and c refer to the volume and the lattice parameters of the unit cell of the crystal. In the often encountered cubic case, one obtains:

$$\overline{m}\,(xyz) = \frac{1}{(2\delta)^3} \int_{x-\delta}^{x+\delta} \int_{y-\delta}^{y+\delta} \int_{z-\delta}^{z+\delta} m\,(xyz)\; dx\; dy\; dz$$

$\overline{m}$ (xyz) is obtained by a series similar to (VI.5) where F_M (Q) is replaced by F'_M (Q):

$$F_M\,(Q) = \left(\frac{\sin 2\pi h\,\delta}{2\pi h\,\delta}\right) \left(\frac{\sin 2\pi k\,\delta}{2\pi k\,\delta}\right) \left(\frac{\sin 2\pi l\,\delta}{2\pi l\,\delta}\right) F_M\,(Q) \qquad \text{(VI.6)}$$

The convergence of the series is faster when F'_M (Q) replaces F_M (Q), and the larger δ, the faster the series converges.

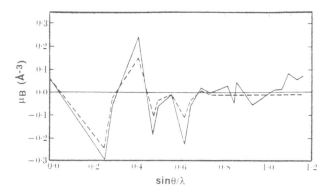

Figure VI.4. — Convergence of the Fourier series for the magnetization density at position (1/2 0 0) of nickel. Full line: not averaged, broken line: averaged[5].

An example of accelerated convergence is given in Fig. VI.4 where the magnetization density at position (1/2 0 0) of the nickel cell is shown as function of $(\sin \theta/\lambda)_{max}$ for an averaged and for a non averaged density[5].

Average over other volumes, adapted to specific problems, may be used. They result in multiplying by other (apodisation) functions which transform $F_M (Q)$ into $\bar{F}_M (Q)$.

VI.3.2.2. Experimental uncertainties

The experimental structure factors $F_{M\ obs} (Q)$ are measured with a certain accuracy. In the procedure represented by equation (VI.5), the data are introduced without any use of these uncertainties: a precise structure factor is treated exactly as a very imprecise structure factor. This part of the information is forgotten.

VI.3.2.3. Probabilities

One can consider a reconstructed map as a model in the real space for which a number of structure factors $F_{M\ obs} (Q)$ has been observed. It is usual, to evaluate the probability of such a model, to compare the agreement of the observed and the calculated structure factors. To do that, a quantity χ^2 is defined such as:

$$\chi^2 = \frac{1}{n} \sum_Q \frac{1}{\sigma^2} | F_{M\ cal} (Q) - F_{M\ obs}(Q)|^2 \qquad (VI.7)$$

where n is the number of independent observations, and σ the experimental uncertainty of each reflection.

For the Fourier inversion, $F_{M\ cal} (Q) = F_{M\ obs} (Q)$, whatever the experimental uncertainties are. This shows that, among all the possible reconstructions which are compatible with the data $(\chi^2 \sim 1$ or $\chi^2 < 1)$, Fourier chooses that one which gives exactly $\chi^2 = 0$ for those reflections which have been measured, and exactly $F_{M\ cal} = 0$ for those which have not been measured. This is the main bias of the method.

VI.4.   The maximum entropy method: an intelligent model-free reconstruction

VI.4.1.   A Bayesian approach: conditional probabilities

The reconstruction of the spin density map from the knowledge of a set of data (the magnetic structure factors) can be considered in terms of probabilities and conditional probabilities. Considering all the possible maps, one tries to evaluate for each of these maps the probability of such a map, knowing that the structure factors are those measured. Such a conditional probability can be written p (map | data).

It is then suitable to use the very general Bayes equality

$$p(A \,|\, B)\, p(B) = p\,(B \,|\, A)\, p(A) \qquad (VI.8)$$

which gives the posterior probability

$$p\,(map \,|\, data) = \frac{p\,(data \,|\, map)\, p\,(map)}{p\,(data)} \qquad (VI.9)$$

In this relation:

• p (data| map), the likelihood, represents the probability of the set of the experimental data if one given density map is supposed to be true, in other words it represents the agreement between the $F_{M\ obs}$ and the $F_{M\ cal}$, as expressed by the χ^2

• p (map), the prior probability, represents an intrinsic probability of the map, without any reference to the data.

• p (data) represents an intrinsic probability of the data, without any reference to the map. This probability is unity once the set of data has been obtained. We are then left with the relation:

$$p(map \mid data) = p\ (data \mid map)\ p(map) \qquad (VI.10)$$

which means that the probability of a map, knowing the set of measured data, is not only represented by the agreement between observed and calculated structure factors but also by the intrinsic probability of the map. Clearly the Fourier inversion completely neglects this last term. We can then say that among all the possible reconstructions which are compatible with the data, the maximum entropy method chooses that one which corresponds to the highest intrinsic probability of the distribution.

VI.4.2. The entropy of a distribution

The intrinsic probability of a map, p (map) may be expressed in term of entropy. This concept of entropy was first introduced by Boltzmann to express the probability of a given configuration of the phase space (6 dimensions space representing positions and velocities) of N particles. He showed that all the configurations have not the same chances of occuring: the most probable being those which maximize the quantity called entropy:

$$S_B = -N \sum_i p_i \ Log\ p_i \qquad (VI.11)$$

where $p_i = N_i/N$ and where N_i is the occupancy of cell number i. This approach, which uses no dynamical law except the conservation of energy, appears to be very efficient as it produced the concept of absolute temperature, the Boltzmann distribution and many other consequences.

The concept of entropy was then generalised by Shannon[6] to the theory of information. The kangaroos's problem[7] illustrates what may represent entropy in such a context: "one knows that 50% of the kangaroos of an uninhabited island are blue eyed (BE) and 40% of them are left handed (LH): the question is to infer, in absence of any other information, the proportion which is and blue eyed and left handed". The solution is not unique and all the answers from 0% to 40% are all possible but have not the same probability. If one restricts to 10 kangaroos with names AB... IJ, Figure VI.4 represents 3 possible repartitions of the 10 kangaroos, all respecting 50% (BE) and 40% (LH) but repartitions (a) and (b) corresponding to a configuration $\begin{pmatrix} 1 & 4 \\ 3 & 2 \end{pmatrix}$ while (c) corresponds to $\begin{pmatrix} 2 & 3 \\ 2 & 3 \end{pmatrix}$. The number of complexions, or number of different possible repartitions of kangaroos, which give the same configuration is easily calculated by combinational analysis and reported in

Table VI.I. It is clear that, in absence of other information, configuration $\begin{pmatrix} 2 & 3 \\ 2 & 3 \end{pmatrix}$ is more probable than the other as it corresponds to more possible repartitions than the other. Let us note that this answer of 20% of the kangaroos and (BE) and (LH) corresponds to the decoupling of the two characters, which is the most probable in absence of any other information.

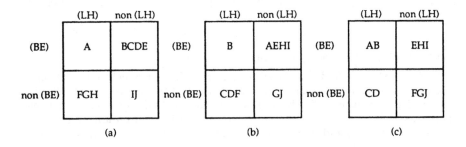

Figure VI.5. — Three possible repartitions of the 10 kangaroos. Repartition (a) and (b) correspond to a configuration $\begin{pmatrix} 1 & 4 \\ 3 & 2 \end{pmatrix}$ while (c) corresponds to $\begin{pmatrix} 2 & 3 \\ 2 & 3 \end{pmatrix}$.

Configuration	Number of complexions
$\begin{pmatrix} 0 & 5 \\ 4 & 1 \end{pmatrix}$	$\dfrac{10!}{5!4!1!} = 1260$
$\begin{pmatrix} 1 & 4 \\ 3 & 2 \end{pmatrix}$	$\dfrac{10!}{4!3!2!1!} = 12600$
$\begin{pmatrix} 2 & 3 \\ 2 & 3 \end{pmatrix}$	$\dfrac{10!}{3!3!2!2!} = 25200$
$\begin{pmatrix} 3 & 2 \\ 1 & 4 \end{pmatrix}$	$\dfrac{10!}{4!3!2!1!} = 12600$
$\begin{pmatrix} 4 & 1 \\ 0 & 5 \end{pmatrix}$	$\dfrac{10!}{5!4!1!} = 1260$

Table VI.1. — Number of complexions for the different configurations of the 10 kangaroos.

To generalize the number of repartitions to M cells, the number of complexions of one given configuration is

$$W = \frac{N!}{(N_1)! \, (N_2)! \, ... \, (N_M)!} \qquad (VI.12)$$

which becomes by application of the Stirling formula to large numbers:

$$W = \frac{N^N}{(N_1)^{N_1} (N_2)^{N_2} ... (N_M)^{N_M}} = \frac{1}{(p_1)^{N_1} (p_2)^{N_2} ... (p_M)^{N_M}} \qquad (VI.13)$$

with $p_i = Ni/N$.

The repartition which maximimes W will also maximise its logarithm, that is the entropy of the distribution:

$$Log\ W = S_B = NS$$

where

$$S = -\sum_i p_i\ Log\ p_i \qquad (VI.14)$$

VI.4.3.   Application to the spin density distribution [8, 9]

In order to be able to define the entropy of the spin (or magnetization) density, this continous function of space is quantized by dividing the unit cell in subcells i in which the density is supposed to be constant: $m_i = m\ (r_i)$. As the spin density may be negative as well as positive, one considers a double distribution of positive quantities: $m_i = m^+\ (r_i)$ and $m_{i+M} = m^-\ (r_i)$ with $m\ (r_i) = m^+\ (r_i) - m^-(\ r_i)$. One then defines the normalized densities:

$$p_i = \frac{m_i}{\sum\limits_{j=1}^{2M} m_j}$$

The entropy of any distribution is then defined by equation (VI.14).

The most probable spin distribution is that which will fit the data and which has the maximum entropy. To construct the maximum entropy map, one starts with a flat distribution, calculates the corresponding structure factors F_{Mcal} and evaluate the agreement with the data through the value of χ^2 defined by equation (VI.7). The distribution will be refined in order to bring down χ^2 to unity, but not to a lower value, and at the same time, to reach the highest possible value for the entropy. The refinement programme, uses the MEMSYS software package subroutine [10].

An example of application of this procedure is shown in   Fig. VI.6 which compares the spin density projection of the high T_c superconductor $YBa_2\ Cu_3\ O_7$ when the data are treated by Fourier inversion and by maximum entropy reconstruction. The comparison speaks for itself: the artefacts which are mainly due to the truncation effects have been suppressed. Moreover, any sizable density appearing locally in the maximum entropy maps must be considered as significant and resulting from the data themselves. That stems from the fact that such reconstructions are as featureless as possible, while still fitting the data within the experimental error bars.

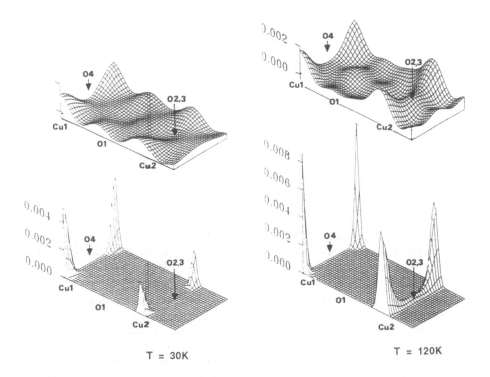

Figure VI.6. — YBa$_2$Cu$_3$O$_7$ below and above the critical temperature T$_c$ = 90 K. The upper spin density reconstructions have been obtained by Fourier inversion, the lower reconstructions by maximum entropy[11].

VI.5.   Modeling the spin density: the multipolar expansion

A natural way to retrieve the spin density distribution, while avoiding the problems due to Fourier inversion, is to model the spin density and to determine the parameters involved by comparing the model to the experimental data.

VI.5.1.   Multipole expansion

A well adapted model results from a multipolar expansion of the density around the nuclei at rest[12]. It consists of a superposition of aspherical atomic densities, each one which described by a series expansion in real sherical harmonic functions $y_l^m(r)$.

$$m(r) = \sum_{atoms} \sum_{l=0}^{\infty} R_l(r) \sum_{m=-l}^{l} P_l^m y_l^m(\hat{r}) \qquad (VI.15)$$

where the P_l^m are the population coefficients and $R_l(r)$ are radial functions of the spin density, for instance of the Slater type:

$$R_l(r) = \frac{\zeta^{(n_l + 3)}}{(n_l + 2)!} r^{(n_l)} e^{-\zeta r}$$ (VI.16)

The magnetic structure factors corresponding to equations (VI.8) and (VI.9) are

$$F_N(Q) = \left(\sum_{atoms} \sum_{l=0}^{\infty} \phi_l(Q) \sum_{m=-l}^{l} P_l^m y_l^m(\hat{Q}) \right) e^{i\,Q\,r}\, e^{-W}$$ (VI.17)

$$\phi_l(Q) = i^l \int_0^{\infty} R_l(r)\, j_l(Qr)\, r^2\, dr$$ (VI.18)

where $j_l(x)$ are the spherical Bessel functions. The thermal motion of atoms is taken into account through the term e^{-W}.

The set of parameters $\left(\zeta, P_l^m\right)$ for each atom characterizes the spin distribution. These parameters are fitted by a least square refinement of the data, that is, in the general case, the set of experimental structure factors. This determines the spin density.

An example of application of this method is given for the spin distribution of the tanol suberate $(C_{13} H_{23} O_2 NO)_2$. This molecule is a binitroxide free radical where the unpaired electrons are localised on the NO groups located at the two ends of the chain molecules. The flipping ratios of reflections (0kl) corresponding to a projection parallel to a where measured[7]. A complete set of data, up to $\sin\theta/\lambda = = 0.45\ Å^{-1}$, comprises 150 reflections. Only the stronger nuclear reflections were measured giving a partial set of 69 magnetic structure factors $F_{M\ obs}$. Figure VI.7 compares two spin density reconstructions obtained by Fourier inversion and by multipole expansion. The last map clearly shows less noise, an enhanced resolution and also values of the density closer to reality than the partial Fourier summation.

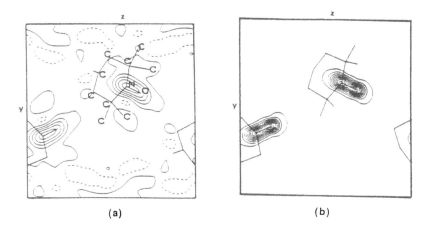

(a)   (b)

Figure VI.7. — Tanol Suberate: the comparison of the spin distribution projected along the a direction (a) by Fourier inversion (b) by multipole expansion[13].

VI.5.2.   Application to acentric structures

In the case of acentric structures, both F_N and F_M are complex quantities

$$\begin{vmatrix} F_N = \dot{F}_N + i\, \overset{''}{F}_N \\ F_M = \dot{F}_M + i\, \overset{''}{F}_M \end{vmatrix} \qquad\qquad (VI.18)$$

Equation (VI.3) contains then 2 unknown quantities $\dot{F}_M$ and $\overset{''}{F}_M$ which cannot be deduced from the single measurement of the flipping ratio R. The method of modeling the spin density allows nevertheless the determination of the parameters $\left(\zeta, P_l^m\right)$, and thus of the spin density, by a least square refinement, comparing directly the experimental and the calculated flipping ratios: R_{mes} (Q) and R_{cal} (Q).

The spin density of diphenyl picryl hydrazil (DPPH) was determined that way[14]. This molecule (Fig. VI.8a) crystallizes, together with one molecule of the solvant C_6H_6 in the monoclinic space group Pc which is acentric.

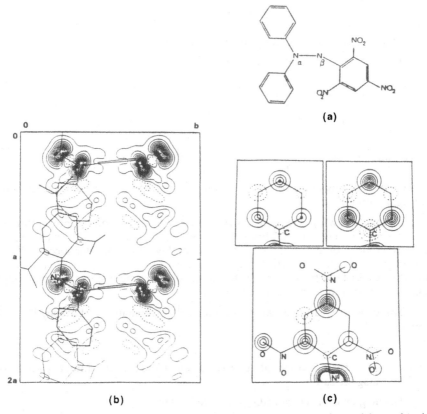

(a)

(b)   (c)

Figure VI.8. — DPPH: (a) the molecule, (b) and (c) the spin density obtained by multipole expansion[14]: (b) projected along the c axis of the crystal, (c) projected onto the plane of each aromatic ring.

The projection of the spin density, parallel to the c axis is represented in Fig. VI.8b. One sees that most of the density is localised and equally distributed on the two central N atoms of the hydrazyl but that some of the density is delocalized on the cycles. Figure VI.8c represents this delocalized density, projected onto the plane of the cycles. The striking point is the alternance of positive and negative densities on the carbon atoms, alternance which is maintained on the N and O atoms of the picryl ring.

VI.6.   Modeling the spin density: the magnetic wave function

The natural model to represent a spin density is the wave function of the magnetic electrons. The structure factor can be expressed by:

$$F_M = <\psi \,|\, e^{i\,Q\,r} \,|\, \psi> = \int \psi^* \, e^{i\,Q\,r} \, \psi \, d^3 r \qquad (VI.19)$$

$|\psi>$ is in the general case a molecular wave function

$$|\psi> = \sum_{atoms} \alpha_j \, |\varphi_j>$$

where the sum runs over all the magnetic atoms.

Two types of terms enter the expression of F_M:

• one center integrals $\int \varphi_j^* \, (r) \; e^{i\,Q\,r} \varphi_j \, (r) \, d^3 r$

which represent the main contribution

• two center integrals $\int \varphi_{j'}^* \, (r) \; e^{i\,Q\,r} \varphi_j \, (r) \, d^3 r$

which are correcting terms corresponding to the overlap between the wave functions of neighbouring atoms.

In the following we will restrict ourselves to the one center integrals, considering the magnetic amplitude scattered by one atom only. We shall express this amplitude in term of magnetic form factor and we shall consider successfully the case of p and d electrons with a magnetic moment which is mainly of spin origin, and the case of f electrons (rare earths and actinides) where spin and orbit couple together to give a total angular momentum.

The general treatment can be found in the Marshall and Lovesey[15] or in Lovesey[16].

VI.6.1.   The form factor of p and d electrons: and almost pure spin magnetization

The magnetic form factor $f(Q)$ can be defined by

$$F_M(Q) = mf\,(Q) = \int \varphi^* \, e^{i\,Q\,r} \varphi \, d^3 r \qquad (VI.20)$$

where m is the magnetic moment and where the one electron atomic wave function φ is expanded in a radial part $R(r)$ and in an angular part:

$$\varphi (r) = \sum_{l=0}^{\infty} R^l (r) \sum_{m=-l}^{l} a_{lm} Y_l^m (\theta,\varphi) \qquad (VI.21)$$

where θ, φ are the angular coordinates of r and where the Y_m^l are the usual spherical harmonics.

In this sum $l = 0, 1, 2 \ldots$ for s, p, d $\ldots$ electrons. In most of the cases only one value of l is concerned by magnetism.

One expands the exponantial of (VI.20) by

$$e^{i\,Q\,r} = 4\pi \sum_{L=0}^{\infty} i^L j_L (Qr) \sum_{M=-L}^{L} Y_L^{M^*} (\theta_Q, \varphi_Q) Y_L^M (\theta,\varphi) \qquad (VI.22)$$

where θ_Q, φ_Q are the angular coordinates of Q and where the $j_L(x)$ are the spherical Bessel functions.

Introducing this expansion in (VI.20), one obtains

$$f(Q) = \sum_{L} < j_L (Q) > \sum_{M=-L}^{L} C_{LM} Y_L^{M^*} (\theta_Q, \varphi_Q) \qquad (VI.23)$$

where the $<j_L(Q)>$ are the radial integrals (sort of Fourier transform) of the magnetic electrons.

$$<j_L (Q)> = \int^{\infty} r^2 R^2 (r) j_L (Qr)\ dr \qquad (VI.24)$$

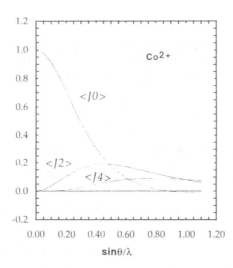

Figure VI.9. — The radial integrals calculated for Co^{2+} [17].

Such radial integrals are displayed in Fig. VI.9 for Co^{2+} [17]. The coefficients C_{LM} are given by:

$$C_{LM} = i^L (2l + 1) \left[4\pi (2L + 1) \right]^{1/2} \begin{pmatrix} 1 & 1 & L \\ 0 & 0 & 0 \end{pmatrix} \sum_{m\,m'} (-1)^m \, a^*_{lm} \, a_{lm'} \begin{pmatrix} 1 & 1 & L \\ -m & m' & M \end{pmatrix} \quad \text{(VI.25)}$$

using the 3j symbols $\begin{pmatrix} a & b & c \\ d & e & f \end{pmatrix}$ which are closely related to the Clebsch Gordan coefficients.

Because of the triangular relations which exist for the 3j symbols:

$$\left\{ \begin{matrix} L \leq 2l \\ -m + m' + M = 0 \end{matrix} \right\}$$

one is restricted to

for p electron (l=1) $f(Q) = <j_0(Q)> + A(\theta_Q, \varphi_Q) <j_2(Q)>$ \hfill (VI.26)

for d electron (l=2) $f(Q) = <j_0(Q)> + A(\theta_Q, \varphi_Q) <j_2(Q)> + B(\theta_Q, \varphi_Q) <j_4(Q)>$ (VI.27)

From the knowledge of the magnetic wave function, the calculation of the form factor is straightforward. But at the opposite the wave function is a very convenient model with adjustable parameters such a_{lm} and some others which could change the shape of the radial part $R(r)$.

When the orbital contribution to the magnetic moment is not completely quenched.

$$m = sm_S + lm_l \quad \text{(VI.28)}$$

$$f(Q) = sf_S(Q) + l\, f_L(Q) \quad \text{(VI.29)}$$

Within a spherical approximation the orbital form factor is expressed as

$$f_l(Q) = <j_0> + <j_2> \quad \text{(VI.30)}$$

wich gives for the total form factor

$$f(Q) = <j_0> + l <j_2> \quad \text{(VI.31)}$$

Let us note that the orbital magnetization being produced by orbital currents, it is more localized than the spin magnetization due to the same magnetic electrons, and therefore its form factor falls down less rapidly in the reciprocal space.

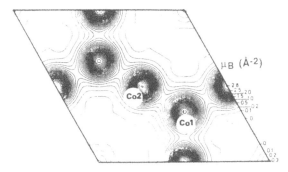

Figure VI.10. — YCo_5: the unit cell and the magnetization density projected along c[18].

As an illustration Fig. VI.10 and Table VI.II show the experimental magnetization obtained on the intermetallic compound YCo_5 and the analysis of the form factor of its two magnetic sites[18].

Site	Localized moment	Spin proportion	Occupation parameter	
Co_I	1.77 (2) μ_B	0.74 (5)	dz^2	0.23 (3)
			dxz, dyz	0.18 (12)
			$dx^2 - y^2, dxy$	0.58
Co_{II}	1.72 (2) μ_B	0.84 (4)	dz^2	0.15 (2)
			dxz	0.24 (4)
			dyz	0.19 (3)
			$dx^2 - y^2$	0.22 (3)
			dxy	0.20
Sum of the localised moments in one cell			8.90 (10) μ_B	
Magnetisation measured for one cell			7.99 (2) μ_B	

Table VI.II. — Analysis of the form factors of the two sites of YCo_5 in terms of spin and orbital contributions and wave function populations (ref. [18]).

VI.6.2.   The form factor of f electrons : rare earths and actinides

In these cases where the spin orbit coupling is large, spin and orbit couple together to give a total angular momentum.

$$J = L + S$$

J is a good quantum number (J = L-S for the first half and L+S for the seconf half), and the magnetic moment can be expressed as

$$m = g_J J$$

A complete formalism of the atomic form factor is exposed in[15-16]. Practical expressions can be found in Lander et al[19].

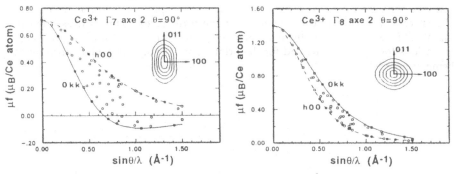

Figure VI.11. — The form factor and the magnetization density of Ce^{3+} calculated for the two cubic states Γ_7 and Γ_8[20].

The important point concerning rare earth and actinides in the presence of strong anisotropies in the magnetization distribution, resulting of the strong orbital contribution. This is illustrated in Fig. VI.11 for cerium[20].

A simplified expression for the form factor is given by the spherical approximation:

$$f(Q) = <j_0 (Q)> + C_2 <j_2 (Q)> \qquad (VI.32)$$

$$\text{with } C_2 = \frac{2}{g_J} - 1 = \frac{J (J + 1) + L (L + 1) - S (S + 1)}{3J (J+1) - L (L + 1) + S (S + 1)} \qquad (VI.33)$$

	L	S	J	$C_2 = \dfrac{2}{g_J} - 1$
f^1	3	1/2	5/2	1.333
f^2	5	1	4	1.500
f^3	6	3/2	9/2	1.750
f^4	6	2	4	2.333
f^5	5	5/2	5/2	6.000
f^6	3	3	0	no moment
f^7	0	7/2	7/2	0
f^8	3	3	6	0.333
f^9	5	5/2	15/2	0.500
f^{10}	6	2	8	0.600
f^{11}	6	3/2	15/2	0.667
f^{12}	5	1	6	0.714
f^{13}	3	1/2	7/2	0.750

Table VI.3. — Coefficient C_2 for the different filling of an f shell.

Table VI.III displays the values of C_2 for the different filling of the f shell. One can note the particular case of 5 electrons where the spin part and the orbital part almost cancel, giving unusual shapes for the form factor. This is illustrated in Fig. VI.12 for $SmCo_5$, where the maximum at $Q \neq 0$ results from the different spatial extensions for the spin part and for the orbital part, the sign of both contributions being opposed[21].

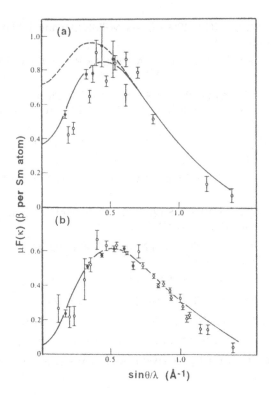

Figure VI.12. — The form factor of Sm measured in Sm Co5 at different temperatures[21].

REFERENCES

(1)   SHULL C.G., YAMADA Y., 1962 - J. Phys. Soc. Jap. 17 Suppl. BIII, 1.

(2)   BOUCHERLE J.X., GIVORD F., LEJAY P., SCHWEIZER J., STUNAULT A., 1989 - Physica B156-B157, 809.

(3)   JAMES R.W., 1954 - The optical principles of the diffraction of X rays. The crystalline state, Vol. 2. Bells and Sons (London).

(4)   SHULL C.G., MOOK H.A., 1966 - Phys. Rev. Lett. 16, 184.

(5)   MOOK H.A., 1966 - Phys. Rev. 148, 495.

(6)   SHANNON C., 1948 - Bell System Tech. J. 27 379, 623.

(7)   GULL S.F., SKILLING J., 1984 - IEEE Proceedings, 131, 646.

(8)   PAPOULAR R., GILLON B., 1990 - Europhysics Lett. 13, 429.

(9)   PAPOULAR R., GILLON B., 1990 - "Neutron Scattering Data Analysis 1990", p. 101, Ed by M.W. JOHNSON, Adam Hilger Publisher.

(10)   GULL S.F., SKILLING J., 1989 - Users' MEMSYS manual, Maximum Entropy Data Consultant Ltd.

(11)   BOUCHERLE J.X., HENRY J.Y., PAPOULAR R.J., ROSSAT-MIGNOD J., SCHWEIZER J., TASSET F., UIMIN G., 1993 - Physica B, to be published.

(12)   GILLON B., SCHWEIZER J., 1989 - "Molecules in Physics, Chemistry and Biology", Vol. 3, p. 111, Edited by J. MARUANI, Kluwer Academic Publishers,.

(13)   BROWN P.J., CAPIOMONT A., GILLON B., SCHWEIZER J., 1979 - J. Mag. Magn. Mat. 14 289.

(14)   BOUCHERLE J.X., GILLON B., MARUANI J., SCHWEIZER J., 1987 - Mol. Phys. 60, 1121.

(15)   MARSHALL W., LOVESEY S.W., 1971 - "Theory of Thermal Neutron Scattering, Clarendon press, (Oxford).

(16)   LOVESEY S.W., 1983 - Theory of Neutron Scattering from condensed Matter, Clarendon Press (Oxford).

(17)   WATSON R.E., FREEMAN A.J., 1961 - Acta Cryst. 14, 27.

(18)   SCHWEIZER J., TASSET R., 1980 - J. Phys. F., 10, 2799.

(19)   LANDER G.H., BRUN T.O., 1970 - J. Chem. Phys. 53, 1387.

(20)   BOUCHERLE J.X., SCHWEIZER J., 1985 - Physica, 130, 337.

(21)   GIVORD D., LAFORET J., SCHWEIZER J., TASSET F., 1979 - J. Appl. Phys. 50, 2008.

CHAPTER VII

EXCITATIONS AND PHASE TRANSITIONS
R.A. COWLEY

VII.1. Introduction

Neutron and X-ray scattering experiments have been responsible for much of our detailed knowledge of phonon and magnon excitations in crystals, and also of the behaviour of the fluctuations associated with phase transitions in crystals. In a short article it is clearly impossible to do justice to the immense amount of important work which has been performed, as there are many books written about only small parts of these topics. Consequently the approach will not be to review the results obtained, but to concentrate on the experimental techniques and the way in which the results are then analysed to provide information about particular systems.

The characteristic feature of excitations is that for particular wavevectors there are well defined frequencies, and the determination of these frequencies enables the phonon and magnon dispersion relations to be determined. Since the characteristic frequencies are of order 1THz or energies are of order 1meV, these frequencies are better determined using neutron scattering than with X-ray scattering techniques. In the next section we shall describe the way in which the dispersion relations are determined experimentally, and some of the problems which can occur. Because very similar techniques are used for both phonons and magnons the development will concentrate on the phonon case and there will be only a brief account of the differences in determining magnons.

Although ideally the excitations are well defined, as given by the harmonic approximation for the phonons, in practice, the interactions between the excitations give rise to significant effects which can in principle be identified by careful experiments. In section VII.3, these effects are discussed, so that the different effects can be looked for and studied with the experiments.

A similar approach is then adopted to the study of phase transitions. In section VII.4, the experimental techniques are described, for both neutron and X-ray scattering. In this case we shall largely, but not exclusively, describe magnetic scattering. The problems of the data analysis will then be described in a final section with illustrations taken from a number of different experiments.

VII.2. Determination of Phonon and Magnon Dispersion Relations

VII.2.1. Theory

The theory of the inelastic scattering of neutrons by phonons and magnons has been described in detail by, for example, Lovesey (1984), and for phonons is given in this series by Zabel (1993). In an ideal experiment a beam of monochromatic neutrons with energy, E, and wavevector, $\mathbf{k}$, is scattered to give an energy E', and wavevector, $\mathbf{k'}$. The scattering vector is then given by

$$Q = k-k' \qquad\qquad\qquad\qquad \text{(VII.1)}$$

and the energy transfer by

$$\hbar\omega = E-E' \qquad\qquad\qquad\qquad \text{(VII.2)}$$

The differential neutron scattering cross-section for a coherent monatomic system is then given by

$$\frac{d^2\sigma}{d\Omega dE'} = b^2 \frac{k'}{k} S(Q,\omega), \qquad\qquad\qquad \text{(VII.3)}$$

where the Van Hove correlation function depends only on Q and ω, and b is the coherent scattering length for the nuclei. The correlation function is then given by:

$$S(Q,\omega) = \frac{1}{2\pi} \sum_{\ell\ell'} \int \left\langle \exp iQ.\left(r_\ell(t) - r_{\ell'}(o)\right) \right\rangle \exp(-i\omega t)\, dt \qquad \text{(VII.4)}$$

The analysis is then continued by writing the atomic position of the ℓth atom at time t in terms of the equilibrium position R_ℓ and the displacement from equilibrium, $u_\ell(t)$, and then expanding the exponentials in eq. VII.4 by assuming the displacements are small. The leading term in the expansion gives the Bragg reflections for which $\omega=0$ and $Q=G$, a reciprocal lattice vector. If the displacements are given in terms of the harmonic normal modes of vibration for the crystal, the one-phonon term is then given by the first terms in the expansion of the exponential as

$$\left(\frac{d^2\sigma}{d\Omega dE'}\right)^{1\,phonon} = \frac{k'}{k} N \frac{(2\pi)^3}{v} \sum_{jG} \left|F\left(Q,q_j\right)\right|^2 \left(n\left(q_j\right)+1/2\pm1/2\right) \delta\left(\omega\pm\omega\left(q_j\right)\right) \delta(Q\pm q - G)$$

$$\text{(VII.5)}$$

where $n(q_j)$ is the occupation number of the phonon with wavevector q and belonging to branch j, $\omega(q_j)$ is its frequency, and the dynamic structure factor is

$$F\left(Q,q_j\right) = \sum_K \left(\frac{\hbar}{2\omega\left(q_j\right)M_K}\right)^{\!1/2} b_K\left(Q.E_K(q_j)\right) \exp(-W_K(Q)) \exp(iG.R_K), \quad \text{(VII.6)}$$

where for the sake of completeness the expression is given for many atoms in the unit cell described by index K of mass M_K, scattering length b_K, and position R_K, while $W_K(Q)$ and $E_K(q_j)$ are the Debye-Waller factors, and eigenvectors respectively.

The difference between the Bragg and one-phonon cross-sections lies in the evaluation of the expectation value of eq. VII.4. In the former case there is no correlation between sites ℓ and ℓ', so that motion of the atoms enters solely in the Debye-Waller factors which describe the way in which one site behaves independently of the other motions. In the one-phonon term a single phonon propagates from site ℓ to ℓ' giving rise to correlations or coherence between the sites, and hence to the inelastic scattering.

VII.2.2.   Triple-Axis Spectrometer

VII.2.2.1. Principles

Neutron scattering is the most effective technique for determining phonon dispersion curves because the wavevector of thermal neutrons is comparable to that of phonons in crystals, while simultaneously the energy transfers are also comparable with the energies of phonons. Consequently both Q and $\hbar\omega$ in eq. VII.1 and VII.2 are readily measured. By far the most successful instrument for these measurements has been the triple-axis crystal spectrometer first built by Brockhouse (1961) in 1958, and illustrated in Fig. VII.1. There are now many similar instruments in operation at reactor sources throughout the world.

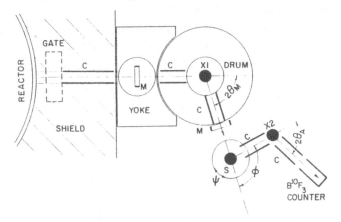

Figure VII.1. — A schematic diagram of a triple axis crystal spectrometer with the monochromator at X1, analyser at X2 and sample at s.

The thermal neutrons are incident on a monochromator which through Bragg's law, gives a monochromatic beam with an energy, E, determined by the monochromator plane spacing and the angle $2\theta_M$. Likewise the analyser and the angle, $2\theta_A$, determine that scattered energy, E', which is detected. The angles ϕ between the incident and scattered beam, and ψ for the crystal orientation give a total of four controllable angles. For scattering in a plane perpendicular to say z, eq. VII.1 and VII.2. have three variables Q_x, Q_y and ω. In principle therefore it is possible to adjust three of the four angles to study the scattering at any point in Q_x, Q_y and ω space, apart of course, from certain constraints based on the limitations of the instrument and the kinematics of the process. The fourth variable is adjustable to provide different experimental conditions such as the resolution. It is these two features which make the triple axis spectrometer so powerful; firstly the experimenter chooses the point in Q_x, Q_y and ω to study, and secondly simultaneously chooses the resolution.

The spectrometer can then be scanned so as to determine the scattered intensity at successive points in (Q,ω) space, and often the scans chosen are ones in which ω is varied while keeping the wavevector transfer Q fixed, constant Q, or ones in which ω

is fixed and Q scanned along a particular important direction, so-called constant ω scans.

The choice of the wavevector, Q for a particular phonon, q, is determined by a number of different factors many of which can be seen from eq. VII.5. and VII.6. Obviously the intensity of the scattering should be a maximum; this implies that |Q| should be a maximum, apart from the Debye-Waller factor, and for |Q| to be large k must also be large which implies an increase in the resolution width for fixed angular collimation, while if E is too large there is a reduction in the number of incident neutrons when E is above the peak of the Maxwellian spectrum of the source. Furthermore the $(Q.E)^2$ factor shows that for longitudinal modes Q and q should be parallel with one another, while for transverse modes Q and q should be perpendicular to one another, as described by Zabel (1993).

A final factor arises from the resolution of the instrument, particularly for transverse modes. In the case of these modes the resolution function is strongly correlated in q and ω as illustrated in Fig. VII.2.

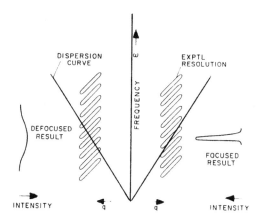

Figure VII.2. — A schematic diagram of a constant Q experiment showing that if the slope of the resolution function and dispersion relation match, a sharp focused peak is obtained, while if they do not a defocused peak is the result.

If the experiment is designed so that the slope of the resolution function is nearly the same as that of the dispersion relation, a sharp focused peak will result. If the slope of the resolution function and dispersion curve are opposite, then a broad defocused result is obtained. Clearly focusing should be achieved whenever possible. One way to achieve focusing is to make a detailed calculation of the resolution function as described by Cooper and Nathans (1967), and most laboratories have computer programmes available for this. In practice focusing conditions can often be chosen by considering, the analyser, and whether a small increase in the length of k' will increase or decrease q and ω. The sign of the slope of the resolution function is then readily obtained, and compared with the dispersion relation. Of course if a very accurate calculation of the line width is needed, there is no substitute for a full analysis using the Cooper and Nathans formalism to obtain the optimum conditions.

VII.2.2.2. Intensity

The integrated intensity in a scan across a dispersion relation is of importance for a detailed analysis of the results. The spectrometer is usually operated with the incident flux on the sample determined by a thin monitor detector between the monochromator and the sample. Because this detector is thin, its efficiency is proportional to $1/k$. The observed intensity is dependent on a large number of factors such as the angular collimations, the efficiencies of the crystals and the counters and many of these are difficult to determine. Absolute intensity measurements are therefore very difficult. Relative intensity measurements are usually performed with the constant Q technique with then either the incident energy E or scattered energy E' held fixed. These result in different relationships between the observed intensities and the theory. In the latter case the whole of the analyser system after the sample is fixed and acts as detector with a constant energy resolution, dE', which is fixed. The integrated intensity including the effect of the monitor detector is then proportional to $k\dfrac{d^2\sigma}{d\Omega dE'}$ which in turn is proportional to the structure factor squared, $|F|^2$. More generally if the analyser angle and collimations are kept fixed the relative intensities are proportional to the van Hove correlation function, $S(Q,\omega)$, eq. VII. 3.

The experiments are however also performed keeping the incident energy, E, fixed and varying E', and indeed the hardware often make this the only possible way of using the instruments. In these circumstances dE' varies across the scan due to the changing $2\theta_A$, even if the collimation remains fixed. Using Bragg's law and the kinematics of the neutron gives

$$dE' = \frac{\hbar^2}{m_N} k'^2 \cot\theta_A \, d\theta_A \qquad\qquad (VII.7)$$

so that if the experiment is performed with fixed angular collimation $d\theta_A$ the intensity becomes proportional to

$$k'^3 \cot\theta_A S(Q,\omega)$$

In Fig. VII.3. we sketch the effect that this difference can have on the observed cross-section by comparing the observed intensity for constant E, and constant E' modes. This difference has in the past often been overlooked, and led to erroneous conclusions. If it is further noted that the analyser reflectivity may also vary with E', and that this variation is usually unknown, the constant E' technique is the one to be preferred. If instead of a constant Q scan, the triple axis spectrometer is controlled so that $Q = Q_0 + B\omega$, the integrated intensity of a peak is then proportional to $\left|1 - B.\nabla\omega(q_j)\right|^{-1}$ where $\nabla\omega(q_j)$ is the slope of the dispersion relation. It is then necessary to know this gradient before the intensity can be reliably interpreted.

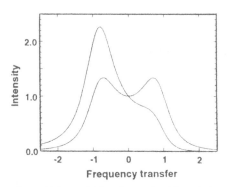

Figure VII.3. — The scattering which would be observed with a triple crystal spectrometer with constant E' = 2.5, symmetrical curve (ω_0 = 1 and $\Gamma = \omega$) and with constant E = 2.5.

VII.2.2.3. Problems

One of the requirements of a triple axis crystal spectrometer is that the monochromator and analyser reflect only those neutrons with the required energies.

Since Bragg's law is $n\lambda = 2d\,\sin\theta$, a monochromator will usually reflect neutrons of energy E, 4E, 9E, etc. The higher order contaminants must be suppressed if the peaks are to be reliably interpreted. One of the best ways is to choose an energy E so that the peak of the Maxwell Boltzmann distribution from the source is of sufficiently high energy that there are few neutrons with energies of 4E and 9E. Similarly the neutrons from a cold source and a curved guide will suppress these neutrons if the curvature of the guide is appropriate.

Another approach is to use a monochromator crystal which has zero structure factor for the 4E Bragg reflection. This is the case for germanium and silicon crystals for which the (222) reflection has zero scattering power. As a result of this there has been considerable effort made to squeeze these crystals so as to give them a sufficiently large mosaic spread ~0.25°, that the reflectivity is large enough for them to be used as monochromators.

An alternative approach is to use a filter to suppress the 4E and 9E neutrons. The two most common filters are powdered beryllium, cooled to 80K, which only transmits neutrons with energies of less than 5 meV but scatters ones with larger energies out of the beam, and pyrolytic graphite which is very effective at suppressing the higher order contaminants when E = 14 meV and to a lesser extent 40 meV. The use of a filter to produce a clean incident beam then implies that the incident energy is fixed throughout any scans, with the consequences discussed above. Nevertheless the introduction of these techniques has been so useful that some form of filter is usually used to suppress the unwanted contaminant neutrons.

Another problem arises from multiple scattering in the sample. The crystals used for inelastic studies are large and it is quite likely that either the incident or scattered beam will be Bragg reflected by the sample in a direction which may well be out of the scattering plane. This Bragg reflected beam is then inelastically scattered into the counter. The geometry for this effect is illustrated in Fig. VII.4; the multiple scattering does not influence the energy or wavevector conservation conditions. It does however alter the intensities so that it enables, for example, a transverse mode to be observed where the $(Q.E)^2$ factor would suggest that only longitudinal modes should be observed. Multiple scattering can be tested for by rotating the sample about Q, which should not alter the primary scattering but does change the multiple scattering. This requires, however, a very flexible sample mounting which may not be feasible for a sample in a large cryostat. An alternative approach is to change the neutron energy continuously while keeping the energy transfer fixed when the intensity of the primary scattering will change slowly and steadily, while the multiple scattering will change erratically. Unfortunately this cannot be used when filters are in use.

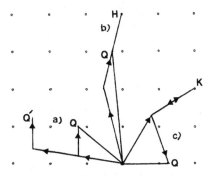

Figure VII.4. — A reciprocal space diagram showing spurious events which can occur with a triple axis spectrometer, (a) shows second order neutrons from the monochromator scattering with a wavevector Q' in addition to Q; (b) shows that Bragg scattering from the lattice point H is incident on the analyser and well as inelastic from Q, while (c) shows that Bragg scattering from the lattice point K, gives inelastic scattering with wavevector transfer KQ.

The final difficulty to be discussed is that of Bragg reflections in the sample, Fig. VII.4. The spectrometer has three crystals which are used to give Bragg and Inelastic and Bragg scattering. It is not surprising that Bragg and Bragg and Inelastic gives a similar intensity. In practice this process gives intense sharp peaks which can usually, be separated from "genuine" peaks with a little experience.

VII.2.3.   Time-of-Flight Spectrometers

Time-of-flight spectrometers are a system of choppers to produce a pulsed monochromatic source of neutron from either a continuous or pulsed neutron source. The advantage of the technique is that there can be many detectors, and that the data is simultaneously recorded for a wide range of scattered neutron velocities, as determined by the time-of-flight, and angles. A typical diagram in reciprocal space is shown in Fig. VII.5.

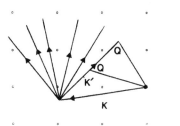

Figure VII.5. — The reciprocal space diagram for time-of-flight instrument with pulsed monochromatic incident neutrons. Note that as the time-of-flight varies, k' varies, and Q and ω vary in each detector.

The difficulty with the technique is that the data is not collected along desired directions in reciprocal space, but rather generally scattered in (Q,ω) space and that the relative counting statistics or resolution in the different detectors cannot be independently adjusted. As a consequence time-of-flight techniques have only occasionally been used for studying excitations. They have proved useful in the special case of systems usually magnetic, for which the interactions are one or two dimensional because in these cases the unique direction is aligned parallel to the incident beam when the data from all of the detectors can be used efficiently and effectively. In these circumstances time-of-flight spectrometers are very successful.

The intensity directly measured with a time-of-flight spectrometer is a function of the time-of-flight for constant intervals of time. Since an energy interval dE' corresponds to a time interval dt if

$$dE' = \frac{-\hbar^3}{m_N^2 L} k'^3 \, dt,$$

where L is the sample to detector distance the observed intensity is proportional to $k'^4 S(Q,\omega)$ or $(E')^2 S(Q,\omega)$ where Q and ω are simultaneously varying as determined by the kinematics of the process, Fig. VII.5. This correction to the intensities is very similar to that shown in Fig. VII.2 for a triple axis spectrometer and its neglect has led to errors in interpretation particularly for instruments for which E is small compared with the energy of the excitations.

Other types of spectrometers have been developed for pulsed sources, which attempt to incorporate some of the advantages of crystal spectrometers with the advantages of pulsed neutron sources. One is the PRISMA spectrometer and in Fig. VII.6 we illustrate the principle. A pulsed white incident beam is incident on the sample and there are a number, ~16, of analysers whose positions and energies are all independently controllable so that the wavevector transfers Q for all the detectors lie on a particular symmetry line. This instrument has some uses, but the multiple detector arrangement is difficult to use fully, due to geometric constraints and collisions, while the background is at present high. More development is required before these pulsed source instruments are fully competitive with the triple axis spectrometer on a continuous source.

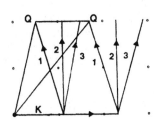

Figure VII.6. — The reciprocal space diagram for a time-of-flight instrument with a white incident beam and analyser crystal. Note all the detectors can be arranged to give Q along a symmetry line in reciprocal space.

VII.2.4. Magnon Scattering

The experimental study of magnons is very similar to that of phonons. The main difference lies in the intensity because the magnon scattering (Lovesey, 1984) is not proportional to Q^2, but depends on the magnetic form factor $|f(Q)|^2$. This decreases with increasing Q, and so magnon scattering is best studied with the wavevector transfer, Q, as small as possible.

Another substantial difference is the dependence on the eigenvector. Instead of the neutron coupling to the eigenvector components parallel to Q, (Q.E(qj)), in magnetism the magnetic dipolar coupling between the dipole moment of the neutron and the sample leads to a coupling to the components of the spins perpendicular to Q. In an experiment it is necessary to distinguish between the magnon and phonon scattering, and this is usually done from their different dependences on Q, or from their different temperature dependences. In a few cases it is useful to use polarised neutrons to study the magnetic scattering, and to identify the different magnetic modes uniquely. Polarising monochromators are, however, much less efficient than non-polarising ones so that these experiments become very difficult and the technique is only used in exceptional circumstances.

VII.3.   Interactions between the Excitations

VII.3.1.   Lineshapes

The harmonic theory of the normal modes of vibration gives a scattering cross-section with a delta function for the energy transfer. The scattering is then a sharp peak centred at the frequency of the normal mode, $\omega(\mathbf{qj})$. In real crystals the interatomic forces are anharmonic and there are interactions between the excitations. The effects of these interactions are described qualitatively in this section, with particular emphasis on their effect on the scattering cross-section. Although throughout we shall only describe the results for phonons, similar effects occur for magnons, and for the interactions between magnons and phonons. A full description of the theory of anharmonic crystals has been given earlier (Cowley, 1963, Cowley, 1968).

In most cases the anharmonic terms have only a small effect on the normal modes of vibration, so they can be treated using perturbation theory, and not surprisingly they lead to a change in the frequency of the normal modes and to a decay of the phonon with a characteristic lifetime. The frequency delta function and Bose Einstein occupation numbers in eq. VII.3 are then replaced by

$$\sigma(\omega) = \left(\frac{n(\omega)+1}{\pi} \right) \frac{4\omega_0^2 \, \Gamma(\omega)}{\left(\omega^2 - \omega_0^2 - 2\omega_0 \, \Delta(\omega) \right)^2 + 4\omega_0^2 \, \Gamma^2 \, (\omega)} \, , \qquad (VII.8)$$

where $\omega_0 = \omega(\mathbf{qj})$ and $\Delta(\omega)$ and $\Gamma(\omega)$ are the shift and width (1/lifetime) of the $(\mathbf{qj})$ normal mode. The width, $\Gamma(\omega)$, arises in lowest order by the decay of the phonon into pairs of phonons. It therefore depends on the two-phonon density of states with the appropriate matrix elements, and hence depends on ω, the frequency transfer in the experiment. Energy conservation is only required between the initial and final state of the neutron and the crystal and not the intermediate one-phonon state. The shift, $\Delta(\omega)$, consists of a similar frequency dependent part from the two-phonon term, but also two other terms which correspond to the changes in the effective harmonic potential due to the thermal expansion and the amplitude of the thermal motion.

Since in the experiments a peak is usually observed with a particular frequency and width, it is not, in practice, possible to determine $\Delta(\omega)$ and $\Gamma(\omega)$ experimentally and so they are frequently approximated by constants, Δ and Γ when if the Δ and $\Gamma \ll \omega_0$

$$\sigma(\omega) \approx \frac{1}{\pi} \left[\frac{(n(\omega)+1)\Gamma}{(\omega - \omega_0 - \Delta)^2 + \Gamma^2} + \frac{n(\omega)\Gamma}{(\omega + \omega_0 + \Delta)^2 + \Gamma^2} \right] ,$$

which gives Lorentzian profiles centred at $\omega_0 + \Delta$ and with a half-width of Γ.

This approximation fails if the damping becomes large because the response is then centred around $\omega = 0$ and causality implies that $\Gamma(\omega)$ is an odd function of ω. It is then often approximated as $\Gamma(\omega)=\gamma\omega$, when the response function (VII.8) is equivalent to that of a classically damped harmonic oscillator. As is easily seen in this case the peak response is not at $\omega_0+\Delta$, and the width not directly related to γ, and so a detailed fit is needed to obtain these parameters from the experimental results.

Detailed measurements of the temperature dependence of the scattering have been made and have enabled the phonon shifts and linewidths to be obtained for a number of different branches as illustrated in Fig. VII.7. Since the shifts and widths are often small compared with the instrumental resolution effects, it is difficult to obtain very accurate measurements.

Anharmonic effects in potassium

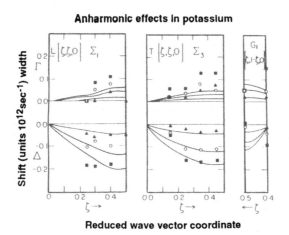

Reduced wave vector coordinate

Figure VII.7. — The anharmonic frequency and linewidth changes as a function of temperature in potassium (Cowley and Buyers, 1969). The triangles show the changes between 100K and 9K, the circles between 200K and 9K and the squares 300K and 9K.

The theory of the propagation of sound in an anharmonic crystal is complicated because at low frequencies there are many phonon collisions in each period and the sound propagates by thermodynamic processes rather than by ballistic transport. The anharmonic terms for Δ and Γ are then different from at high frequencies so that the velocity of sound at low frequencies as measured by ultrasonics can be quite different from those determined at higher frequencies, above the phonon inverse collision time. This is illustrated in Fig. VII.8, similar measurements have been made in a variety of other materials (Macdonald et al, 1988).

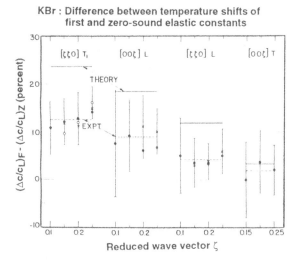

Figure VII.8. — Differences between the thermodynamic (first) second and ballistic (zero) sound in KBr at 400K and 90K (Cowley, 1967). The different points show different experimental configurations.

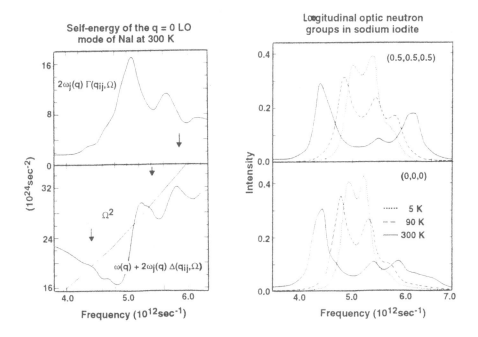

Figure VII.9. — The Self-Energy of the LO q=o mode in NaI at 300K. The lower part shows the real part. The structure gives rise to the group shown in part b.

Finally it is at times necessary to use the full frequency dependence of $\Gamma(\omega)$ and $\Delta(\omega)$, particularly if they vary rapidly within the linewidth of the scattering. This is the case for the longitudinal optic modes of alkali halides and Fig. VII.9 shows calculations of the predicted profiles showing that the frequency dependent effects can give rise to multiply peaked structures very different from the simple Lorentzian profile usually adopted. These predictions have been carefully tested by infra-red measurements (Bruce, 1972, Hisano et al, 1972) and by neutron scattering data (Cowley at al, 1983).

These effects are all consequences of the way in which the phonons propagate through the crystal, and so consequently the shape of the scattering cross-section is identical at different points in reciprocal space, Q, with the same phonon wavevector, q. The same profile is obtained for Q in different Brillouin zones, ie different reciprocal lattice vectors, G.

VII.3.2. Interference Terms

The effect of the coupling between the interactions, which is described in this section, alters the scattering cross-section in a different way from that described in section VII.3.1. The result is that it is no longer possible to unambiguously assign a particular frequency and lifetime to a phonon. These effects arise when one phonon is anharmonically coupled so that it changes into another phonon or into a pair of phonons. The scattering cross-section, eq. VII.4, shows that the one-phonon scattering arises from the neutron exciting a phonon (q_j) at the point ℓ', and the phonon propagating to the point ℓ at time t. In an anharmonic crystal the phonon (q_j) can be created at ℓ', but the phonon at ℓ can be $(q_{j'})$. The wavevector is conserved because anharmonic interactions have translational symmetry, and symmetry also shows that this can only occur if j and j' belong to the same irreducible representations of the little group of q.

A detailed derivation of the cross-section (Cowley, 1965) then shows that the coupling terms are proportional to $F(Q \mid q_j) F(Q \mid q_{j'})^*$, and that in the neighbourhood of $\omega(q_j)$, the cross-section has two terms. One of these has a very similar form to eq. VII.8 and alters the intensity of the scattering, but the other is asymmetric about $\omega(q_j)$ and has the form

$$\sigma'(\omega) \approx \frac{n(\omega)+1}{\pi} \frac{2\omega_0\left(\omega^2 - \omega_0^2 - 2\omega_0\,\Delta(\omega)\right)}{\left(\omega^2 - \omega_0^2 + 2\omega_0\,\Delta(\omega)\right)^2 + 4\left(\omega_0\,\Gamma(\omega)\right)^2} \qquad \text{(VII.9)}$$

As illustrated in Fig. VII.10, this type of term alters the lineshape and hence can lead to a different frequency being assigned to a peak. Since the size of the effect depends on the relative sizes of the structure factors, $F(Q \mid q_j)$ and $F(Q \mid q_{j'})$, the lineshape observed and the corresponding frequency then differs from zone to zone. These effects can be distinguished from the lineshape anomalies produced by the frequency dependence of $\Delta(\omega)$ and $F(\omega)$ because the effects resulting from the latter are identical in every zone, whereas structure resulting from the interference effects depends explicitly on the coupling between the phonons and the neutrons, and so varies from zone to zone.

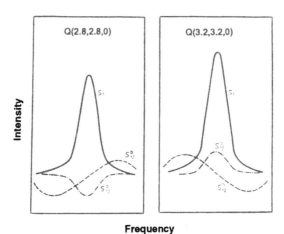

Frequency

Figure VII.10. — The contribution to the intensity of the one phonon peaks, S_1, from interference with the two phonon background, S_{12}. Note the asymmetric contribution S_{12}^b which changes sign with position in Q.

In the preceding paragraph, the coupling between two different phonons was discussed. The same type of effects occur if a phonon decays into two phonons which then reach position ℓ, to interact with the neutron. In this case the coupling with the neutron is determined by the product of the one and two-phonon structure factors, and so the effects also vary with Q. There is then no clear distinction between one, two and three-phonon process in an anharmonic crystal, but there is still a unique separation between Bragg scattering and inelastic scattering. The one and two-phonon interference effects can be distinguished in centrosymmetric crystals from the two one-phonon effects by studying the scattering for $Q_1 = G + q$ and $Q_2 = G\text{-}q$. The two one-phonon effects, discussed first in this section give the same asymmetry at Q_1 and Q_2, but the one and two-phonon terms are of opposite signs. Consequently measurements at Q_1 and Q_2, which are then repeated for different G vectors, enable the origin of the different asymmetries to be found experimentally.

There have been several studies of these effects in alkali halides (Cowley et al, 1969) and in metals (Buyers et al, 1979). They were first however observed (Buyers and Smith, 1966) in X-ray scattering where they produce asymmetries in the integrated intensity which prohibit the analysis of the data to give accurate measurements of the phonon frequencies.

These asymmetric contributions to the scattering cross-section are well known in other branches of physics. In nuclear physics they are known as Breit-Wigner resonances and in atomic physics as Fano resonances.

VII.4.   Phase Transitions

VII.4.1.   Theory of Phase Transitions

The simplest theory of phase transitions is the mean field or molecular field theory which neglects the effect of fluctuations on the behaviour by replacing the value, of say the magnetisation on a site ℓ, $S(\ell)$, by its average value. It is now realised and confirmed experimentally that the fluctuations neglected by this approximation do play an important role and can very strongly modify the behaviour. The power of X-ray and neutron scattering techniques is that they enable these fluctuations to be studied experimentally, and our present understanding is in large measure due to the success of these measurements.

The theory of phase transitions is described in may texts (for example Stanley, 1971, Binney at al, 1992), and concentrates on the behaviour close to the phase transition temperature, T_c. The order parameter is the magnitude of the parameter which is zero in the high symmetry phase and non-zero in the low symmetry phase. In a magnetic system with a continuous phase transition and the order parameter described by a wavevector q_s, then

$$\langle S(q_s) \rangle = M_0 |t|^\beta \quad t < 0 \qquad\qquad (VII.10)$$

$$\langle S(q_s) \rangle = 0 \qquad t > 0$$

where t is the reduced temperature $(T-T_c)/T_c$ and β is the critical exponent. Mean field theory gives $\beta = 0.5$, but in practice it is usually smaller but the same for all phase transitions having the same symmetry, number of components for the order parameter, and dimensionality. This is known as the universality of exponents.

The fluctuations in the order parameter decay with distance, and the scaling theory of phase transactions postulates that there is only one length scale at any temperature, and that the inverse of this correlation length is given by

$$K = K_0^+ t^\upsilon \qquad t > 0 \qquad\qquad (VII.11)$$

$$K = K_0^- |t|^\upsilon, \qquad t < 0$$

where the amplitudes K_0^+ and K_0^- depend on the interactions, but their ratio K_0^+ / K_0^- is a universal constant, like the exponent υ.

Likewise the susceptibility for the ordering field, $\chi(o)$, behaves as

$$\chi(o) = C^+ t^{-\gamma} \qquad t > 0 \qquad\qquad (VII.12)$$

$$\chi(o) = C^- t^{-\gamma} \qquad t < 0$$

where the ratio C^+/C^- and γ are more universal quantities. The scaling theory then predicts that in general $\chi(q)$ depends only on K and has the general form

$$\chi(q) = \chi(o) X^\pm (q/K) \qquad\qquad (VII.13)$$

where $X^\pm (y)$ is known as the scaling function, and $X(o) = 1$. It is frequently assumed that

$$X(y) = (1 + y^2)^{-1}, \tag{VII.14}$$

except at T_c when $y \to \infty$ and

$$X(y) = y^{-2+\eta}, \tag{VII.15}$$

and consistency then leads to a relationship between the critical exponents:

$$\gamma = (2 - \eta)\upsilon \tag{VII.16}$$

Scaling theory can also be applied to the dynamical effects of the fluctuations. If the fluctuations have a characteristic frequency it will depend on the wavevector $\mathbf{q}$ and the temperature, t. The scaling theory assumes that for small t, the frequency is not separately a function of $\mathbf{q}$ and t but can be written in the scaling form,

$$\omega_c(t,\mathbf{q}) = q^z \Omega(K/q), \tag{VII.17}$$

where z is the dynamic critical exponent and $\Omega(x)$ is another scaling function, whose behaviour depends on the detailed nature of the interactions and the symmetry of the system. The objective of experiments is to determine the universal features of the phase transition: the exponents β, γ, υ and z, the amplitude ratios K_0^+ / K_0^-, C^+ / C^-, and the structures of the scaling functions.

VII.4.2. Neutron and X-ray Scattering and Phase Transitions

Although the theory for phase transitions can be developed equally well for structural and magnetic systems, we choose the magnetic case because most of the important experiments have been performed on magnetic systems. The differential scattering of unpolarised neutrons by a system with only one type of magnetic ion is given by (Lovesey, 1984)

$$\frac{d^2\sigma}{d\Omega dE'} = \left|\frac{k'}{k}\right| F_M^2 \sum_{\alpha\beta} \left(\delta_{\alpha\beta} - \frac{Q_\alpha Q_\beta}{Q^2}\right)$$
$$\frac{1}{2\pi} \sum_{\ell\ell'} \int_{-\infty}^{\infty} \langle S_\alpha(\ell,t)\, S_\beta(\ell',0)\rangle \cdot expi\big(Q.(R(\ell) - R(\ell')) - \omega t\big)\, dt, \tag{VII.18}$$

where F_M^2 is the square of the magnetic scattering length of each ion, and α and β are sums over the coordinate axes: x, y, z. The angular factor arises from the dipolar interaction between the magnetic moments in the sample and the neutron. $S(\ell,t)$ is the spin on site ℓ at time t and position $R(\ell)$, and t is now again time.

As with the nuclear scattering it is convenient to rewrite the cross-section in terms of a Bragg and a diffuse component:
where

$$\frac{d^2\sigma}{d\Omega dE'} = \left|\frac{k'}{k}\right| F_M^2 \left(S_B(Q) + S_D(Q,\omega)\right),$$

$$S_B(Q) = N\sum_\alpha \left(1 - \frac{Q_\alpha^2}{|Q|^2}\right) M_0^2\, t^{2\beta} \sum_G \delta(Q - q_S - G)\, \delta(\omega) \tag{VII.19}$$

while if $S(Q) = \int_{-\infty}^{\infty} S_D(Q,\omega)\, d\omega$ (Collins, 1989).

$$S(Q) = \frac{kT}{\hbar} \sum_{\alpha} \left(1 - \frac{Q_{\alpha}^2}{|Q|^2}\right) \chi(q) \sum_{G} \delta(Q - q_s - q - G) \qquad \text{(VII.20)}$$

if $kT \gg \hbar\omega_c$, and $\chi(q)$ is the wavevector dependent susceptibility.

These results show that the Bragg scattering allows the order parameter to be determined including its wavevector, q_s, its amplitude M_0 and its exponent, β. The frequency integrated diffuse scattering gives a measure of the susceptibility, and hence enables the behaviour of K, χ_0 and the exponents γ and υ, to be measured. The dynamic effects and characteristic frequency can also be obtained from the energy resolved diffuse scattering.

VII.4.3. Experimental Considerations

Universal critical behaviour occurs when the correlation length of the fluctuations is considerably longer than the lattice spacing, a, say 3a. This means that the inverse correlation length is less than 0.05 reciprocal lattice units $(2\pi/a)$. Since the exponents can only be fully trusted if they are determined over two or more decades, this implies that we need to measure inverse correlation lengths of 0.005 reciprocal lattice units or smaller. This needs very good experimental resolution. The second requirement is that the temperature be sufficiently close to T_c. For three dimensional systems $K \sim 1/3a$ is typically reached when $|t| < 0.05$ and for two-dimensional systems when $|t| < 0.3$. Hence to reliably measure the inverse correlation length over an adequate range of temperature requires a temperature control of about $0.0005\, T_c$. The experiments are therefore technically demanding in that they require the best resolution, and very good temperature control. Furthermore, since the theory is only asymptotically correct as $T \to T_c$, the most crucial part of the data is often the most difficult to obtain.

Because the experiments require a very detailed study of the scattering in a very small region of reciprocal space as a function of temperature, they are most efficiently performed with a continuous source of neutrons and by using a double or triple axis crystal spectrometer. The time-of-flight techniques are not competitive because the data in most of the detectors and in most of the time channels is not useful. Experiments to determine the dynamics of the critical fluctuations are usually performed using a triple axis spectrometer operated to perform constant Q or constant ω scans as described in section VII.2. The considerations discussed there, concerning the choice of keeping $2\theta_M$ or $2\theta_A$ fixed and the use of filters are equally important in these experiments. Many of the experiments require however the energy integrated scattering, eq. VII.20. This can always be reliably obtained by measuring $S(Q,\omega)$ with a triple axis instrument and integrating over constant Q scans. This is very time consuming however, and a more efficient technique is to use a double axis spectrometer, so that the detector then collects simultaneously scattered neutrons of all energies. Unfortunately a change in the neutron energy causes a change in the wavevector transfer, Q. If the characteristic frequency of the fluctuation is ω_c, this change is given by $\delta Q = k' \hbar\omega_c / (2E'))$, and this must be kept small if a reliable energy integration is to be made. Fortunately $\omega_c \to 0$ as $T \to T_c$, and since the errors produced in neutron energy gain and loss tend to cancel the results are more accurate than might have been expected (Tucciaroni et al, 1971). Nevertheless the neutron energy should be kept as large as possible for these measurements. This problem can be largely eliminated for one and two-dimensional systems because the scattering varies only slowly for wavevectors perpendicular to the sheets or lines. By arranging δQ in these directions the energy integration is performed reliably, and the so called quasi-elastic approximation is satisfied.

The quasi-elastic approximation is always valid when the motion of the order parameter is very slow as occurs in order - disorder transitions in alloys. It is also satisfied in X-ray scattering experiments because δQ is then very small due to the high energy of the X-rays. Due to this and to the fact that X-ray beams are much more intense than neutron beams, X-ray scattering is usually to be preferred for measurements of structural phase transitions where it offers the possibility of much higher resolution than neutron scattering.

Although neutron and X-ray scattering, in principle, provide very detailed information about phase transitions, there are a number of experimental difficulties. There are several difficulties, for example, in determining the temperature dependence of the order parameter from the Bragg reflections. Firstly due to the thermal expansion and possible thermal effects in the crystal, the integrated intensity should be measured at each temperature with the collimation sufficiently relaxed to ensure that all the Bragg scattering is collected even if there is some misalignment of the system. At low temperatures when the intensity is large, the scattering can be substantially reduced by extinction, and corrections for extinction are notoriously difficult to make. This reduction of intensity at low temperatures tends to give a decrease in the measured β exponent. In principle, this effect can be reduced by using smaller crystals for the low temperature part of the measurement or by measuring for a variety of different reciprocal lattice vectors. If all the points do not then lie on a common curve extinction is probably the cause.

A third difficulty is multiple Bragg scattering. This is particularly the case when q_s, the ordering wavevector, is a reciprocal lattice vector of the high temperature structure but one that has zero intensity, as is the case of the (100) lattice point in MnF_2 at the antiferromagnetic transition. Multiple Bragg scattering then gives rise to additional scattering which may be temperature dependent near T_c due to relief of extinction or changes of lattice parameter close to T_c. A fourth difficulty is that close to T_c, the diffuse scattering is large and temperature dependent while the Bragg scattering is weak. Reliable measurements of the Bragg scattering can then only be made if a correction is applied to subtract the diffuse scattering. This effect is minimised if very good resolution is used, because the intensity of the diffuse scattering is roughly proportional to the resolution volume, while the Bragg scattering is not. Unfortunately there is no satisfactory way of correcting the Bragg scattering for the diffuse scattering. If the diffuse scattering is measured above T_c, then an approximate correction can be obtained by assuming that for the same $|t|$, the correlation length and amplitude can be deduced by scaling with the theoretically known amplitude ratios. This correction is clearly unsatisfactory in that it assumes the theoretical results are correct, and furthermore of necessity assumes a form for the scaling function and this is often not known reliably from either theory or experiment.

Finally the exponent deduced is very dependent on the temperature T_c, and due to the diffuse scattering it is often very difficult to determine the temperature at which the Bragg scattering becomes zero. Bruce (1981) showed that the total scattering observed close to T_c has a temperature dependence like the energy $|t|^{1-\alpha}$, where α is the specific heat exponent. Bruce then suggests that the temperature derivative of the intensity will locate T_c, and that the temperature dependence of the intensity enables the specific heat exponent, α to be determined. This suggestion has not, however, often been used in practice. More usually T_c is located by the maximum in the diffuse intensity $\chi(q)$ for wavevectors q close to the ordering wavevector. The temperature of the maximum in the intensity is then extrapolated as $q \rightarrow q_s$ to give the transition temperature, T_c. Since at small q these measurements may be contaminated by the Bragg reflections the extrapolation requires care. One of the most successful ways of locating T_c is if the lattice parameter or domain structure changes at T_c. The first can be determined from lattice parameter measurements, and the latter gives rise to a change in the extinction and hence intensity of the Bragg reflections of the crystal.

The biggest problem in determining the diffuse scattering is to be able to correct reliably for the resolution function. In the previous section we showed that we needed to be able to measure the inverse correlation length K to values as small as 10^{-3} A^{-1}. The resolution in a neutron scattering experiment with collimation in the scattering plane ~0.2° and perpendicular to the plane 2.0° is typically 0.005 A^{-1} and 0.05 A^{-1} respectively. In the case of X-ray scattering measurements the in-plane collimation may be 0.005° when the resolution is then better than 0.0005 A^{-1}, but these experiments are still most often performed with an angular collimation of about 1° perpendicular to the scattering plane. The reason for the relaxed collimation perpendicular to the scattering plane is in part at least due to the difficulty of aligning and performing the experiments if very tight collimation is used throughout. Nevertheless it is clear that corrections for the experimental resolution must be applied if reliable measurements of K are to be obtained close to T_c.

The resolution function of a neutron (Cooper and Nathans, 1967) or X-ray (Cowley, 1987b) spectrometer have been calculated. We do not give the details here, because in practice it is better to measure the resolution function by using an appropriate Bragg reflection. This has the advantage that it includes the sample mosaic spread. The resolution is then assumed to be unchanged for small changes in the wavevector transfer, Q. This approximation fails for small angle scattering, very small Q, when a detailed calculation is used for each Q and ω setting of the spectrometer (Mitchell et al, 1984).

Experimental results are analysed by assuming a parameterized functional form for convoluting with the measured resolution function, and adjusting the parameters to give a good fit to the experimental data. Ideally the convolution for a dynamic experiment is a four dimensional integral which must be performed numerically, but in practice this is too time consuming. More frequently the resolution function is assumed to be of Gaussian form, and often the out-of-plane resolution can be treated analytically. For example, if $\chi(q)$ has an isotropic Lorentzian form

$$\chi(\mathbf{q}) = A / \left(K^2 + q_x^2 + q_y^2 + q_z^2\right) \qquad \text{(VII.21)}$$

and the out of plane, z, resolution is a triangular form of width δ, the intensity as a function of q_x and q_y is

$$I(q_x, q_y) = \frac{2A}{\lambda\delta}\tan^{-1}\frac{\delta}{\lambda} - \frac{A}{\delta^2}\log\left(1 + \frac{\delta^2}{\lambda^2}\right)$$ (VII.22)

where $\lambda^2 = K^2 + q_x^2 + q_y^2$.

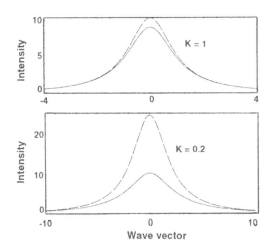

Figure VII.11. — The critical scattering gives by the Lorentzian form (dotted curve), and when convoluted with the vertical resolution. The units of K are in terms of the FWHM of the vertical resolution.

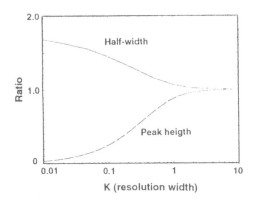

Figure VII.12. — The width and peak height of the curves calculated with the resolution correction as compared with those of a Lorentzian as a function of K in terms of the resolution width.

The result of eq. VII.22 is compared with the Lorentzian form in Figs. VII.11 and VII.12, and these show that whereas eq. VII.21 for q=o diverges as A/K^2, eq. VII.22 diverges only as A/K, and that the half-width of eq.VII.22 is not given by K. The advantage of using eq. VII.22 is that it considerably speeds up the numerical convolution. Although the Cooper-Nathans form for the resolution gives a Gaussian form whereas the derivation of eq. VII.21 has used a triangular form, the actual resolution function is intermediate between these two forms. Fortunately similar calculations to those shown in Figs. VII.11 and VII.12 but with a Gaussian resolution function give half widths that differ by only 1%.

The above procedure enables the parameters in the diffuse scattering to be determined with an accuracy of about 10% of the horizontal resolution width. Normally measurements of K with a better precision than 10% of the resolution width are untrustworthy because of uncertainties in the functional forms of the resolution function and $S_D(q,\omega)$. Indeed by far the largest problem is determining whether the theoretical forms for $S_D(q,\omega)$ are accurate, because in the absence of a detailed knowledge of these functional forms, K for example, cannot be extracted reliably. We shall elaborate further on this problem in section VII.5.2.

VII.5.   Studies of Phase Transitions

VII.5.1.   Real Systems

The study of phase transitions by neutron and X-ray scattering techniques has produced a wealth of important information as reviewed previously (Cowley, 1987, Collins, 1989). It is not intended to repeat or up-date these reviews here, but to illustrate the problems which occur in analysing the data by giving a few specific examples. The theory outlined in section VII.4 was based on the assumption that the fluctuations were isotropic in the wavevector q. Although this may be correct in systems with cubic symmetry, most real systems have considerable anisotropies. The simplest example of these occurs for the magnetic materials, like MnF_2 or CoF_2, which are tetragonal where the susceptibility for the fluctuations has the tetragonal structure:

$$\chi(q) = \frac{A}{K^2 + q_x^2 + q_y^2 + cq_z^2} \qquad (VII.23)$$

and the parameter c must be determined if the correlation length is to be obtained accurately. This is usually determined by measuring the scattering as a function of q_x with $q_y=q_z=0$ and then as a function of q_z with q_x and $q_y=o$.

A more difficult situation occurs when there is more than one component to the order parameter, as occurs in $SrTiO_3$ where the oxygen rotations can be around the x, y, z axes. In this case we need to introduce the susceptibility matrix $\chi_{\alpha\beta}(q)$, which is given by the inverse of the three dimensional matrix with diagonal elements, $\alpha\alpha$, $K^2 + q^2 + c_1q_\alpha^2$, and off-diagonal elements $c_2q_\alpha q_\beta$. There are then three constants which determine the fluctuations, and furthermore measurements made around different reciprocal lattice points give different elements of the susceptibility matrix. Clearly it is much more difficult to unambiguously determine the correlation length as shown by the work of Andrews (1986).

The complications become even more severe in systems with dipolar interactions as illustrated by the case of the ferroelectric material KD_2PO_4 (Shalyo et al, 1970). The dipolar interactions are of long range and suppress the fluctuations with wavevectors along the ferroelectric, axis. The susceptibility is given by the form

$$\chi(\mathbf{q}) = A \bigg/ \left(K^2 + q^2 + C\left(\frac{q_z^2}{q^2}\right)\right) \tag{VII.24}$$

showing that when $q_z=0$ $\chi(0)=A/K^2$ but if $q\to 0$ along q_z, $\chi(0)=A/(K^2+C)$. Fig. VII.13 shows the critical scattering measured for DKDP illustrating the effect of this singular term. The results are even more surprising for the fluctuations in the (q_x,q_y) plane because there is a piezoelectric coupling between the ferroelectric fluctuations and the acoustic modes governed by the elastic constant, C_{66}. Consequently the symmetry of the scattering is modified by the symmetry of the elastic waves, and then further modified by the interference between the ferroelectric fluctuations and the acoustic modes as discussed in section VII.3.2. The result for the scattering is shown in Fig. VII.14 and clearly a very complex scattering pattern results which makes a detailed convolution with the resolution function almost impossible.

These examples illustrate that the form of the scattering close to T_c can be very complex in some systems, and that it is necessary to carefully consider the nature of the fluctuations before attempting a deconvolution. Nevertheless despite these considerations, in many cases the scattering is closely isotropic, and the deconvolution can be performed reliably.

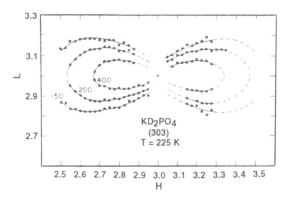

Figure VII.13. — The critical scattering observed in the (HOL) plane of KD_2PO_4 for $T-T_c = 0.4K$ (Skalyo et al, 1970).

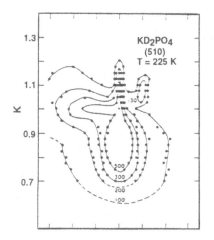
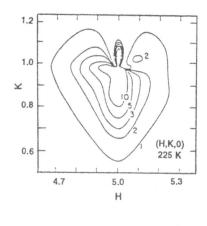

Figure VII.14. — The critical scattering observed in the (HKO) plane of KD_2PO_4 at $T-T_c = 0.4K$ (a) experiment (Skalyo et al, 1970) (b) theory (Cowley, 1976).

VII.5.2. The Scaling Function

Most of the analysis of critical scattering data has been performed by assuming that the scaling function $X(y)$, eq. VII.13, is a Lorentzian, eq. VII.14. Since $X(o)=1$ and is even in y, the Lorentzian form is the leading term in a Taylor series expansion of the inverse of $X(y)$. Nevertheless close to T_c, the resolution is comparable to K, and data is often taken for $q/K \sim 50$, when it is by no means clear that using only the leading term in the expansion is adequate. Unfortunately the nature of $X(y)$ is for most systems unknown, and so it is difficult to assess the errors involved with the use of a Lorentzian form. There is one system, however, for which $X(y)$ has been calculated in detail and we shall now review the results obtained for this, the two-dimensional Ising model.

The two-dimensional Ising model can be studied experimentally through the behaviour of the magnetic ordering in K_2CoF_4. Measurements by Ikeda and Hirakawa (1974) showed that the order parameter behaved as predicted by Onsager (1944). More recently there has been a detailed study of the critical fluctuations (Cowley et al, 1984) and as shown in Fig. VII.15, the fits to the data for $T>T_c$ by the Lorentzian profile gave a good account of the data and yielded values of K which agreed with the theory. The results obtained below T_c are more interesting in that Tracy and McCoy (1975) and Tarko and Fisher (1975) had shown that the ubiquitous Lorentzian form was a very poor description, and that

$$X(y) = \frac{\left(1+\phi^2 y^2\right)^{\eta/2}}{\left(1-\lambda+\lambda\left(1+y^2\right)^{1/2}\right)^2} \qquad (VII.25)$$

where the parameters ϕ and λ are not universal. As shown in Fig. VII.15 both the Lorentzian form and the Tarko-Fisher form give a good description of the scattering, but the resulting K values are very different, Fig. VII.16.

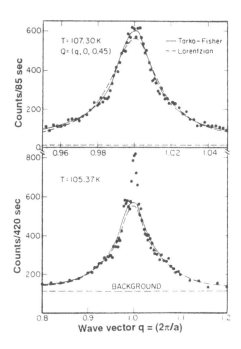

Figure VII.15. — Fits to the critical scattering from K_2CoF_4 with Lorentzian and Tarko-Fisher forms. The peak in the centre for T = 105.37K is a Bragg component which was not fitted (Cowley et al, 1984).

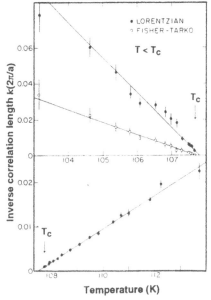

Figure VII.16. — The inverse correlation length, K resulting from the fits shown in Fig. VII.15. Note that the Lorentzian and Tarko-Fisher forms give very different results for $T<T_c$.

The results for K deduced from the Tarko-Fisher fits are in excellent accord with the theoretical predictions, but those from the Lorentzian fits are very different leading to an incorrect amplitude. This illustrates one of the biggest difficulties in interpreting the experimental results. Usually we do not know the detailed form of the scaling function, and when incorrect forms are convoluted with the resolution function they may give a good description of the data but quite misleading values of the parameters.

There are now many instances where a Lorentzian profile has been found to be inadequate. One of these is in analysing data from experiments in which a magnetic field is applied to a random antiferromagnet. In this case if the sample is cooled in a field, long range order is not established and instead of the magnetic Bragg peak, the scattering has a Lorentzian squared profile.

$$I(q) = \frac{A}{\left(k^2 + q\right)^2} \tag{VII.26}$$

as illustrated in Fig. VII.17. In contrast when the material is cooled in the absence of the field and the field then applied long range order persists until the sample is heated above a well defined metastability temperature above which the behaviour is always ergodic. Fig. VII.17 shows the scattering observed just below the instability line for both field cooling and zero field cooling experiments (Cowley et al, 1989). The behaviour of these random field systems now introduces two new features, namely the Lorentzian squared profile and the concept of metastability. Unfortunately there is not a satisfactory understanding of these systems and the nature of their fluctuations so that a detailed interpretation of the data is controversial because the functional forms for the scattering are unknown.

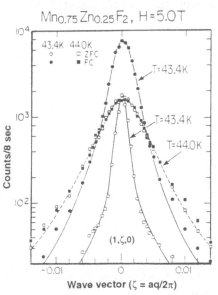

Figure VII.17. — Scattering from $Mn_{0.75} Zn_{0.25} F_2$ in a magnetic field of 5T, showing the difference between the field cooled and zero field cooled results at 43.4K but ergodic behaviour at 44.0K (Cowley et al, 1989).

VII.5.3.   Is Scaling Theory Correct ?

The basic concept of the modern theory of phase transitions is the scaling theory which states that the static susceptibility of the fluctuations is determined by a single length scale, 1/K, at each temperature and for the dynamical susceptibility there is a single frequency scale at each temperature. Nearly all theories implicitly depend on these scaling concepts, and most experiments are interpreted in terms of them. It is, however, worthwhile to consider experiments which may question these hypotheses.

The first experiments in this class were performed on the structural phase transitions of $SrTiO_3$ and $KMnF_3$. In both cases there is a phase transition in which the cubic high temperature structure distorts to a tetragonal structure while the oxygen or fluorine octahedra rotate around the unique axis. Neutron scattering measurements above T_c at first observed the normal mode of vibration associated with the rotations and found that its frequency decreased as the temperature approached T_c, approximately as

$$\omega(\mathbf{q}_s)^2 = A(T - T_c).$$ (VII.27)

This frequency then provided the frequency scale which varied with temperature as expected for the scaling theories. This behaviour of a soft mode above T_c is often observed when the high temperature phase is fully ordered. In contrast in magnetic systems where the high temperature phase is disordered the susceptibility is diffusive or quasi-elastic in character.

More detailed measurements on $SrTiO_3$ and $KMnF_3$ showed however that even above T_c there were two components: one is quasi-elastic and the other characteristic of the soft mode as shown in Fig. VII.18 (Shapiro et al, 1974).

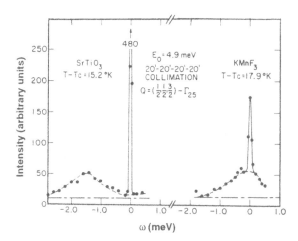

Figure VII.18. — Neutron scattering from $SrTiO_3$ and $KMnF_3$ showing an inelastic peak and a quasi-elastic peak (Shapiro et al, 1974).

Many experiments have now been performed to measure the frequency width of the quasi-elastic peak including a neutron backscattering experiment with a resolution width of $2 \times 10^8 H_z$, but no appreciable width has been observed. The data can be described by a susceptibility of the form

$$\chi^{-1}(q,\omega) = \omega(qj)^2 - \omega^2 - 2i\omega\gamma - \frac{i\omega\zeta^2}{1-i\omega\zeta} \qquad (VII.28)$$

when there are clearly two frequencies involved, $\omega(qj)^2$ and $1/\zeta$ the frequency of the additional relaxation term. As the temperature approaches T_c the proportion of the scattering in the quasi-elastic peak steadily increases while the intensity in the phonon peak tends to saturate. Despite much theoretical effort, as reviewed by Bruce and Cowley (1981), there is no convincing explanation of the quasi-elastic scattering or central peak. It is very difficult to explain it by an intrinsic process involving phonon collisions because of its small frequency, but it is almost equally unsatisfactory to describe it in terms of defects, because the nature of the defects is unknown, and unknown defects, with unknown couplings and relaxation times can explain anything. Experiment on "pure" $SrTiO_3$ show that the central peak occurs for temperatures 20K above T_c=105K, and when defects are put into $SrTiO_3$ the intensity of the quasi- elastic scattering increases but not nearly as rapidly as the concentration of the defects. The experiments do therefore question the existence of a single characteristic frequency especially as the central peak phenomena has now been observed in many different transitions of both a magnetic and structural nature.

More recently the existence of a single length scale has been questioned by X-ray scattering studies of the transitions in $SrTiO_3$ and $RbCaF_3$ (Andrews, 1986, Ryan et al, 1986, McMorrow et al, 1990). These experiments were performed with much

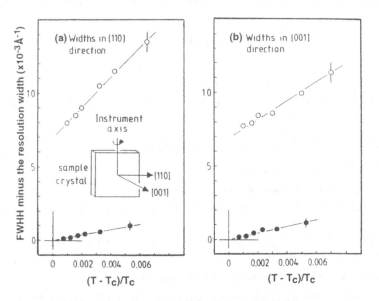

Figure VII.19. — X-ray scattering from $RbCaF_3$ showing two length scales close to T_c (Ryan et al, 1986).

better wavevector resolution than the neutron scattering ones, and when the data was analysed it was found that there were two components; a large K component found in low resolution experiments, and a small K component found in the high resolution experiments as shown in Fig. VII.19. There measurements suggest that there is not a single length scale for describing the transition, but two in contrast to the scaling theory. Since both the frequency dependent results and the length scale results were obtained on similar systems, it is obviously possible that they both arise from the same origin. Since however the X-ray wavevector central peak is observed only close to T_c, a significant part of the X-ray broad scattering arises from the quasi-elastic neutron component. The possibility that these results may be more general has recently been enhanced by the observation of similar effects in both the X-ray and neutron scattering from the magnetic phase transition in holmium (Thurston et al, 1993). There have as yet been very few studies of the critical phenomena at phase transitions with good enough resolution to be capable of distinguishing the two length scales. Clearly further work is needed to determine the extent to which this is a general phenomena requiring a reworking of the theories of phase transitions, or whether it is an unusual quirk of these particular materials and their defects. The results do however clearly show theories can only really be tested if the resolution is sufficiently good that the experiments directly test the theoretical forms as well as fit particular parameters. Too few of the experiments done so far enable us to do this.

ACKNOWLEDGEMENTS

My knowledge of neutron scattering, and phase transitions has benefitted form many discussions with R. J. Birgeneau, B. N. Brockhouse, A. D. Bruce, W. Cochran and G. Shirane. This manuscript was expertly produced by J. L. Andrews, and financial support was provided by the Science and Engineering Research Council.

REFERENCES

ANDREWS S.R., 1986 - J. Phys. C19, 3721.

BINNEY J.J., DOWRICK N.J., FISHER A.J., NEWMAN M.E.J., 1992 - The Theory of Critical Phenomena - Oxford University Press.

BROCKHOUSE B.N., 1961 - Inelastic Scattering of Slow Neutrons form Solids and Liquids - International Atomic Energy Agency (Vienna).

BRUCE A.D., 1972 - J. Phys. C6, 174.

BRUCE A.D., 1981 - J. Phys. C14, 193.

BRUCE A.D., COWLEY R.A., 1981 - Structural Phase Transitions - Taylor and Francis (London).

BUYERS W.J.L., DOLLING G., JACUCCI G., KLEIN M.L. GLYDE H.R., 1979 - Phys. Rev. 20, 4859.

BUYERS W.J.L., SMITH T., 1966 - Phys. Rev. 150, 758.

COLLINS M.F., 1989 - Magnetic Critical Scattering - Oxford University Press.

COOPER M.J., NATHANS R., 1967 - Acta Cryst., A23, 357.

COWLEY E.R., SATIJA S., YOUNGBLOOD R., 1983 - Phys. Rev. B28, 993.

COWLEY R.A., 1963 - Adv. Phys. 12, 421.

COWLEY R.A., 1965 - Phil. Mag. 11, 673.

COWLEY R.A., 1967 - Proc. Phys. Soc. 90, 1127.

COWLEY R.A., 1968 - Rep. Prog. in Phys. 31, 123.

COWLEY R.A., BUYERS W.J.L., 1969 - Phys. Rev. 180, 755.

COWLEY R.A., SVENSSON E.C., BUYERS W.J.L., 1969 - Phys. Rev. Lett. 23, 525.

COWLEY R.A., 1976 - Phys. Rev. Lett., 36, 744.

COWLEY R.A., HAGEN M., BELANGER D.P., 1984 - J. Phys. C17, 3763.

COWLEY R.A., 1987 - Phase Transitions in Neutron Scattering Part C - Academic Press (New York).

COWLEY R.A., 1987b - Acta Cryst., A43, 825.

COWLEY R.A., SHIRANE G., YOSHIZAWA H., UEMURA U.J., BIRGENEAU R.J., 1989 -Zeit Phys. B75, 303.

HISANO K. PLACIDO F., BRUCE A.D., HOLAH G.D., 1972 - J. Phys. C5, 2511.

IKEDA H., HIRAKAWA K., 1974 - Solid State Commun 14, 529.

LOVESEY S.W., 1984 - Theory of Neutron Scattering from Condensed Matter, Oxford Univ. Press.

MACDONALD J.E., SAUNDERS G.A., CLAUSEN K.N., 1988 - J. Phys. C, 21, Ll.

MCMORROW D.F., HASNAYA N., SHIMOMURA S., FUJI Y., KISHOMOTO S., IWASAKI H., 1990 - Solid State Commun 76, 443 (1990).

MITCHELL P.W., COWLEY R.A., HIGGINS S.A., 1984 - Acta Cryst. A40, 152.

ONSAGER L., 1944 - Phys. Rev. 65, 117.

RYAN T.W., NELMES R. J. COWLEY R.A., GIBAUD A., 1986 - Phys. Rev. Lett. 56, 2704.

SHAPIRO S.M., AXE J.D., SHIRANE G., RISTE T., 1974 - Phys. Rev. B6, 4332.

SKALYO J., FRAZER B.C., SHIRANE G., 1970 - Phys. Rev. B1, 278.

STANLEY H.E., 1971, - Introduction to Phase Transitions and Critical Phenomena - Oxford University Press.

TARKO H.B., FISHER M.E., 1975 - Phys. Rev. B11, 1217.

THURSTON T.R. HELGESEN G., GIBBS D., HILL J.P., GAULIN B.D., SHIRANE G., 1993, - Phys. Rev. Lett. 70, 315.

TRACY C.A., MCCOY B., 1975 - Phys Rev. B11, 1217.

TUCCIARONE A. LAU H.Y., CORLISS L.M., DELAPALME A., HASTINGS J.M., 1971 -Phys. Rev. B4, 3206.

ZABEL H., 1993 in "Neutron and Synchrotron Radiation for Condensed Matter Studies; Theory, Instruments and Methods", Vol. 1 p. 285, edited by J. Baruchel, J.L. Hodeau, M.S. Lehmann, J.R. Regnard, C. Schlenker, Les Editions de Physique, Springer-Verlag.

CHAPTER VIII

NON-PERIODIC SYSTEMS (AMORPHOGRAPHY)
Adrian C. WRIGHT

VIII.1. Introduction

An amorphous solid is one in which the structure is not only non-periodic but also lacks both extended symmetry and any form of long range order, in contrast to the quasicrystalline materials discussed in the next Chapter which, although non-periodic and lacking extended symmetry, still retain long range order and hence give rise to sharp Bragg-like peaks in their diffraction pattern. This Chapter will describe the general principles involved in studying the structure of inorganic amorphous solids by neutron and X-ray diffraction, using examples taken from a wide range of materials, and indicate how their various structures are interrelated. A much more detailed account, together with a comprehensive list of references, can be found in Chapter 8 of "Experimental Techniques of Glass Science", as cited in the bibliography (Wright, 1993), and two earlier reviews by the present author (Wright, 1974; Wright and Leadbetter, 1976).

Amorphous solids not only comprise glasses, obtained by melt quenching[1], but may be prepared by a variety of techniques from each of the three states of matter, as summarised in Table VIII.1. Only inorganic materials will be considered here, since amorphous organic polymers are included elsewhere in the HERCULES course (Volume III). The lack of a periodic structure combined with the fact that amorphous solids are in a state of metastable thermodynamic equilibrium means that the structure of any given amorphous solid depends on its detailed preparation. This has important consequences in the study of amorphous solids in that any samples must be carefully prepared and adequately characterised for the results obtained to be of any significant use.

A knowledge of the structure of amorphous solids is an important prerequisite for a full understanding of their physical properties at a microscopic level. As with crystals X-ray and neutron diffraction are the major direct structural probes but, due to the inherent differences between the two classes of materials, the problems encountered in diffraction studies of amorphous solids are somewhat different from those experienced in crystallography. The existence of a periodic structure means that, given good diffraction data over a reasonable region of reciprocal space, it is in practice possible to determine the structure of simple crystalline solids absolutely. The same is not true for amorphous solids. Because they are normally isotropic on a macroscopic scale the maximum that can be obtained from a diffraction experiment on an amorphous solid is a one-dimensional correlation function, from which the regeneration of the underlying three-dimensional structure can never be unique. It is for this reason that modelling plays such an important role in structural studies of

---

[1] In this Chapter the term "glass" will be reserved for materials covered by the A.S.T.M. definition of a glass as "an inorganic product of fusion which has cooled to a rigid condition without crystallising".

amorphous solids and why the choice between possible models involves a wide range of experimental techniques and not just X-ray and/or neutron diffraction.

Table VIII.1. — Preparation routes to the amorphous solid state.

Precursor phase	Preparation technique	Examples
Gas	Thermal evaporation	Se, $As_{1-x}S_x$, Si, $Si_{1-x}O_x$
	Sputtering	Ge
	Vapour-phase hydrolysis	SiO_2
	Vapour-phase pyrolysis	SiO_2
	Glow-discharge decomposition	$Si_{1-x}H_x$
Liquid	Melt quenching	Se, SiO_2, $Pd_{1-x}Si_x$
	Gel desiccation (Sol-gel techniques)	SiO_2
	Precipitation	As_2S_3, $Ca_3(PO_4)_2$
	Electrolytic deposition	Ge
Solid	Photon irradiation	As_4Se_4
	Particle irradiation	Ge, SiO_2, $U_{1-x}Fe_x$
	Pressure amorphisation	$AlPO_4$, Fe_2SiO_4, H_2O, ZnP_2
	Shock wave amorphisation	SiO_2
	Solid state diffusion	$Cu_{1-x}Hf_x$
	Mechanical alloying	$Co_{1-x}B_x$, $Pd_{1-x}Si_x$
	Hydrogenation	$CeNi_2$
	Oxidation	SiO_2
	Decomposition	
	Dehydration	

Amorphous solids may be divided into two broad categories:

VIII.1.1. Amorphous Network Structures

Amorphous solids, in which the bonding is directional and has significant covalent character, have network structures traditionally described in terms of Zachariasen's (1932) random network theory. Structural units similar to those in the corresponding crystalline materials are linked together randomly to form a continuous three-dimensional network (c.f. Fig. VIII.1). In multicomponent systems the added components may act as either further network formers or network modifiers.

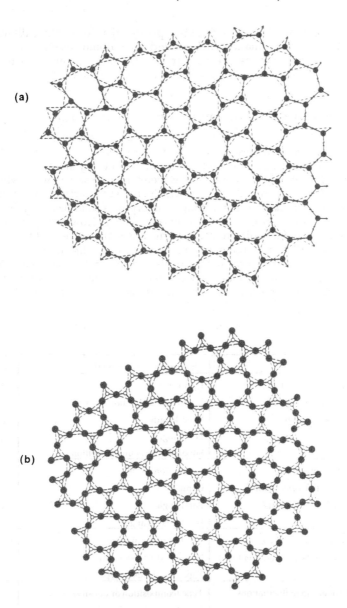

Figure VIII.1. — Schematic two-dimensional representation of (A) an amorphous elemental semiconductor and (B) an A_2X_3 network glass

VIII.1.2. Random Packing Structures

Where the bonding is essentially non-directional, as in metallic or ionic systems, a better first order model is a random packing of spheres. In such structures the important concepts are those of radius ratio, packing density and hole filling and the nature of the first co-ordination shell around each atomic species. For structural models, the co-ordination geometry is frequently analysed in terms of Voronoi polyhedra or the related simplicial graph (Finney, 1970; Zallen, 1983).

VIII.2. Quantification of amorphous solid structures

Although the concepts of a random network and a random sphere packing are easy to describe qualitatively there is no way to define the resulting structures quantitatively except by specifying the co-ordinates of every atom present, which is clearly impossible for any real material. It is therefore necessary to work with a series of statistically averaged parameters which are summarised in Table VIII.2. Traditionally the structural order within an amorphous solid has been divided into three ranges: short, intermediate (medium) and long. However, particularly for systems with well defined directional bonding, it is more convenient to use the four ranges in Table VIII.2, since not only do they relate more directly to the analysis of X-ray and neutron diffraction data but also the boundary between short and intermediate range order is not clearly defined in the literature, different workers including a varying number of the range II parameters under the heading of short range order.

Table VIII.2. — Ranges of order for an amorphous solid $A_{1-x}X_x$

Range	Order	Characteristic parameters	Symbol
I	Structural unit (s.u.)	Identity	AX_n
	or co-ordination	Internal co-ordination number	$n_{A(X)}$
	polyhedron	Bond length	r_{A-X}
		Bond angle	β_A
		Superstructural unit definition	-
II	Interconnection /	Connection mode	C
	relative orientation	Connectivity	c
	of adjacent units	Bond angle	β_X
		Bond torsion angle	α
III	Network topology /	Shortest path ring size	m
	order beyond adjacent unit	Network dimensionality	D
		Topological cluster type	-
IV	Longer range fluctuations	Type (composition or density)	-
		Morphology (droplet, spinodal	-
		decomposition, etc)	
		Radius of gyration	R_g
		Inter-region separation	R

VIII.2.1. Range I - the Structural Unit / First Co-ordination Shell

The most important fact to establish for an amorphous solid is the identity of the structural unit(s) present or for random packing structures the predominant co-ordination polyhedra. The structural parameters in range I specify the detailed geometry of the structural unit/co-ordination polyhedra, including the detailed distribution of bond lengths and angles.

VIII.2.2. Range II - the Interconnection / Relative Orientation of Adjacent Structural Units

The range II order involves the relative orientation and where appropriate the interconnection of adjacent structural units or co-ordination polyhedra. The number of parameters required depends on the number (if any) of shared atoms. For example the interconnection of two corner-sharing SiO_4 tetrahedra requires one (Si-O-Si) bond angle and two torsion angles as defined in Fig. VIII.2.

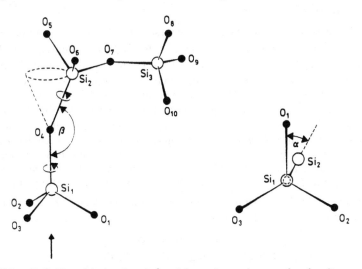

Figure VIII.2. — Definition of the bond angle β and the torsion angles α_1 and α_2 for vitreous silica.

VIII.2.3. Range III - the Network Topology / Order Beyond the Adjacent Structural Unit

As indicated in Fig. VIII.1, a useful concept in discussing the so-called intermediate range order for an amorphous network solid is that of an underlying topological network (dashed triangles) which can be decorated using various atomic motifs to represent different amorphous solids. The topology of the network can only be fully specified by a connectivity matrix or a near neighbour table, which again is impossible for a real material, and hence it is usual to use the ring statistics by shortest path analysis.

VIII.2.4. Range IV - Long Range Fluctuations

Although amorphous solids lack long range order there may nevertheless be longer range fluctuations in density and/or composition arising from phase separation, etc. The characteristic distances involved in such fluctuations are such that they must be studied using small angle scattering techniques (Vol. I, Chapter X).

VIII.2.5. Experimental Parameters

In general the parameters just outlined are not those which are obtained from experiment, diffraction or otherwise. The isotropic nature of most amorphous solids means that in real space the result of a single diffraction experiment is a weighted sum of component correlation functions

$$t_{ij}(r) = 4\pi r \rho_{ij}(r) \qquad \text{(VIII.1)}$$

in which $\rho_{ij}(r)$ is on average the number density of atoms of element j a distance r from the ith atom in the composition unit. For a sample containing n elements there are $n(n+1)/2$ independent component correlation functions.

VIII.3. Theoretical outine

As indicated in the core part of the course (Vol. I), the scattering of X-rays or neutrons is characterised by a momentum transfer, $\hbar Q$, and an energy transfer, $\hbar\omega$, and may be expressed in terms of a generalised scattering law $S(Q,\omega)$, which is the double Fourier transform of the Van-Hove (1954) space-time correlation function, $G(r,t)$. In a conventional total diffraction experiment the detector records both the elastic and the inelastic scattering arising from thermal vibrations and hence performs an integration over ω. As a result the real space quantity obtained is the instantaneous space-time correlation function $G(r,0)$ which may be envisaged as the average of a large number of "snapshots" of the system, one for each neutron or X-ray photon scattered. It is for this reason that this type of measurement can be used to study liquids where the local structure is constantly fluctuating with time.

For an isotropic amorphous solid, the final corrected X-ray or neutron diffraction pattern takes the general form

$$I(Q) = I^S(Q) + i(Q) \qquad \text{(VIII.2)}$$

in which Q is the magnitude of scattering vector for elastic scattering

$$Q = \frac{4\pi}{\lambda} \sin\theta \qquad \text{(VIII.3)}$$

λ being the incident wavelength and 2θ the scattering angle. $I^S(Q)$ is the self (independent) scattering and $i(Q)$ the distinct (interference) scattering. The required structural information contained within $i(Q)$ may be extracted by means of a Fourier sine transformation of the interference function $Qi(Q)$ as outlined for neutrons in Fig. VIII.3. The total neutron correlation function is given by

$$T^N(r) = T^o(r) + D^N(r) \qquad \text{(VIII.4)}$$

in which $D^N(r)$ is the corresponding differential correlation function

$$D^N(r) = \frac{2}{\pi} \int_0^\infty Qi^N(Q)M(Q)\,\sin rQ\,dQ \qquad \text{(VIII.5)}$$

and

$$T^o(r) = 4\pi r \rho^o \left(\sum_i \overline{b_i} \right)^2 \qquad \text{(VIII.6)}$$

M(Q) is a modification function to allow for the fact that data can only be obtained for Q less than or equal to some maximum value Q_{max} and is zero for $Q>Q_{max}$. The i summation is taken over the atoms in one composition unit (c.u.), $\overline{b}_i$ is the neutron scattering length for atom i and $\rho°$ is the composition unit number density.

T(r) from a neutron diffraction experiment

$$I(Q) = I^S(Q) + i(Q)$$
$$\text{self} \quad \text{distinct}$$

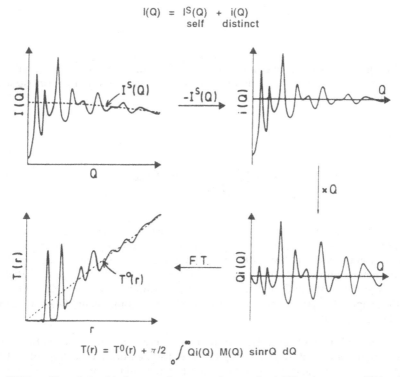

$$T(r) = T^0(r) + \tau/2 \int_0^\infty Qi(Q) \, M(Q) \, \sin rQ \, dQ$$

Figure VIII.3. — The relationship between the corrected, normalised diffraction pattern I(Q) and the real space correlation function T(r). The data are for vitreous silica.

A simple Fourier sine transformation of the X-ray interference function along the lines of eq. (VIII.5) leads to the electronic correlation function of Finbak (1949), which suffers from poor real space resolution since the peaks reflect the finite size of the electron clouds surrounding each atom/ion. For this reason it is conventional to divide $Qi^X(Q)$ by a "sharpening function" before Fourier transformation. For systems containing more than one atomic species the usual sharpening function is $f_e^2(Q)$, $f_e(Q)$ being the average form factor per electron. This yields the X-ray differential correlation function

$$D^X(r) = \frac{2}{\pi} \int_0^\infty \frac{Qi^X(Q)}{f_e^2(Q)} M(Q) \, \sin rQ \, dQ \qquad (VIII.7)$$

The expression for $T^X(r)$ is analogous to eq. (VIII.4) for $T^N(r)$ and in $T^o(r)$ [eq. (VIII.6)] the atomic number Z_i replaces $\overline{b_i}$.

The fact that data can only be obtained for $Q \le Q_{max}$ means that the relationship between $T^N(r)$ or $T^X(r)$ and the component correlation functions $t_{ij}(r)$ is one of convolution

$$T(r) = \sum_i \sum_j t'_{ij}(r)$$

$$t'_{ij}(r) = \int_0^\infty t_{ij}(r') \left[P'_{ij}(r-r') - P'_{ij}(r+r') \right] dr'$$

$$(VIII.8)$$

The j summation is taken over atom types, r' is a dummy convolution variable and the prime indicates N or X for neutrons or X-rays respectively. The corresponding peak functions

$$P^N_{ij}(r) = \frac{\overline{b_i} \ \overline{b_j}}{\pi} \int_0^\infty M(Q) \cos rQ \ dQ$$

$$P^X_{ij}(r) = \int_0^\infty \frac{f_i(Q) \ f_j(Q)}{f_e^2(Q)} M(Q) \cos rQ \ dQ$$

$$(VIII.9)$$

define the experimental resolution in real space, $f_i(Q)$ and $f_j(Q)$ being atomic scattering factors. The reduced neutron peak function is given for the common modification functions in Fig. VIII.4. Note that the experimental real space resolution (width of the central maximum) is inversely proportional to Q_{max} and that, for X-rays, the fact that the Q dependence of $f(Q)$ is different for each element means that the shape of $P^X_{ij}(r)$ changes for each independent component. (See, for example, Fig. 14 of Wright, 1993).

In principal $Qi(Q)$ and $T(r)$ contain precisely the same information, albeit expressed in a different form, although in practice the information content of $T(r)$ is slightly reduced by the use of a modification function other than a step function. The advantage of $T(r)$ is that the information for a particular interatomic distance is concentrated around the appropriate value of r whereas it is spread throughout reciprocal space. Both functions are one-dimensional representations of a three-dimensional structure and are a gross average over the whole irradiated volume. In addition, as shown by eq. (VIII.8), the experimental correlation function is broadened by $P'_{ij}(r)$ and a single diffraction experiment on a multi-element sample yields only a weighted sum of the individual components $t'_{ij}(r)$.

As an alternative to the above formalism, based on the component correlation functions, $t_{ij}(r)$, it is possible to analyse the scattering for a binary system in terms of atom - atom, atom - atom fraction and atom fraction - atom fraction correlation functions (Bhatia and Thornton, 1970). This approach, which yields information concerning chemical short range ordering, is employed particularly for amorphous metallic alloys and is discussed by Sadoc and Wagner (1983). Ionic systems, on the other hand, may formally be analysed using atom - atom, atom - charge and charge - charge correlation functions but, as shown by Wright and Wagner (1988), with a change of normalisation constant these correlation functions are equivalent to those of Bhatia and Thornton.

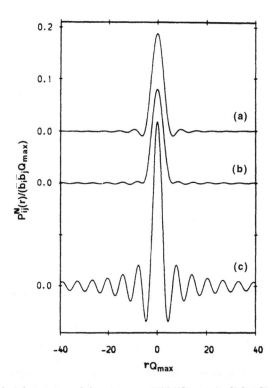

Figure VIII.4. — Reduced neutron peak functions: (A) $M(Q) = \sin(\Delta rQ)/(\Delta rQ)$, $\Delta r = \pi/Q_{max}$;

(B) $M(Q) = \exp(-BQ^2)$, $B = \ln 10/Q^2_{max}$

(C) $M(Q) = 1$.

VIII.4. Experimental techniques

An X-ray or neutron diffraction experiment involves a determination of the scattered intensity as a function of the scattering vector Q for elastic scattering [c.f. eq. (VIII.3)]. In order to achieve the required variation in Q it is possible to scan the scattering angle, 2θ, at fixed incident wavelength, λ, (conventional technique) or to make measurements as a function of λ at fixed 2θ.

VIII.4.1. Neutrons

The two techniques are compared for neutrons in Fig. VIII.5 and examples of the type of conventional and time-of-flight diffractometers used for amorphous solids are given in Wright (1993).

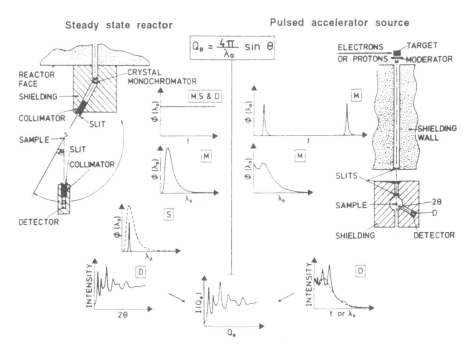

Figure VIII.5. — A schematic comparison of reactor and pulsed source techniques for measuring the neutron diffraction pattern from an amorphous sample.

The main difference between the real instruments and the schematic diagrams of Fig. VIII.5 is that the former have multiple counters to increase the count rate and, for the time-of-flight spectrometer, to provide coverage of the full range of Q. The great advantage of a pulsed accelerator source over a steady-state reactor is that the former is undermoderated, which gives rise to a strong epithermal component in the incident spectrum, $\phi(\lambda)$. These short wavelength neutrons allow data to be obtained to much higher values of Q, resulting in a corresponding increase in real space resolution, and as an example a time-of-flight interference function for vitreous silica is shown in Fig. VIII.6. Older time-of-flight spectrometers, such as LAD at ISIS employ large time-focussed banks of detectors at scattering angles in excess of 90° to increase count rates. However, theoretical studies of the form of the Placzek (1952) corrections for departures from the static approximation for both conventional twin axis and time-of-flight diffractometers, indicate that to avoid distortion of $Qi^N(Q)$ due to departures from the static approximation, measurements should be performed at as low a scattering angle as possible, consistent with adequate Q resolution, and for this reason the latest pulsed source instruments (such as SANDALS at ISIS and GLAD at IPNS) are being designed to operate at scattering angles of less than ~45°.

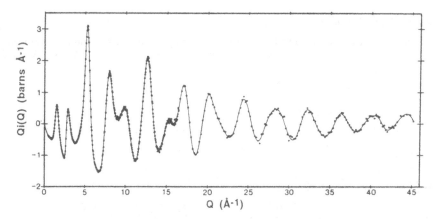

Figure VIII.6. — The interference function for vitreous silica obtained from time-of-flight data (●, experimental points; —— cubic spline fit (Grimley et al., 1990)).

VIII.4.2. X-rays

Accurate quantitative X-ray diffraction measurements on amorphous solids employ a modified powder diffractometer. The situation for X-rays is complicated by the presence of incoherent Compton scattering which at high values of Q can be of much higher intensity than the required coherent contribution. Since neither the wavelength distribution nor the integrated intensity of the Compton scattering from a given sample can be satisfactorily calculated it is necessary to remove this contribution by means of a diffracted beam monochromator or to measure the energy distribution of the scattering at each angle and then extract the desired coherent scattering using peak fitting techniques.

The increasing use of synchrotron radiation sources in X-ray diffraction studies of amorphous solids is one of the most important recent developments. Provided X-rays of sufficiently short wavelength (high Q_{max}) can be generated using a wiggler, this type of source has several advantages over a standard laboratory X-ray generator and when used with an appropriate diffractometer optimised for amorphous materials, is capable of yielding high quality data. An effective method of removing the Compton scattering is that due to Warren and Mavel (1965), which has subsequently been exploited with synchrotron radiation (Bushnell-Wye et al., 1992). The continuous spectrum of a synchrotron radiation source allows the selection of the optimum wavelength for anomalous dispersion experiments and the highly collimated beams produced make it possible to perform grazing incidence experiments to investigate very thin films or surface layers.

The dispersion technique is also possible with X-rays using a semiconductor detector. The advantage of this method is that for complex sample environments (such as cryostats, furnaces and high-pressure cells) only two or three small windows are needed, a fixed angle apart. Similarly, all values of Q are examined simultaneously, making the technique ideally suited to following phase changes or studying kinetic effects. Outside such special applications, however, X-ray variable-wavelength experiments are unlikely to prove important for work on amorphous materials owing to the difficulty of accurately determining the incident spectrum and making corrections for absorption and Compton scattering.

VIII.4.3. Data Reduction

Any measurement involves a certain number of corrections to the raw data but the good experiment minimises these corrections or puts them in a form in which they are easily handled. Normally (c.f. Wright, 1974) corrections are included for:

VIII.4.3.1.   Counter paralysis time

This correction is becoming increasingly important with modern high intensity X-ray and neutron sources.

VIII.4.3.2.   Instrumental background

A correction is required both for the empty spectrometer background and also for the scattering from any sample container used.

VIII.4.3.3.   Absorption, self-shielding and multiple scattering

These corrections are closely coupled and can only be treated independently if one or other of them is small.

VIII.4.3.4.   X-ray effects

Corrections specific to X-ray diffraction include those for polarisation and for the residual Compton scattering not removed by the diffracted beam monochromator.

VIII.4.3.5.   Neutron static approximation distortions (Placzek corrections)

For total neutron diffraction, a correction is required for the fact that the detector integration over ω is performed at constant 2θ rather than at constant Q and, in the case of time-of-flight diffraction, that different energy transfers arise from neutrons with different incident energies, E_0.

VIII.4.4. Normalisation

Following correction it is necessary to normalise the data to absolute units. This may be achieved either by the use of self consistent [Krogh-Moe (1956) -Norman (1957)] integration techniques or, for neutrons, by measuring the (incoherent) scattering from a standard vanadium sample.

VIII.4.5. Fourier Transformation

Probably one of the most controversial aspects of the analysis of X-ray and neutron diffraction data for amorphous solids is the Fourier transformation of the interference function to give the correlation function T(r). This may be achieved using either Filon's (1929) quadrature or a fast Fourier transformation algorithm, the latter being more economical on computer time but much less flexible. The data may also be smoothed before Fourier transformation but this has very little effect on the resulting transform since the frequency range of the noise removed mainly corresponds to distances in excess of those of interest.

VIII.4.6. Assessment of Accuracy

The accuracy of experimental data can most easily be gauged by the behaviour of the transform at low r, below the first true peak. If the transform is well behaved in this region then the data are of reasonable quality. Great care is needed in making this assessment, however, as some authors either plot the radial distribution function $rT(r)$, which has the effect of reducing the amplitude of error ripples at low r, or use these false oscillations as a criterion for "massaging" their data before publication and do not include the original unadulterated transform. In the absence of other information, such data must be treated with the utmost suspicion. Similar techniques are sometimes used to "remove" the effects of terminating the data at finite Q_{max}. Information theory, however, indicates that it is impossible to replace the unmeasured data without making some assumption about the material under investigation so that the resulting correlation function merely becomes one possible model which fits the results rather than an unbiased Fourier transform. Similar objections can be raised to the maximum entropy technique, which also has the additional problem that it is impossible to know what to do to the correlation function from a structural model in order to be able to directly compare it with the experimental maximum entropy version.

VIII.4.7. Separation of Individual Component Correlation Functions

Various methods exist for the separation of individual component correlation functions or linear combinations of subsets of these components and examples of their use occur later in this Chapter. Unfortunately, for many amorphous solids of interest, none of these techniques is feasible, but even in such cases valuable extra information can be obtained from a combination of neutron and X-ray diffraction, for which the component weighting factors are different ($\bar{b}_i \bar{b}_j$ and $\sim Z_i Z_j$ respectively), or by a systematic variation of the composition in systems with more than one component. A particularly striking example of the complementarity of neutron and X-ray diffraction is afforded by Fig. VIII.7, which compares the neutron and X-ray correlation functions for vitreous As_2O_3 and illustrates the use of neutron diffraction data to investigate light atoms in the presence of elements of much higher atomic number (c.f. Chapters V and VI, Vol. I). Important supplementary information in range I can also be obtained from EXAFS spectroscopy (c.f. Chapter XV, Vol. I).

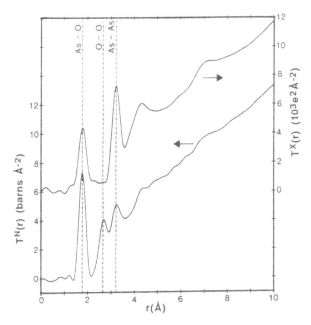

Figure VIII.7. — A comparison of the neutron and X-ray correlation functions for vitreous As_2O_3. The vertical dashed lines indicate the positions of the first As-O, O-O and As-As distances (Clare et al., 1989).

IV.4.7.1. Isotopic substitution

The scattering length of a given element A, with more than one isotope, may be varied by altering the relative isotopic abundances. Samples may thus be prepared with different values of $\overline{b}_A$. For a few elements (H, Li, Ti, Cr, Ni, Sm, Dy and W) isotopes exist with both positive and negative scattering lengths, the latter corresponding to scattering without the normal phase change of π. This gives rise to a special form of isotopic substitution known as the null technique (Wright et al., 1984) in which a combination of isotopes is used such that element A has a zero scattering length and hence does not contribute to the measured interference or correlation functions. A measurement of the scattering from such a sample thus yields the X-X component directly, where X indicates any element present other than A. Examples of the use of isotopic substitution can be found in Gaskell (1983) and Sadoc and Wagner (1983) and data for melt spun Dy_7Ni_3, obtained using the (double) null technique, are shown later in Fig. VIII.13.

VIII.4.7.2. Anomalous dispersion

The anomalous dispersion technique (Krogh-Moe, 1966; Warburton et al., 1987) is based on the fact that near an absorption edge the X-ray form factor has both real and imaginary parts which are wavelength dependent

$$f(Q) = f°(Q) + \Delta f'(\lambda) + i\Delta f''(\lambda) \qquad (VIII.10)$$

The utilisation of anomalous dispersion to separate component correlation functions has, however, only really become feasible with the advent of suitable diffractometers on synchrotron X-ray sources, which allow the wavelength to be continuously varied to obtain the optimum values of $\Delta f'(\lambda)$ and $\Delta f''(\lambda)$. Anomalous dispersion experiments are also possible with neutrons (Wright et al., 1982) for which a variation in scattering length occurs close to an absorption resonance and is much larger for the isotope in question than is the case for X-ray scattering. At present the neutron anomalous dispersion technique is limited to very few elements, but this number should increase with the development of the next generation of amorphous diffractometers on spallation pulsed neutron sources which make measurements at smaller scattering angles and much shorter neutron wavelengths.

VIII.4.7.3.   Neutron magnetic diffraction

The spins of the magnetic atoms/ions in paramagnetic amorphous solids (i.e. many transition metal and rare earth atoms/ions) can frequently be aligned either by cooling below a magnetic ordering transition or by applying a high magnetic field at low temperatures. In either case the difference between the diffraction patterns of the ordered and paramagnetic states yields the magnetic interference function $Qi^M(Q)$. For a system in which there is no correlation between the magnetic moment directions and the interatomic vectors, $Qi^M(Q)$ may be Fourier transformed to give a magnetic correlation function (Wright, 1980) which depends on both the relative positions and the magnetic moments of the magnetic ions and hence it is possible to extract information concerning both the atomic and magnetic structure. The former application is particularly useful when studying complex systems containing a single magnetic species A since it yields the A-A distances alone. Measurements on magnetic systems are in general limited to $Q \leqslant 10$ Å^{-1} due to the fall-off in the magnetic scattering with increasing Q, which limits real space resolution, but on the other hand the correlation functions obtained are simpler since they refer only to the magnetic species.

VIII.4.7.4.   Isomorphous substitution

Isomorphous substitution involves the replacement of one element with another of different scattering amplitude along with the assumption that the structure is otherwise unchanged. In order for this to be the case the chemistry of the two elements must be identical for the system in question and also the bonding to the other elements present, particularly in respect of bond lengths and angles. In general isomorphous substitution, like doping, is not nearly so useful for amorphous solids as in crystalline materials, since there is no longer the constraint of a lattice and hence an atom/ion has a much greater tendency to modify its immediate environment to give the optimum (minimum energy) local structure. The most likely candidates for isomorphous substitution are the lanthanide rare earth ions, since their chemistry and ionic radius are relatively insensitive to changes in the 4f electron configuration. Other possibilities include transition metal ions such as the substitution of Hf^{4+} for Zr^{4+} in fluorozirconate glasses.

VIII.4.7.5.   The difference technique

In any multicomponent system the information obtainable from diffraction experiments is greatly increased if a systematic study is carried out as a function of composition. A common approach in such cases is to use the difference technique in which the correlation function for the base amorphous solid is subtracted from those of the more complex material and the assumption made that the resulting difference correlation functions apply only to the added component(s). Unless the concentration of the added component(s) is small however, this assumption is unlikely to be valid except perhaps for the first coordination shell.

VIII.5. Methods of interpretation

The data obtained from an experimental investigation of an amorphous solid may range from a single total correlation function T(r) to a complete determination of all the individual components $t'_{ij}(r)$. The object in each case is to extract the maximum information on the structure of the material in question. Many amorphous solids contain a well defined structural unit which gives rise to one or more relatively sharp peaks in the correlation function at low r, from which it is possible to extract information on bond distances and angles by means of peak fitting techniques. The way in which the individual structural units pack together is then usually investigated through structural modelling.

The neutron correlation function for vitreous silica obtained from a combination of the data in Fig. VIII.6 and those obtained with various twin-axis spectrometers is shown in Fig. VIII.8, marked with the approximate limits of ranges I to III. The improvement in real space resolution afforded by the increased Q range of

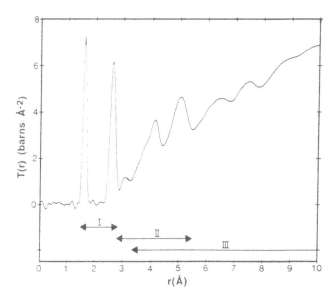

Figure VIII.8. — The correlation function for vitreous silica (Grimley et al., 1990). The roman numerals indicate the extent of the ranges of order defined in Table VIII.2.

the time-of-flight data over that of the twin-axis experiments, may be seen by comparison with the correlation function in Fig. VIII.3 which is also for vitreous silica but only utilises the twin-axis data. The general analysis procedures used to extract structural information from experimental diffraction data will now be outlined under the headings of the various ranges of order using both vitreous silica and other amorphous solids as examples.

VIII.5.1. Range I Order

The modern method of extracting range I parameters from diffraction data is via peak fitting techniques. Such fits may be performed either in reciprocal or real space but the latter has the advantage that the Fourier transform has the effect of at least partially decoupling the parameters for different peaks. Assuming a Gaussian peak in $t_{ij}(r)$ about a mean interatomic distance r_{ij} with an r.m.s. deviation of $\left(\overline{u_{ij}^2}\right)^{1/2}$, the contribution to the neutron interference function is given by

$$Qi_{ij}(Q) = n_{ij}\overline{b_i}\ \overline{b_j}\ \frac{\sin r_{ij}Q}{r_{ij}}\ e^{-Q^2\overline{u_{ij}^2}/2} \qquad (VIII.11)$$

in which n_{ij} is the co-ordination number of j type atoms about the ith atom in the composition unit. In the equivalent X-ray equation, $\overline{b}$ is replaced by the X-ray form factor $f(Q)$. The real space fitting procedure employed by the present author (Grimley et al., 1990) involves folding this idealised peak with the peak function $P'_{ij}(r)$ and then optimising the parameters r_{ij}, $\left(\overline{u_{ij}^2}\right)^{1/2}$ and n_{ij} using a χ^2 minimisation routine. Peak fits for vitreous SiO_2 with and without a modification function are illustrated in Fig. VIII.9.

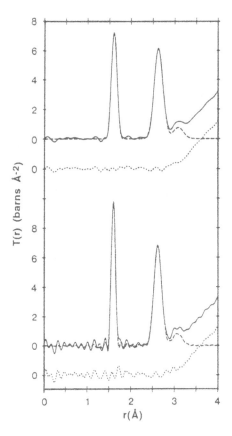

Figure VIII.9. — A fit to the first two peaks in T(r) for vitreous silica (c.f. Fig. VIII.8.) with (upper curves) and without (lower curves) a modification function (——, experiment; - - - fit; ...., residual) (Grimley et al., 1990).

VIII.5.2. Range II Order

In addition to interatomic distances and bond angles the quantification of the order for an amorphous network solid in range II involves the distribution of bond torsion angles $A(\alpha)$ which cannot be obtained directly from diffraction data without assuming some model. For materials such as SiO_2 the difference between the structures of the crystal and glass are that, in the latter, there is a broad distribution of Si-O-Si and bond torsion angles whereas in general the former is characterised by one, or at most a few, discrete values. One simple model for the interconnection of structural units in a random network assumes a uniform distribution of torsion angles except where excluded by steric effects. This model has been used by both Mozzi and Warren (1969) and later by Galeener (1985) to predict Mozzi and Warren's Si-O-Si bond angle distribution. The contribution to Mozzi and Warren's X-ray correlation function from range I and II order, according to their model, is shown in Fig. VIII.10.

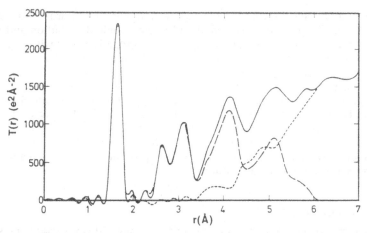

Figure VIII.10. — The contribution to the X-ray correlation function of vitreous silica from the range I and II order (——, experiment; - - -, fit; ...., residual) (Mozzi and Warren, 1969).

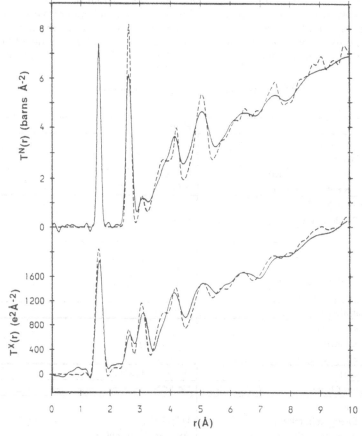

Figure VIII.11. — Neutron (upper curves) and X-ray (lower curves) correlation functions for the Bell and Dean (1972) model of vitreous silica (dashed lines) compared to experiment (solid lines).

The structural models discussed in the next section on range III order must also, by definition, contain some inherent bond and torsion angle distribution and indeed often fail in this respect, making a comparison of range III order with experiment pointless. An example is the Bell and Dean (1972) model for vitreous silica. The correlation function for the Bell and Dean model, after energy minimisation and thermal broadening, is compared to neutron and X-ray data in Fig. VIII.11. The range II order gives rise to the peaks at 4.1 Å (mainly Si-O interactions) and 5.0 Å (O-O) and it can be seen that the shoulder on the low r side of the 4.1 Å peak is much too pronounced compared to experiment, indicating that the range II order in the models is incorrect. This is a common failing of models of vitreous silica and a similar discrepancy between model and experiment occurs for almost all the other models of this material discussed in the literature.

VIII.5.3. Range III Order

The normal method of investigating order in range III is via modelling studies in which an attempt is made to build a "typical region" of the structure, which can be used both to compare to diffraction data and in the calculation of other properties. Various types of model are used, depending on the material under investigation, and the most important of these are summarised in Table VIII.3. In the limited space available, only a few carefully chosen models can be considered in detail here, but further examples are discussed in Wright (1993) and by Gaskell (1983 and 1991).

Table VIII.3. — Structural models of amorphous solids.

Model	Example(s)
Random Network	
(a) Hand built	As, As_2O_3, B_2O_3, Ge, SiO_2
(b) Computer generated	Ge, SiO_2
(c) Geometric transformation	B_2O_3, H_2O, P, P_2O_5
Polymer Models	
(a) Random coil	S, Se
(b) Bundled coil	
Random Sphere Packing	
(a) Hand built	$Ni_{1-x}P_x$, Pd_4Si
(b) Computer generated	Fe_4P
(c) Percus-Yevick	$Dy_{1-x}Ni_x$
Molecular Model	CCl_4, P_4Se_3, $As_{1-x}S_x$
Crystal Based Model	
(a) Limited range of order	$GeSe_2$, SiO_2
(b) Strained crystal	SiO_2
Layer Model	$As_{1-x}S_x$, $Ge_{1-x}Se_x$
Amorphous Cluster	Ge, SiO_2
Monte Carlo	
(a) Energy minimisation	BeF_2, Ge
(b) χ^2 minimisation	$As_{1-x}Se_x$, SiO_2, $ZnCl_2$
Molecular Dynamics	BeF_2, SiO_2, Na_2O-SiO_2

Diffraction data are frequently held to be insensitive to the exact nature of the structure of an amorphous solid but, if used correctly, modern diffraction experiments are able to yield accurate data with high real space resolution which in practice provide a stringent test for the various structural models proposed in the literature. The main problem with the analysis of diffraction data from amorphous solids is one of uniqueness in that even if a perfect fit is obtained with experiment there is no guarantee that there are not other models which would fit equally well. Agreement with diffraction data is thus *a necessary but not sufficient criterion* for any valid structural model. What can be stated with absolute certainty, however, both for diffraction and for the results of any other experimental technique, is that *any model which is not consistent with the data is wrong and must be rejected*. Indeed many models may be rejected on the basis of quite simple measurements, the average density ρ^0 being an obvious and most important example.

Too often diffraction data are published along with the author's favoured model, totally ignoring other published models and/or the results of other complementary techniques which would invalidate the model in question. For the reasons outlined in the previous paragraph, it is essential to consider, and if possible eliminate, all reasonable alternative models, particularly those already proposed in the literature for the material under investigation. An example of the approach required is afforded by Fig. VIII.12, which compares neutron diffraction data for amorphous Ge with all the then published structural models for this material. It can be seen that in fact none of the models is consistent with the experimental data within their known uncertainty, as indicated by the behaviour of the transform below the first true peak.

The amount of structure in the experimental interference and correlation functions, for a given material, and hence their information content, varies considerably but tends to be greater for amorphous network solids than for their random packing based counterparts. In particular, for the latter, even the peaks due to the first co-ordination shell(s) tend not to be isolated which means that it is much more important in such cases to try to use the techniques of section VIII.4.7 to separate the individual component correlation functions. The earliest random sphere packing models comprised hand constructed random close packings of single sized spheres, the most well known example being that due to Finney (1970). Sphere packings can also be computer generated for both hard and soft spheres with a wide range of packing densities and in the case of binary systems with varying radius ratio. An alternative approach is to use the Percus-Yevick hard sphere analytical expressions for either single sized spheres or a binary system.

For a simple binary hard sphere system $A_{1-x}X_x$ there should be all three types of interatomic contact (A-A, A-X and X-X) and the A-X first neighbour distance should be the mean of the A-A and X-X distances. For most metallic glasses, however, there appear to be no mutual contacts between the smaller atoms (ie the metalloid atoms in transition metal-metalloid glasses and the transition metal atoms

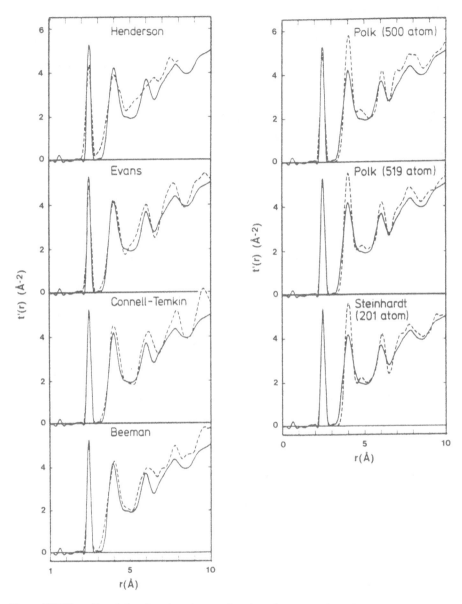

Figure VIII.12. — Correlation functions for a series of random network models of amorphous Ge (dashed lines) compared to neutron data (solid lines, Q_{max} = 23.2 Å^{-1}) (Etherington et al., 1982).

in rare earth-transition metal glasses) as may be seen from Fig. VIII.13 which compares the three component correlation functions for melt spun Dy_7Ni_3, obtained using the double null isotopic substitution technique, with those calculated for a binary hard sphere liquid using the Percus-Yevick equation.

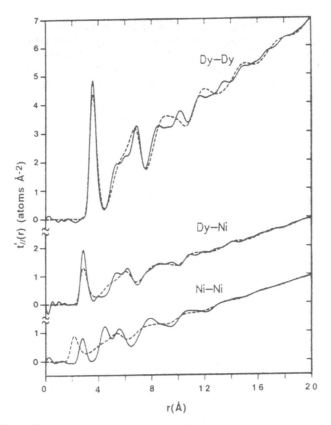

Figure VIII.13. — The three component correlation functions for melt-spun Dy_7Ni_3 (solid line), compared to the results of a Percus-Yevick calculation for a binary hard sphere liquid (dashed line) (Hannon et al., 1991).

The partial interference functions $Qi_{ij}(Q)$ were multiplied by a Debye-Waller factor with a value of 0.147 Å for the rms bond length variation so as to include the effects of thermal vibrations. Fourier transforms were then performed, using the same Q_{max} and modification function as for the experimental data. The hard sphere diameters used in the calculation were adjusted so as to obtain coincidence between the calculated and experimental first peaks of $t^N_{DyDy}(r)$ and $t^N_{DyNi}(r)$, yielding final values for the Dy and Ni atom diameters of 3.347 and 1.938 Å respectively. The composition and number density (x_{Ni} = 0.3; ρ^0 = 0.03938 atoms Å^{-3}) were not adjusted. If the atoms are treated as hard spheres then it must be concluded that the Ni atoms are close to each other but not touching and hence a more ordered structural model than a simple binary random close packing of hard spheres is required. A much more detailed discussion of amorphous metallic structures can be found in Gaskell (1983) and Sadoc and Wagner (1983).

The final two examples reflect the increasing use of computer modelling and simulation techniques in structural studies of amorphous solids. The first involves a novel, and highly successful Monte Carlo technique for generating random network models of amorphous Si and Ge ("Sillium" models) that has been evolved by Wooten and Weaire (1987). This starts with the ambient crystalline network and then introduces disorder via a series of topological transformations which are tested against a Maxwell-Boltzmann factor as in the case of conventional Monte Carlo simulations. Correlation functions for three models due to Wooten and Weaire are compared to neutron diffraction data in Fig. VIII.14.

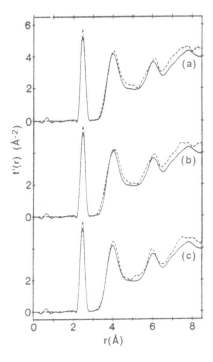

Figure VIII.14. — A series of periodic boundary models of amorphous Ge generated by Wooten and Weaire (1987, dashed lines) compared to experiment (solid lines). (A) model relaxed with Keating potential, (B) as model A but relaxed with Weber potential and (C) model containing 4-membered rings.

Models A and B are topologically equivalent but have been relaxed with, A, a Keating potential and, B, a generalisation of Weber's adiabatic bond-charge model. Model C, on the other hand, contains 4-membered rings which were specifically excluded from models A and B. The complete shortest path ring statistics for both model topologies are given in Fig. VIII.15. It can be seen that all three models have a bond angle distribution very much closer to that of the real material than the Polk and Boudreaux (1973) model (Fig. VIII.12, centre right) and in general are in very good agreement with experiment, except that the average model number densities are too high. The additional structure in t'(r) at higher r is almost certainly statistical in nature due to the small model size (216 atoms).

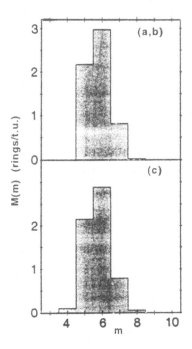

Figure VIII.15. — Ring statistics for the amorphous Ge models in Fig. VIII.14.

    Molecular dynamics simulations have been performed for a number of glasses (Soules, 1990), mainly using simple two-body ionic potentials, an example being the simulation of vitreous BeF_2 by Brawer (Wright et al., 1989) compared to neutron data in Fig. VIII.16. The great majority of these simulations published in the literature claim to be in (good) agreement with diffraction data in that they yield the appropriate structural unit. However, when the resulting correlation functions are correctly compared to experiment it is usually found that the structural units are much too distorted as demonstrated in Fig. VIII.16 by the broad distribution of intratetrahedral F-F distances and hence F-Be-F bond angles. (The width of the F-Be-F bond angle distribution is ~30°). The main reason for the tetrahedral distortion is the absence of any covalent character in the potential used for the simulation and the corresponding lack of bond angle restoring forces, and consequently the most recent state of the art simulations have been performed with three body potentials. A comparison of the third peak in the correlation function shows that the Be-F-Be bond angle distribution is also incorrect while the structure in the simulation correlation function at higher r is much less than that for the real material. The increase in the mean Be-F-Be bond angle and the narrower distribution is probably the result of Be^{2+} ion repulsion shifting the distribution towards 180°.

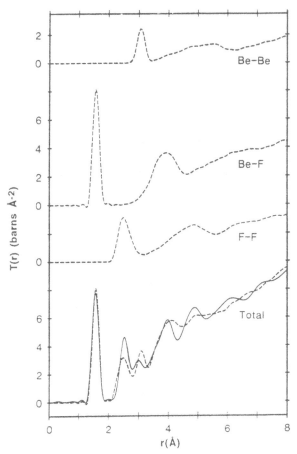

Figure VIII.16. — The neutron correlation function for a molecular dynamics simulation of vitreous BeF$_2$ (dashed lines), compared to experiment (solid line) (Wright et al., 1989).

VIII.6. General discussion

Many authors term good agreement with experiment as getting the peaks in either Qi(Q) or T(r) in the right place. Only very poor models fail to achieve this. What is necessary is also to get peak shapes and areas correct, which involves including the effects of thermal vibration for static models (e.g. computer relaxed random networks) and in real space folding the model component correlation functions with the correct component peak functions P$'_{ij}$(r) for the experiment in question. Otherwise the comparison between model and experiment is meaningless. Note that P$'_{ij}$(r) has satellite features on either side of the central maximum which depend on the exact form of M(Q) and themselves lead to features in T(r). Thus a Gaussian approximation to P$'_{ij}$(r) is simply not adequate, especially when the width is arbitrarily chosen to give the best agreement with experiment. The quantitative comparison of computer simulations and other structural models with experiment is discussed by Wright (1993).

Resolution effects are also important in reciprocal space and typically involve both model and experimental interference functions. In calculating the interference function for a real-space model it is necessary to truncate the model correlation function at some value r_{max}. This leads to a broadening of the model interference function and the introduction of satellite features in reciprocal space in exactly the same way as for an experimental correlation function in real space. Thus in order to compare model and experiment in reciprocal space it is necessary to fold the *experimental* interference function with the reciprocal space peak function $P(Q)$ appropriate to the model calculation. Similarly the model interference function must be broadened by the reciprocal space resolution function for the instrument used in the experiment.

A controversial aspect of the modelling of amorphous solids is the use of periodic boundary conditions. A periodic boundary model is in reality a triclinic P1 crystal, although in practice the unit cell is usually pseudo-cubic (pseudo-orthorhombic in the better models which use different repeat distances a_0, b_0 and c_0 along the x, y and z axes). In reciprocal space the application of periodic boundary conditions means that, with the exception of thermal diffuse scattering, the diffraction pattern is only sampled at discrete values of Q (Q_{hkl}, the Bragg peaks) and there is no scattering below the Bragg cut-off at $Q = 2\pi/a_0$, a_0 being the largest unit cell dimension (assuming orthogonal axes). This should be compared to the case of a cluster model where the finite model size leads to a broadening of the interference function relative to that for an infinite network (c.f. particle size broadening for polycrystalline materials).

The spacing of the Bragg peaks at low Q for model C in Fig. VIII.14 relative to the experimental neutron diffraction pattern for amorphous Ge is shown in Fig. VIII.17. Each Bragg peak is represented by a vertical line in the appropriate position, the height of which is proportional to the peak intensity for the static model $\left[\left(\overline{u_{Ge}^2} \right)^{1/2} = 0 \right]$. The prominent Bragg peak at 1.87 Å^{-1} is close to the 111 Bragg reflection for crystalline Ge, suggesting that the model may contain a small amount of residual crystallinity. A similar prominent Bragg peak is also observed for Model A. Conventionally such a diffraction pattern is artificially broadened, eg by summing the δ-function Bragg peaks into relatively broad histogram columns, in which case it is imperative that before intercomparison the experimental data are treated similarly. However, such a broad spacing of Bragg peaks is unlikely to lead to an accurate prediction of the shape of the first diffraction peak for an amorphous network solid and hence an important objective is to increase the size of the unit cell for periodic boundary models.

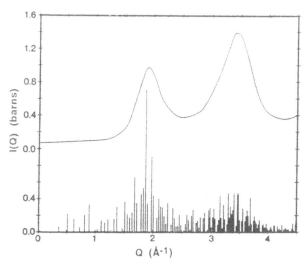

Figure VIII.17. — The positions and intensities of the Bragg peaks from model C in Fig. VIII.14 (vertical bars) relative to the experimental neutron diffraction pattern for amorphous Ge (solid line) (Wright et al., 1991).

VIII.7. Conclusions

The examples discussed in this chapter and in Wright (1993) demonstrate the role of X-ray and neutron diffraction techniques in investigations of the structure of inorganic amorphous solids. The best modern diffraction experiments are capable of providing accurate data with high real-space resolution, which if used correctly are an extremely fine filter for the various models proposed in the literature. The greatest barrier to progress in understanding the structure of amorphous solids lies not so much with the diffraction technique itself but in the development of modelling procedures in which model parameters can be varied in a systematic way until agreement with experiment is obtained in both real and reciprocal space within the known experimental uncertainties. This situation is unique to amorphous materials and has no parallel in corresponding studies of the crystalline state.

BIBLIOGRAPHY

CUSACK N.E., 1987 - "The Physics of Structurally Disordered Matter: An Introduction", Adam Hilger, Bristol.

ELLIOTT S.R., 1990 - "Physics of Amorphous Materials", 2nd Edn., Longman, Harlow.

WARREN B.E., 1969 - "X-ray Diffraction", Addison-Wesley, Reading.

WRIGHT A.C. 1993 - "Neutron and X-ray Amorphography", Chapter 8 of: SIMMONS C.J. and EL-BAYOUMI O.H., editors, "Experimental Techniques of Glass Science", American Ceramic Society, Westerville, pp 205-314.

ZALLEN R., 1983 - "The Physics of Amorphous Solids", Wiley, New York.

REFERENCES

BHATIA A.B., THORNTON D.E., 1970 - Phys. Rev. B2, 3004.

BELL R.J., DEAN P., 1972 - Philos. Mag. 25, 1381.

BUSHNELL-WYE G., FINNEY J.L., TURNER J., HUXLEY D.W., DORE J.C., 1992 - Rev. Sci. Instrum. 63, 1153.

CLARE A.G., WRIGHT A.C., SINCLAIR R.N., GALEENER F.L., GEISSBERGER A.E., 1989 - J. Non-Cryst. Solids 111, 123.

ETHERINGTON G., WRIGHT A.C., WENZEL J.T., DORE J.C., CLARKE J.H., SINCLAIR R.N., 1982 - J. Non-Cryst. Solids 48, 265.

FILON L.N.G., 1929 - Proc. Roy. Soc. (Edinburgh) 49, 38.

FINBAK C., 1949 - Acta Chem. Scand. 3, 1279.

FINNEY J.L., 1970 - Proc. Roy. Soc. (London) A319, 479.

GALEENER F.L., 1985 - Philos. Mag. B51, L1.

GASKELL P.H., 1983 - "Models for the Structure of Amorphous Metals", In: "Glassy Metals II", Eds BECK H. and GÜNTHERODT H.-J., Springer-Verlag, Berlin, 5.

GASKELL P.H., 1991 - "Models for the Structure of Amorphous Solids", In: "Materials Science and Technology" Vol. 9, Ed. ZARZYCKI J., VCH, Weinheim, 175.

GRIMLEY D.I., WRIGHT A.C., SINCLAIR, R.N., 1990 - J. Non-Cryst. Solids 119, 49.

HANNON A.C., WRIGHT A.C., SINCLAIR R.N., 1991 - Mater. Sci. Eng. A134, 883.

KROGH-MOE J., 1956 - Acta Crystallogr. 9, 951.

KROGH-MOE J., 1966 - Acta Chem. Scand. 20, 2890.

MOZZI R.L., WARREN B.E., 1969 - J. Appl. Crystallogr. 2, 164.

NORMAN N., 1957 - Acta Crystallogr. 10, 370.

PLACZEK G., 1952 - Phys. Rev. 86, 377.

POLK D.E., BOUDREAUX D.S., 1973 - Phys. Rev. Lett. 31, 92.

SADOC J.F., WAGNER C.N.J., 1983 - "X-Ray and Neutron Diffraction Experiments on Metallic Glasses", In: "Glassy Metals II", Eds BECK H. and GÜNTHERODT H.-J., Springer-Verlag, Berlin, 51.

SOULES T.F., 1990 - J. Non-Cryst. Solids 123, 48.

VAN HOVE L., 1954 - Phys. Rev. 95, 249.

WARBURTON W.K., LUDWIG K.F.Jr., WILSON L., BIENENSTROCK A., 1987 - Mat. Res. Soc. Symp. Proc. 57, 211.

WARREN B.E., MAVEL G., 1965 - Rev. Sci. Instrum. 36, 196.

WOOTEN F., WEAIRE D., 1987 - Solid State Phys. 40, 1.

WRIGHT A.C., 1974 - Adv. Struct. Res. Diffr. Methods 5, 1.

WRIGHT A.C., LEADBETTER A.J., 1976 - Phys. Chem. Glasses 17, 122.

WRIGHT A.C., 1980 - J. Non-Cryst. Solids 40, 325.

WRIGHT A.C., ETHERINGTON G., DESA J.A.E., SINCLAIR R.N., 1982 - J. Phys. Coll. 43 C9, 31.

WRIGHT A.C., HANNON A.C., SINCLAIR R.N., JOHNSON W.L., ATZMON M., 1984 - J. Phys. F 14, L201.

WRIGHT A.C., WAGNER C.N.J., 1988 - J. Non-Cryst. Solids 106, 85.

WRIGHT A.C., CLARE A.G., ETHERINGTON G., SINCLAIR R.N., BRAWER S.A., WEBER M.J., 1989 - J. Non-Cryst. Solids 111, 139.

WRIGHT A.C., HULME R.A., GRIMLEY D.I., SINCLAIR R.N., MARTIN S.W., PRICE D.L., GALEENER F.L., 1991 - J. Non-Cryst. Solids 129, 213.

WRIGHT A.C., 1993 - J. Non-Cryst. Solids 159, 264.

ZACHARIASEN W.H., 1932 - J. Amer. Chem. Soc. 54, 3841.

CHAPTER IX

QUASICRYSTALS
C. JANOT

IX.1. Introduction

The discovery of new solids exhibiting symmetries forbidden for ordinary crystals was first reported by Shechtman, Blech, Gratias and Cahn (1984) in Al-Mn and Al-Mn-Si alloys. In the intervening eight years, hundreds of other compounds have been observed with quasicrystalline phases. Classically forbidden symmetries, namely 5-fold, 8-fold, 10-fold and 12-fold, have been reported. Most of the quasicrystalline materials which are known now are Al-based binary or ternary metallic alloys, or analogous alloys with Ga or Ti playing the role of Al.

During the early years of the field, there were some speculations that quasicrystals might be inherently disordered and unstable. The speculations proved wrong, though. There are now at least a dozen known compounds which are thought to be thermodynamically stable. In a few cases, the phase diagram has been worked out to some extent (Faudot 1991, Bancel 1991). Also, many of the newer, thermodynamically stable quasicrystals have translational correlation lengths and faceting morphology (with icosahedral symmetry) which rival the best conventional metallic crystals.

At the time of the experimental discovery of icosahedral alloys, Levine and Steinhardt (1984) were independently formulating their hypothesis of a new class of solids which they dubbed "quasiperiodic crystals", or "quasicrystals". They proposed that the new alloys might be laboratory examples of quasicrystals and they outlined some basic mathematical and structural principles.

Quasicrystallography has not yet achieve the level of structure refinement which is currently reported in regular crystallography. Also, the description of the structure, though very easy in its high-dimensional periodic image, is not that straightforward when one deals with the 3-dimensional physical atomic arrangements. This has stimulated many attempts toward atomic modelling, directly in the real space, by approaching the quasiperiodic structures with periodic approximants crystals. So-called "random tiling" models where also introduced in order to give an easy answer to quasicrystal growth problems and also to support the idea that quasicrystals phases may be favoured for entropic reasons (Henley 1991).

In the present paper, the basic principles of quasiperiodicity will be introduced and illustrated with an example of structure determination. (For a complete overview of the subject, see Janot 1992).

IX.2.. Quasicrystallography: the basic principles

Quasicrystals are new types of solids which defy previous standard classifications. They are neither periodically ordered like ordinary crystals, nor are they disordered or amorphous solids. They have a well defined, discrete group symmetry, like crystals, but one which is explicitly incompatible with 3-dimensional periodic translational order (e.g., exhibiting five-, eight-, ten- or twelve-fold symmetry axes). Instead, quasicrystals possess a novel kind of translational order known as *quasiperiodicity*.

Atomic order is best defined in terms of the Fourier transform of the mass density of the solid. In an ordinary crystal, this transform can be written as a Fourier series:

$$\rho(\mathbf{r}) = \frac{1}{V} \sum_{G} \rho(G) \exp(iG \cdot \mathbf{r})$$

(IX.1)

The set of wavevectors **G** define a discrete **reciprocal lattice** in which each wavevector in the sum can be written as an integer linear combination of **three** "basis" vectors $\mathbf{a}_i^*$:

$$G = h\mathbf{a}_1^* + k\mathbf{a}_2^* + \ell\mathbf{a}_3^*$$

(IX.2)

The $\mathbf{a}_i^*$'s are said to span the reciprocal lattice. In a quasicrystal, the Fourier transform of the mass density in also a Fourier series and the wavevectors in the Fourier sum

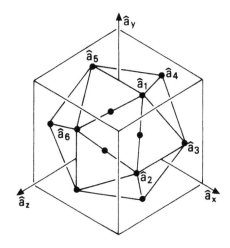

Figure IX.1. — Vertex reciprocal lattice vectors $\mathbf{a}_i^*$ for a regular icosahedron. Cubic axes are also shown on which the $\mathbf{a}_i^*$ vectors have components of the form $(\pm 1, \pm \tau, 0)$.

also form a discrete reciprocal lattice. However, the number of integer linearly independent basis vectors required to "span" the reciprocal space exceeds the spatial dimension and the point symmetry is incompatible with periodic translational order. For example, six basis vectors are required to span the reciprocal space for three-dimensional quasicrystals with icosahedral symmetry:

$$G = n_1 a_1^* + n_2 a_2^* + n_3 a_3^* + n_4 a_4^* + n_5 a_5^* + n_6 a_6^* \qquad \text{(IX.3)}$$

where the a_i^* can be selected to point along the 5-fold axes of an icosahedron. Of course the vectors G, as given by equation (IX.3) must be expressed in a 3-dimensional space. This is illustrated in Fig. IX.1. In the cubic coordinate system, the a_i^* vectors are of the form $(\pm 1, \pm \tau, 0)$ [and permutations] with $\tau = (1 + \sqrt{5})/2 = 2 \cos 36° = 1.618034 \ldots$ the golden mean. Thus, all the vector G have cubic coordinates of the form $(h + \tau h', k + \tau k', \ell + \tau \ell')$ with h, h', k, k', ℓ, ℓ' integral numbers (selected within extinction rules). This may be the simplest possible definition of a quasicrystal. The expression of the reciprocal vectors G in terms of their cubic coordinates shows that, due to irrationality of τ the quasiperiodic structures may exhibit some unusual properties, when referred to regular crystals:

(i)   The G vectors form a dense pseudo-continuous set, and Bragg peaks are expected to show up "everywhere" in the reciprocal space (h + τh' are not integral numbers and their fractional parts fill densely the interval [0,1]). As an example, electron diffraction patterns of an icosahedral quasicrystal are shown in Fig. IX.2.

(ii)   There is not a "smallest" basic G value as a consequence of property (i); this will be a difficulty when indexing experimental diffraction patterns.

(iii) Quasiperiodic structures will obey some inflation rules due again to the peculiar properties of the golden mean τ; for instance, if all h, h', k, k', ℓ, ℓ' integers are allowed by extinction rules, τ inflated structures are identical within rescaling (note that $\tau^{n+1} = \tau^n + \tau^{n-1}$).

(iv)   The dense reciprocal space of a quasicrystal may be given a periodic image in a high dimensional space; for instance, Eq. (IX.3) may be considered as the description of a 6-dimensional periodic reciprocal lattice whose Fourier transform would generate a 6-dimensional periodic mass density distribution. Projection and cut operation will then relate the physical 3-dimensional description to its 6-dimensional image. This is advocated further in the next section.

IX.3.   The hyperspace description

The most important mathematical notion for present thinking on quasicrystals is certainly the hyperspace concept (Janner et al. 1979, Bak 1985, Katz et al. 1985). One application of this idea is in atomic modelling. Diffraction data for icosahedral quasicrystals is now successfully interpreted in terms of a three-dimensional slice through a periodic mass density function in six dimensions. Points of high density in three dimensions (i.e., atoms) are associated with three-dimensional "atomic surfaces" of high density in the six-dimensional space. A three-dimensional slice through the six-dimensional surface shows up as a point in three-dimensional real space.

Figure IX.2. — Selected area electron diffraction
patterns of the AlFeCu icosahedral phase
showing the orientations of the different axes
when rotating the sample around a 2-fold axis.
(courtesy of Marc Audier).

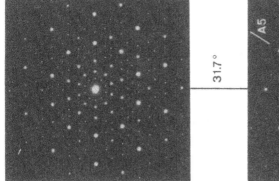

37.4°

20.9°

31.7°

A5

A3

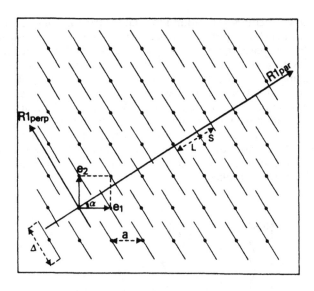

Figure IX.3. — A quasiperiodic chain as generated by a one-dimensional irrational cut of the decorated two-dimensional square lattice. The atomic surfaces A_{perp} reduce here to single straight line segments, whose length is related to the atomic density.

This rather hard-to-visualize description of the quasiperiodic structure is best illustrated by the toy model in Fig. IX.3, in which a one-dimensional quasicrystal is produced by slicing a two-dimensional square lattice. First, the figure illustrates "atomic surfaces", identified as line segments placed in each unit cell, which become atomic points when cut by the "physical" space $R1_{par}$. The slope of $R1_{par}$ is at an angle incommensurate with any lattice plane and results in a one-dimensional chain of atoms separated by short and long intervals occurring in a quasiperiodic sequence.

As suggested by the label $R1_{par}$, the physical space is also called parallel space while the complementary space is consequently dubbed perpendicular space $R1_{perp}$. The atomic surfaces A_{perp}, which become "volume" when real 3-dimensional quasicrystal are concerned, must obey obvious requisites:

(i)   They have no "thickness" in R_{par} if they have to generate point-like atoms in a physical slicing.

(ii) They have the symmetry of the structure e.g. icosahedral, with the same spherical harmonics (Elcoro *et al.* 1992).

(iii) They cannot intercept nor be closer to each other than a threshold distance related to physical atom closeness (the so-called hard core condition).

(iv) They allow translational invariances of the quasiperiodic structure, parallel to both R_{par} and R_{perp} and must be piecewise connected objects (closeness condition).

Such a slicing procedure may be felt awkward or at least, rather artificial. This is actually a simple extension of what may be best understood when referred to simple modulated structures. An n-dim modulated structure can be viewed as the intersection of an (n+1)-dim periodic structure with the n-dim physical space. To illustrate this, let us consider for instance the simple case of an incommensurate structure resulting from the sinusoidal modulation of a periodic chain, periods of the chain and of the modulation function being irrationally related. Position (or origin, or phase) of the modulation function with respect to the chain may be chosen arbitrary. When this phase is changed, atom positions in the chain are "rearranged" into a new configuration having the same total equilibrium free energy as before. A continuous shift of the modulation function with respect to the chain generates an infinite number of "indistinguishable" incommensurate structures which can be visualized simultaneously by piling them up along an axis perpendicular to the chain (Fig. IX.4) (Currat *et al.* 1988).

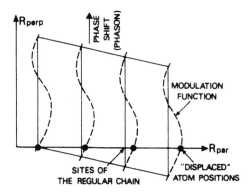

Figure IX.4. — Continuous phase shifts of the modulation function generates "indistinguishable" structures. All these structures can be visualized when piled up together in a "perpendicular" space. This is the fundation of the cut, or embedding, methods (Currat et al. 1988).

In the then defined perpendicular space, a given point of the chain has "equivalent" positions distributed on a periodic profile which is nothing else than the modulation function. The incommensurate structure, in all its "indistinguishable varieties", then appears as a cut of a periodic high-dim (2-dim here) structure by the physical space. The perpendicular space, as describing phases of the modulation function, is called the phase space, or phason space. Any translation in the high-dim space with non-zero component only in the perpendicular space is called a "phason mode" or phason for short. The term phonon is restricted to translations parallel to the physical space. Though all these concepts can easily be extended to quasicrystals, there are some important qualitative differences between incommensurate phases and quasicrystals. Most of the incommensurate structures correspond to 1-dim or 2-dim modulations and the overall symmetry is rather low. The associated high-dim space has then dimension 4 or 5 and the cut that generates the physical space has low symmetry. The relative locations of the atoms are distributed in space according to a modulation law which is a continuous function of the perpendicular space coordinates; the atomic surfaces in the high-dim space are then continuous surfaces

(Fig. IX.4). The existence of an average underlying periodic lattice also gives sense to the notions of average unit cell, average atomic distances, etc.

Conversely, for quasicrystals the high-dim space may have very large dimensions. For instance, the associated high-dim space for icosahedral quasicrystals has minimal dimension six. Moreover, the overall symmetry is the highest possible in 3-dim and there is no underlying periodic lattice. The most dramatic difference comes from the atomic surfaces in the high-dim space, which are not continuous any more, but appear as piecewise discontinuous objects, mostly parallel to the complementary space (see Fig. IX.3). In reciprocal space, the Fourier components of the hyperspace square lattice are modulated by the Fourier transform $A^*_{perp}(G_{perp})$ of the atomic surface A_{perp}, which is a decaying function of the perpendicular component G_{perp} of the Bragg vectors G. As a well known mathematical property of the Fourier transform, the Fourier pattern of the sliced quasiperiodic structure is simply obtained by projection of the transformed high dimensional image on R^*_{par}, the reciprocal space associated to R_{par}. This is shown in Fig. IX.5, which also illustrates the indexing scheme and confirms the dense character of the Bragg pattern. This will be advocated further in the next section.

IX.4. The basics for a diffraction approach

Experimentally, the hyperspace scheme can be used to specify a quasiperiodic structure using diffraction data. Once the point group of symmetry is determined (e.g. icosahedral), Bragg peaks which have been measured are indexed with the high-dimensional Miller indices (e.g. six integral numbers for icosahedral quasicrystals). Figure IX.5B illustrates this indexing for the 1-dim quasicrystals of Fig. IX.3. The position of each Bragg peak in the measured diffraction pattern is indexed with expressions of the sort $h + \tau h'$, h and h' being the Miller integer indices of the high-dimensional image peak. This allows to "lift" the diffraction pattern into its high-dimensional periodic image (Fig. IX.5A). The high-dim periodic mass density is obtained, in principle, via Fourier transform procedures. A final relevant slicing generates the 3-dim quasiperiodic systems. This is quasicrystallography.

In the example shown in Fig. IX.3, the density distribution in R2 can be written:

$$\rho(r_2) = \delta_2\left(r_2 - r_2^{lat}\right) * A_{perp} \qquad (IX.4)$$

r_2^{lat} denotes the vertex positions in the 2-dim square lattice and * describes a convolution product; the $\rho(r_{par})$ density distribution of the 1-dim QC is the cut of $\rho(r_2)$ by $R1_{par}$. The length (size) Δ of the A_{perp} segments is equal to $n_1 a^2$ in which n_1 is the average number of atoms per unit length of the QC chain and "a" the 2-dim lattice parameter. In Fig. IX.3 the slope of $R1_{par}$ in R2 has been chosen to generate a so-called Fibonacci chain $(\cos \alpha / \sin \alpha = \tau)$; in this case Δ is equal to $a(\cos \alpha + \sin \alpha)$ with α the angle of $R1_{par}$ in R2.

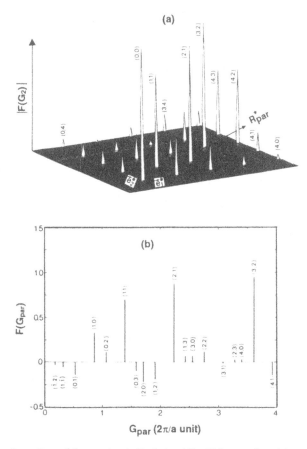

Figure IX.5. — Fourier pattern of the quasiperiodic chain of Fig. IX.3, as projected from the reciprocal "lattice" of the two-dimensional structure. Diffraction data usually gives the one-dim pattern (5B) which, after proper indexing, can be "lifted" in the hyperspace Fourier pattern (5A). In principle, the hyperspace structure (Fig. IX.3) is then deduced via inverse Fourier transform procedure and a cut by the physical space give the final real quasiperiodic structure.

The Fourier components of $\rho(r_2)$ are easily obtained through the expression:

$$F(Q_2) = \delta_2\left(Q_2 - Q_2^{lat}\right) \cdot G\left(Q_{perp}\right) / a^2 \qquad (IX.5)$$

in which Q_2^{lat} are vectors of the reciprocal lattice associated to the 2-dim square lattice; Q_2 is measured in the reciprocal space $R2^*$ and has components Q_{perp} and Q_{par}; $G(Q_{perp}) = \Delta = \left[\sin\dfrac{Q_{perp} \cdot \Delta}{2}\right] / \dfrac{Q_{perp} \cdot \Delta}{2}$ are the Fourier components of the "window function" A_{perp}. As Q_2^{lat} are 2-dim vectors, $F(Q_2)$ are indexed with two integers (see Fig. IX.5) and so are the Fourier component $F(Q_{par})$ of the 1-dim QC density:

$$F\left(Q_{par}\right) = \delta_1\left(Q_{par} - Q_{par}^{lat}\right) \cdot G\left(Q_{perp}\right) / a^2 \qquad (IX.6)$$

as obtained from the projection of $F(Q_2)$ on $R1_{par}^*$ (Fig. IX.5).

Everything so far is easily generalized to 3-dim QC embedded in n-dim space.

The kinematical diffraction is proportional to $|F(Q_{par})|^2$. Thus the diffraction pattern in $R1_{par}^*$ is a very dense set of peaks as expected but terms with large values of Q_{perp} give no significant contributions. This is obvious when considering the analytical form of $G(Q_{perp})$ as given above, which varies as a $\sin x/x$ function, with x proportional to Q_{perp}. Such a function has been well known to exhibit a strong oscillatory decay. Thus, at variance to periodic crystals, the Fourier amplitudes for QC are going to exhibit non uniform behaviours with Q_{par}, depending also on the Q_{perp} values.

It is easy to see that the full sequence of large/short tiles of the 1-dim QC can be generated within the single unit cell of the 2-dim square lattice. Thus, the unit cell and the decoration of this unit cell by A_{perp} elements are the only things which have to be derived from diffraction data. But this is not that easy. Beyond the usual drawbacks of crystallography (truncation effects, background noises, phase reconstruction techniques, etc.) quasicrystallography has to deal with specific problems (Janot 1992). Actually, refinement procedures cannot apply to the many parameters that would be needed to specify all the details (size and shape) of the atomic surfaces. No diffraction experiments will never be enough precise for rendering the boundaries of the atomic objects in R_{perp} by inverse Fourier transform. Some of the details are gained when contrast variation measurements are feasible. As for regular crystals, the derivation of a quasicrystal structure from diffraction data is equivalent to the reconstruction of the phases of the structure factors $F(Q)$. The structure factors are the Fourier component of the density distribution $\rho(r)$. They cannot be obtained directly from the observed Bragg peak intensities $I(Q)$ which, in turn are the Fourier components of the so-called Patterson functions (PF). A correspondence scheme can be summarized as follows:

$$
\begin{array}{ccc}
I(Q) & \leftrightarrow \quad F(Q) \cdot \ F(Q)^* & \leftarrow F(Q) \\
& \uparrow & \uparrow \\
& \downarrow & \downarrow \\
PF & \leftarrow & \rho(r)
\end{array}
\qquad (IX.7)
$$

The $I(Q)$ quantities are integrated intensities of the measured Bragg peaks.

Contrast variation can be used, either with neutron scattering or anomalous X-ray diffraction, to separate partial structure factors of a complex structure. The diffracted intensity for a binary alloy A-B is:

$$I(Q) = |b_A F_A + b_B F_B|^2 \qquad (IX.8)$$

with $b_{A,B}$ neutron scattering length or X-ray atomic scattering factor for A,B atoms and $F_{A,B} = \sum_{A,B} \rho_{A,B} \exp(iQ \cdot r_{A,B})$ the corresponding partial structure factors, or:

$$I(Q) = b_A^2 |F_A|^2 + b_B^2 |F_B|^2 + 2 b_A b_B |F_A| |F_B| \cos \Delta\varphi$$

with Δφ the phase difference between F_A and F_B. The unknown quantities are $|F_A|$, $|F_B|$ and Δφ. They can be determined for each diffraction peak by measuring at least three renormalized independent intensities $I(Q, b_A/b_B)$, b_A/b_B being changed thanks to isotopic (or isomorphous) substitution, or anomalous dispersion effects. This has been successfully applied to QC structure (Dubois *et al*. 1986, Janot *et al*. 1986, Janot *et al*. 1989, De Boissieu *et al*. 1990, 1991). But modelling is always required. Such modellings are necessarily achieved via some arbitrary choice of a finite number of parameters that are intended to characterize the boundaries of the atomic surfaces. These parameters may be a sphere radius in the simplest naive approach. The best empirical way to design crudely the tomic surface may be a fit to data with weighed sums of spherical harmonics of the icosahedral symmetry (Elcoro *et al*., 1992). However, there are severe constraints for the choice of the atomic surface shapes in the high-dimensional images. Composition and density along with the elimination of too short unphysical pair distances restrict freedom to less than 10% in sizes of the atomic surfaces. The connectivity constraints (Katz, 1990) when applied to icosahedral symmetry impose that the boundaries of the atomic surfaces are two-fold planes (Cornier-Quiquandon *et al*. 1991).

Finally, one can also inject semi-artificial constraints which may account for the chemical and local structural features of the material. For instance, Duneau and Oguey (1989) have proposed a model in which the atomic surfaces have been designed to force the existence of a maximum of Mackay icosahedral clusters (Mackay 1962) in the 3-dimensional structure.

Beyond natural constraints (composition, density, unphysical short pairs) models can only show differences corresponding to the "skin" of the atomic surfaces and this may be a very little part of the whole structure. For instance, about 5% only of the atom positions in the 3-dimensional structure are affected when a spherical atomic surface is replaced by a triacontahedron of the same volume. Of course, 5% of atomic positions are already an enormous factor when properties are concerned.

Conclusively, approaching the structure of quasicrystals via high dimensional crystallography techniques is basically very simple. Practical difficulties arise when details of the atomic surfaces are to be specified. One may wonder whether it would be easier to work out the structure of quasicrystals by simple regular refinement of giant cell crystal structures. This has been actually attempted. But it does not provide a more accurate resolution of the problem. Moreover, with the best quasicrysals (equilibrium compounds of the AlPdMn or AlFeCu systems) this crystal approach requires unphysically large unit cells containing more than 10 000 atoms.

IX.5.   A recent example of a quasicrystallography study

As an example of quasicrystallography, it may be interesting to report on the recent study of the structure of the perfect quasicrystal AlPdMn (Boudard *et al*. 1992). The AlPdMn system is a very favourable case for diffraction studies. Beyond the perfection of quasiperiodicity and the possibility to grow stable single grains, it is also possible to play with contrast variation effects (contrast on Mn sites with neutrons, and neutrons versus X-rays or anomalous diffraction for contrast on Pd sites). Indexing the diffraction peaks in neutron and X-ray patterns, shows that the reciprocal lattice of the 6-dimensional image is body center, with indices all odd or all even. Bragg reflections with odd indices are weak, which suggest that the reciprocal space may be best described by a primitive lattice with a set of superstructure reflections having half integer indices. Thus, the Bravais lattice in 6-dim direct space can be pictured as a primitive cubic (parameter a = 6.451 Å), with two families of non-equivalent lattice sites having even or odd parity. (This parity refers to that of the sum of the six corresponding indices).

Sizes and shapes of the atomic surfaces of the high-dimensional image must be induced from 6-dimensional Patterson (PF) functions as calculated from the properly indexed Bragg peaks. Figure IX.6 reproduces planar slice of 6-dimensional PF examples, containing two five-fold axes taken in the physical 3-dimensional space for

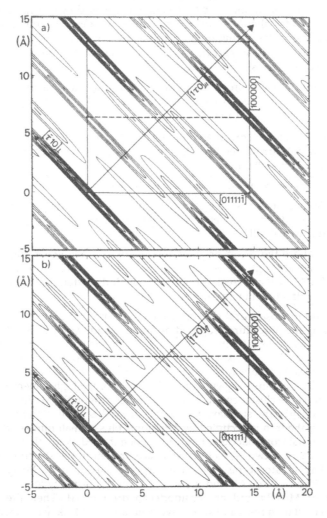

Figure IX.6. — Rational cut of the 6-dim Patterson function, in a plane containing a 5-fold axis in both the physical and the perpendicular space, as deduced from single crystal X-ray (a) and neutron (b) diffraction data. The primitive unit cell is outlined.

one and in the complementary space for the other. The PF features appear as traces, elongated on the perpendicular space axis. Their actual width in the physical space is mainly due to truncation effects of the Fourier transform procedure. These remarks hold true for any 2-dimensional slice of the 6-dimensional PF, which means that the PF features are basically 3-dimensional bulk "objects" in the perpendicular space, with point-like intersection with the physical space. These "objects" appear as located around lattice sites and body centres.

Information about the size of the different atomic surfaces may result from models using balls and spherical shells, even if, obviously, the atomic surfaces have to be faceted volumes. Actually, at small values of Q_{per} (i.e smaller that 0.5 Å^{-1}), the Fourier transform of an atomic surface is mainly influenced by its size and does not depend on its precise shape. A fitting procedure conducted to reproduce the PF maps allows a reasonable definition of these volumes. Within density and chemical composition constraints, the best fit to low Q_{per} data was obtained with 6 atomic surfaces defined as follows:

(i)   at even lattice sites, a core of Mn (radius 0.83 a, where a is the parameter of the 6-dim cube) surrounded by an intermediate shell of Pd (extending up to 1.26 a) and an outer shell of Al (up to 1.55 a),

(ii)   at odd lattice sites, a core of Mn (radius 0.52 a, τ time smaller than the one on the even node) surrounded by a shell of Al (up to 1.64 a, i.e larger than the even one) but without Palladium,

(iii) at odd body centres, a ball of Pd (radius 0.71 a),

(iv)   at even body centres a small ball of Al (radius 0.3 a) or an empty volume (the fit did not show differences between these two hypotheses since the Al volume involved is very small).

Now, what are the 'basic' or 'hard core' features when generating the 3-dim structure as a cut through the 6-dim model ? As for the AlLiCu quasicrystal (de Boissieu *et al.* 1991) we may seek for local icosahedral clusters, or alternatively we can describe the structure in terms of dense atomic planes.

The nature of icosahedral clusters which exist in the 3-dim structure, can be found following the procedure proposed by Duneau and Oguey (1989). Comparing the spheres sizes of interest with Duneau-Oguey atomic surfaces, it can be shown that the external shell of a Mackay icosahedra is actually present in the structure. The internal small icosahedron is replaced by pieces of an Al dodecahedron. Two kinds of external shells are found, depending on the parity of the high dimensional lattice site where is sited the cluster centre. In the 6-dim description the Mn inner core have a radius smaller than the standard triacontahedron; this implies that the resulting 3-dim large icosahedron is occupied by Mn atoms plus a small number of Al or Pd atoms. The external icosidodecahedron is made of either Al atoms alone or of Al+Pd atoms. To summarize, two types of clusters are present in the structure, namely a pseudo-Mackay cluster type 1, with a large icosahedron of Mn+Al and an icosidodecahedron of Pd+Al, and a pseudo-Mackay cluster type 2 with a large icosahedron of Mn+Pd and an icosidodecahedron of Al. Then the structure propagates via inflation rules and successive generation of Mackay icosahedra.

The existence of dense planes in the the 3-dim structure may be evidenced from planar projections of the structure. When looking at such a projection at glancing angle, series of lines corresponding to atomic planes are clearly visible. The densest planes are characterized by large spacing. They are found perpendicular to a 5-fold axis. Two-fold planes are also visible but they are closer to each other and contain a smaller density of points. Two dimensional cuts of the 3-dim structure, perpendicular to a 5-fold axis and made at different level along this axis are pictured in Fig. IX.7 A systematic search of the very dense 5-fold planes shows that only a finite number of different atomic planes has to be considered. Each type of plans is characterized by reasonably well defined average local order and average chemical composition. Some planes are corrugated and may be described as a dense layer sandwiched between two somewhat less dense layers 0.5 Å apart.

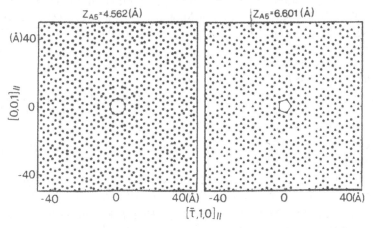

Figure IX.7. —Two examples of dense atomic planes perpendicular to a 5-fold axis. The level Z_{A5} of each plane is shown. Atomic species are indicated: * Al atoms, ▲ Pd atoms, • Mn atoms. Some 5-fold rings, corresponding to external shells of pseudo-Mackay icosahedra have been outlined.

IX.6.   Conclusion

Perfect and stable quasicrystals can now be prepared with several aluminum based systems. They exhibit diffraction patterns with Bragg peaks whose width is only resolution limited.

The basics for structural studies of quasicrystals are also well established. They have periodic images in higher-dimensional spaces which can be approached by models deduced from experimental Patterson functions. Unfortunately, uniqueness of the solutions cannot be guaranteed up to details of the atomic arrangements. Even if a "hard core" including more than 90% of the structure must belong to any model, it may be that properties are strongly influenced by the 5-10% which vary from model to model.

This is probably of importance if one is to understand the peculiar physical behaviours of quasicrystals. In spite of being alloys made of metals they have the electrical or thermal conductivities of dopped semiconductors, and the mechanical properties of ceramics but ceramics with low friction coefficients and that can be coated on a metal surface. These properties are already being used in technological applications (tribology problems, non-stick frying pans, thermal screening, surface protection, etc.) and open the way to a very promising future.

REFERENCES

BAK P., 1985 - Symmetry, stability and elastic properties of icosahedral incommensurate crystals, *Phys. Rev.* **B32**, 5764-5772.

BANCEL P.A., 1991 - Order and disorder in icosahedral alloys. In *Quasicrystals: the State of the Art*, (ed. D.P. di Vincenzo and P.J. Steinhardt) pp. 17-54. World Scientific, Singapore.

BOUDARD M., DE BOISSIEU M., JANOT C., HEGER G., BEELI C., NISSEN H.U., VINCENT H., IBBERSON R., AUDIER M. and DUBOIS J.M., 1992 - Neutron and X-ray single crystal study of the AlPdMn icosahedral phase. *J. Phys: Cond. Matter* **4**, 10149-10168.

CAHN J.W., SHECHTMAN D. and GRATIAS D., 1986 - Indexing of icosahedral quasiperiodic crystals *J. Mater. Res.* **1**, 13-26.

CORNIER-QUIQUANDON M., QUIVY H., LEFEBVRE S., ELKAIM E., HEGER G., KATZ A. and GRATIAS D., 1991 - A tentative methodology for structure determination in quasicrystals. *Phys. Rev.* B **44**, 2071-2092.

CURRAT R. and JANSSEN T., 1988 - Excitations in incommensurate crystal phases, *Solid State Physics* **41**, 201-302.

DE BOISSIEU M., JANOT C. and DUBOIS J.M., 1990 - Quasicrystal structure: cold water on the Penrose tiling scheme, *J. Phys.: Cond. Matter* **2**, 2499-2517.

DE BOISSIEU M., JANOT C., DUBOIS J.M., AUDIER M. and DUBOST B., 1991 - Atomic structure of the icosahedral AL-Li-Cu quasicrystals, *J. Phys.: Condens. Matter* **3**, 1-25.

DUBOIS J.M., JANOT C. and PANNETIER J., 1986 - Preliminary diffraction study of icosahedral quasicrystals using isomorphous substitution, *Physics Letters* A **115**, 177-181; JANOT C., PANNETIER J., DUBOIS J.M. and FRUCHART R. 1986 - *Phys. Letters* A **119**, 309-313.

DUNEAU M. and OGUEY C., 1989 - Ideal AlMnSi quasicrystal: a structural model with icosahedral clusters, *J. Phys. (France)* **50**, 135-146.

ELCORO L., PEREZ-MATO J.M. and HADARIAGA G., 1993 - Spherical harmonics analysis for quasicrystal structures, *J. of Non Crystalline Solids*. In press.

FAUDOT F., QUIVY A., CALVAYRAC Y., GRATIAS D. and HARMELIN M., 1991 - About the Al-Cu-Fe icosahedral phase formation, *Mat. Science and Engineering* A **133**, 383-390.

HENLEY C.L., 1991 - Random tiling models. In *"Quasicrystals: the State of the Art"* (Ed D.P. Di Vincenzo and P.J. Steinhardt) pp. 429-518. World Scientific, Singapore.

JANNER A. and JANSSEN T., 1979 - Superspace groups, *Physica* A **99**, 47-76.

JANOT C., 1992 - *"Quasicrystals: a primer "*. 320 pages. Oxford University Press.

KATZ A. and DUNEAU M., 1985 - Quasiperiodic patterns, *Phys. Rev. Lett.* **54**, 2688-2691.

KATZ A., 1990 - Hardcore and closeness conditions in quasicrystals. In *Number theory and Physics* (eds. J.M. Luck, P. Moussa, M. Waldschmidt and C. Itzykom) pp. 100-123. Springer Verlag.

LEVINE D. and STEINHARDT P.J., 1984 - Quasicrystals: a new class of ordered structures, *Phys. Rev. Lett.* **53**, 2477-2480.

SHECHTMAN D., BLECH I., GRATIAS D. and CAHN J.W., 1984 - Metallic phase with long-range orientational order and no translational symmetry, *Phys. Rev. Lett.* **53**, 1951-1953.

CHAPTER X

STEREO CHEMISTRY AND ELECTRONIC STRUCTURE

XAFS SPECTROSCOPY: DATA-ANALYSIS AND APPLICATIONS

D.C. KONINGSBERGER[1]

X.   Introduction

The structural and electronic characterization of a material provides a basic understanding of its properties. Traditionally, diffraction techniques (XRD, neutron diffraction, LEED) are being used for most of the structural investigations and reliable structures can be determined for materials that exhibit a long-range structural order (like single crystals or polycrystalline material). X-ray Absorption Fine Structure (XAFS) spectroscopy is a powerful technique to characterize all forms of matter irrespective of their degree of crystallinity. EXAFS (Extended X-ray Absorption Fine Structure spectroscopy probes the local structure of a material. The local structure of highly disordered solids, amorphous materials and liquids can be unraveled with EXAFS. In addition, the chemical state and the electronic properties can be determined from the X-ray Absorption Near Edge Structure (XANES) which extends within 40 eV of the X-ray absorption edge. One of the major advantages of XAFS is its atomic selectivity which enables the investigation of the local structure of each different constituent of a sample. As shown by Fontaine (1993), the recent availability of high-brightness synchrotron radiation sources has resulted in a prosperous development XAFS spectroscopy.

Fontaine has extensively discussed in chapter XV of volume I of this edition the underlying physical principles of XAFS. Here the basic concepts of the analysis of XAFS data will be given. These concepts are very general and therefore important for the application of EXAFS in the fields of coordination chemistry, catalysis, bio-inorganic chemistry, biology, surface physics and chemistry. EXAFS have been widely used to determine the stereo chemistry of inorganic compounds. To illustrate the importance of XAFS spectroscopy applications in the field of inorganic chemistry are extremely useful. Inorganic materials are the building stones of an important class of compounds called catalysts. Most industrial catalysts consist of metal, metal-oxides or metal-sulfides particles dispersed on high-area supports exposing most of the particle surface to reactants. Highly dispersed supported metal catalysts are used for chemical processes like hydrogenation, hydrotreating, hydrocracking and catalytic reforming. The particles are stabilized by metal-support interactions that hinder particle migration, coalescence, and loss of catalytic activity by loss of metal surface area. Metal-support interactions also affect the activities of catalysts, and extensive research has been done to elucidate the effects. However, only little is understood about the nature of the metal-support interface or the role of the support in influencing the reactivity of supported metal particles. Metal-oxides are mostly

---

[1]   The unpublished results in this chapter are taken from M. Vaarkamp, thesis 1993 and F.W.H. Kampers, thesis 1988, Eindhoven University of Technology, The Netherlands.

used for oxidation reactions, while metal-sulfides are detrimental for the removal of sulfur from oil feedstocks. Metal-oxides and sulfides can be dispersed as monolayers onto the support. The understanding of particle size and support effects are limited by the lack of precise methods for structural characterization of supported catalytic particles and the particle-support interface, combined with the nonuniformity of the available samples. Extremely high dispersed material consisting of particles built up by only a small number of atoms can not be studied by techniques like X-ray diffraction or Transmission Electron Microscopy. EXAFS can be tuned to study the different constituents of a catalytic system. It can be used to study in detail the particle-support interface. It is possible with *in-situ* techniques to follow the change in electronic properties and the structure due to the absorption of reactants on the surface. The X-ray absorption edge can give additional information about local chemistry and electronic structure (XANES, see also Fontaine chapter XV volume 1 of this edition).

This chapter will show how important XAFS spectroscopy is for the study of catalytic materials mostly consisting of inorganic compounds. The very important property of XAFS spectroscopy which is that experiments can be performed *in-situ* will be demonstrated. In section X.2 important aspects of the analysis of EXAFS data will be discussed. An important example of the determination with the EXAFS technique of the stereo chemistry in catalytic materials will be discussed in section X.3. EXAFS has been used to determine the structure of the metal-support interface of extremely small iridium clusters supported on MgO. This information is very important for a fundamental understanding of catalytic properties. Section X.4 deals with the determination of the electronic structure of small platinum metal particles. The electronic properties of small metal particles are of interest from fundamental standpoint of view. More information is needed to interprete and understand XPS data published in the literature. Moreover, the determination of the electronic structure is important for an understanding of the properties of catalytic materials. It will be shown that the white-line intensity of the L_{II} and L_{III} absorption edges can be used to measure the local d-band density of states of transition metal atoms.

X.2.   EXAFS Data-Analysis

X.2.1.   EXAFS Equation

An accurate theoretical description of EXAFS spectra includes curved wave effects and an energy dependent self-energy (McKale et al., 1988; Vaarkamp (1) et al., 1993). The curved wave EXAFS formula (Ashley et al., 1975; Lee et al., 1975, Stern et al., 1975) can be reduced to the plane-wave EXAFS formula (eq. X.1 and X.2) without loss of accuracy, if the model function used in the data-analysis procedure is based upon the use of a reference absorber-backscatterer coordination (obtained experimentally or theoretically) having approximately the same distance as the coordination in the unknown material

$$\chi(k) = \sum_{j=1}^{Shells} A_j(k)\sin\left[2kR_j + \varphi_j(k)\right]. \tag{X.1}$$

with R_j being the coordination distance and $\varphi_j(k)$ being the phase function determined by both the absorber and backscatterer. The amplitude $A_j(k)$ can be expressed as:

$$A_j(k) = \frac{S_0^2 e^{-2R_j/\lambda}}{kR_j^2} \bullet N_j F_j(k) \bullet e^{-2\sigma_j^2 k^2}. \qquad (X.2)$$

damping   scattering   disorder

power

The first group of terms describes the damping of the electron wave: the exponential term $e^{-2R_j/\lambda}$ accounts for the finite lifetime of the excited state and $S_0^2(k)$ is an amplitude reduction factor which accounts for the photo-electron energy loss due to many body effects and shake-up/shake-off processes in the absorber atom (not all the energy of the impinging photon is transferred to the photo-electron). The second group of terms is characteristic for the scattering power experienced by the reflected electron wave: $F_j(k)$ is the back-scattering amplitude of the neighbouring atoms and N_j is the average coordination number in the jth shell. The third term contains the Debye-Waller factor σ_j which represents the root mean square fluctuation in R_j caused by thermal motion of the atoms and the structural disorder present in the material.

Equations X.1 and X.2 are being used to analyze EXAFS data. The final goal is to determine from the experimental data the coordination distance R_j, the coordination number N_j and the Debye-Waller factor σ_j for each individual coordination shell. EXAFS can distinguish between different types of neighbouring atoms. As an example the phase shift of the Pt-Pt and Pt-O absorber-backscatterer pair and the backscattering amplitude of O and Pt are plotted in Fig. X.1a and X.1b, respectively. It can be seen that both the Pt-Pt and the Pt-O phases are a non-linear function of k. The backscattering amplitude of oxygen (typical low Z scatterer) is decaying rapidly with increasing values of k, whereas the amplitude of platinum (typical high Z scatterer) is still significant at high values of k.

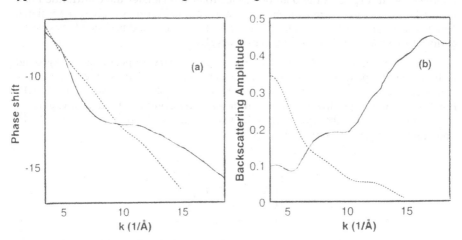

Figure X.1. — a) Phase shift $(\varphi(k))$ and b) Backscattering amplitude $(F(k))$ of Pt-Pt (solid line) and Pt-O (dotted line). Data were obtained from first shell EXAFS data of Pt foil and $Na_2Pt(OH)_6$.

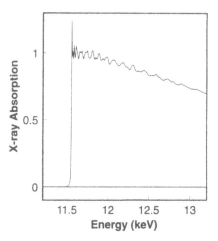

Figure X.2. — Isolated X-ray absorption spectrum of a 4 μ Pt-foil measured at 77K on beamline 9.2 SRS Daresbury.

X.2.2.   Extraction of the EXAFS Function

The first step in the analysis of an XAFS spectrum is to separate the XAFS oscillations from the experimentally measured X-ray absorption coefficient. The necessary steps for the extraction of the EXAFS function are: (i) pre-edge subtraction, (ii) determination of inner potential value, (iii) post-edge background subtraction, (iv) normalisation (Sayers et al., 1988). A standard procedure to remove the pre-edge absorption involves fitting the spectrum in the energy range sufficiently below the absorption edge (typically - 200 to - 40 eV) with a polynomial. The fitted polynomial is extrapolated past the edge and subtracted from the entire spectrum. This step removes the absorption due to all other electrons and isolates the contribution of a particular absorption edge from the total absorption. Figure X.2 gives the isolated X-ray absorption of a 4μ thick platinum foil measured at liquid nitrogen temperature at the EXAFS station 9.2 of the SRS, Daresbury, UK.

The next step consist of removing the smoothly varying post-edge background μ_0 by fitting the post-edge region (typically starting at 20 eV above the edge) with a spline function and subtracting it from the data (Cook et al., 1981). The EXAFS function is then normalized on a per atom basis by dividing the data through the edge jump μ_0 at the absorption edge (Sayers et al, 1988). The normalized EXAFS function is converted to k-space by using:

$$k = \left[\left(\frac{8\pi^2 m}{h^2} \right) (h\nu - E_0) \right]^{1/2} . \tag{X.3}$$

In the initial stage of analysis, the value of E_0 is chosen at the maximum derivative point on the edge. At later stages, E_0 is varied and adjusted to its proper value. The normalized EXAFS spectrum of the platinum foil is shown in Fig. X.3a.

X.2.3.   Fourier Filtering: Separation of Coordination Shells

As discussed above the EXAFS function shown in Fig. X.3a is a superposition of a number of damped sinusoidal functions, each corresponding to a particular

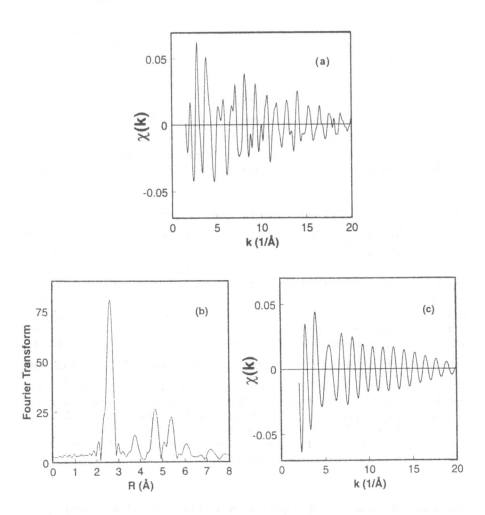

Figure X.3. — a) EXAFS spectrum of Pt foil obtained after background subtraction and normalisation. b) Fourier transform; k^3, Δk: 2.2 - 20.3 Å^{-1}. c) Isolated first shell EXAFS; ΔR: 1.1 - 3.1 Å.

coordination shell. Fourier transformation of $\chi(k)$ yields peaks in r-space corresponding to individual coordination shells around the absorbing atom (Sayers et al., 1988):

$$FT(r) = \frac{1}{\sqrt{2\pi}} \int_{k_{min}}^{k_{max}} k^n \chi(k) e^{-2ikr} dk. \qquad (X.4)$$

The peaks are shifted to lower r-values from the real interatomic distances due to the influence of the phase shift function. The weight factor k^n can be used to compensate for the decay in amplitude of the spectrum with increasing values of k or to emphasize a particular part of the EXAFS spectrum. The Fourier transform of the platinum foil EXAFS spectrum is shown in Fig. X.3b.

The EXAFS function of a particular coordination shell can now be isolated by applying a window in r-space and performing an inverse Fourier transform to k-space (Sayers et al., 1988):

$$k^n \chi_j(k) = \frac{1}{\sqrt{2\pi}} \int_{R_{min}}^{R_{max}} FT(r)\, e^{-2ikr} dr. \tag{X.5}$$

This step produces a modulated sinusoidal function with amplitude $A_j(k)$ and argument $\Phi_j(k) = 2kR_j + \varphi_j(k)$, as given in Eq. X.1. The isolated EXAFS function for the first shell of platinum foil is shown in Fig. X.3c. $A(k)$ and $\Phi(k)$ for a particular coordination shell can now be obtained in tabular form from an EXAFS experiment.

X.2.4.   Non-linear Least Square Multiple Shell Fitting

To extract the coordination parameters R, N and σ from the EXAFS data the quantities $\varphi(k)$, S_0^2, $e^{(-2R/\lambda)}$ and $F(k)$ have to be known. It is now generally accepted that the phase shift $\varphi(k)$ and $F(k)$ are transferable from the reference compound to the compound with unknown structure. The argument $\Phi_r(k)$ and amplitude $A_r(k)$ of an absorber-backscatterer pair can be extracted by Fourier filtering of an EXAFS spectrum (as shown in section X.2.3) from a reference compound (with a well defined known structure) in which the contributions are well separated. The phase shift can be obtained by $\varphi(k) = \varphi_r(k) = \Phi_r(k) - 2kR_r$. To determine $F(k)$ one still has to know input values for S_0^2 and $e^{(-2R/\lambda)}$. To circumvent this requirement we assume that not only $\varphi(k)$ and $F(k)$ but also S_0^2 and $e^{(-2R/\lambda)}$ are transferable from one compound to another. The validity of this assumption has been shown for compounds with the same absorber-backscatterer pair (Citrin et al., 1976; Bunker et al., 1983) but even for absorbers or backscatterers which are neighbours in the periodic table (Teo et al., 1997; Lengeler, 1986). This leads to the definition of a modified backscattering amplitude $F'(k)$ (Vaarkamp (2), 1993):

$$F'(k) = S_0^2 e^{(-2R/\lambda)} e^{-2k^2\sigma^2} F(k). \tag{X.6}$$

This modified backscattering amplitude contains the Debye-Waller factor of the reference compound. Hence, the Debye-Waller factor ($\Delta\sigma^2$) obtained from the data-analysis is relative to the Debye-Waller factor of the reference compound. This modified backscattering amplitude $F'(k)$ can now be derived from the measured amplitude $A_r(k)$ of the Fourier filtered EXAFS of the reference compound by:

$$F'(k) = F'_r(k) = \frac{kR_r^2}{N_r} \bullet A_r(k). \tag{X.7}$$

The function to be minimized with a non-linear least square refinement becomes (Vaarkamp (2), 1993):

$$\chi_{exp}(k) - \sum_{j=1}^{Shells} \frac{N_j}{k'_j R_j^2} e^{-2k'^2_j \Delta\sigma^2_j} F'_j(k'_j) sin\left[2k'_j R_j + \varphi_j(k'_j)\right]. \tag{X.8}$$

Here k' is the photo electron wave vector corrected for the difference in inner potential (ΔE_0) between the sample and the reference compound:

$$k' = \sqrt{k^2 + \frac{4\pi m}{h} \Delta E_0}. \tag{X.9}$$

Generally, $F'(k')$ and $\varphi(k')$ are not known at the values of k' where $\chi_{exp}(k)$ has been measured due to different experimental setup or a shift in k' value due to a change in ΔE_0. As ΔE_0 is a fitting parameter, the values of k', and consequently $F'(k')$ and $\varphi(k')$ are subject to change during a refinement.

Overlapping coordination shells have to be isolated by inverse Fourier transformation and a multiple shell non-linear least square fitting routine has to be applied. The structural parameters N, R and $\Delta\sigma^2$ have to be refined for every contribution (shell). In deriving eq. X.1 one assumes the presence of a Gaussian pair-distribution function and small disorder. If the pair-distribution function is not Gaussian and if the system is highly disordered, the standard data analysis procedure described here is not valid (Crozier et al., 1988).

X.2.5. The Use of Phase Shifts and Backscattering Amplitudes Obtained from Theory

As shown in the previous section, the phase shift $\varphi(k)$ and the backscattering amplitude $F(k)$ can be extracted by Fourier filtering from the EXAFS spectrum of a compound with known structure. They can also be calculated from first principles. Both methods can introduce (systematic) errors in the analysis of unknown compounds. Extraction of backscattering amplitudes and phase shifts from reference compounds suffers from Fourier filtering truncation errors. The accuracy of calculated backscattering amplitudes and phase shifts is hampered by the approximations necessary to make calculations possible. A number of first principle methods to calculate backscattering amplitudes and phase shifts have been developed (Pendry et al., 1974; Teo et al., 1979; McKale et al., 1986; Binsted et al., 1987; Mustre de Leon et al., 1991). Computer programs as EXCURV and FEFF are being widely used, while the tables of phase shifts and backscattering amplitudes as compiled by Teo and Lee, and McKale are applied as well.

One of the goals of the International Committee on Standards and Criteria in X-ray Absorption Spectroscopy (Lytle et al., 1988) is to assess quantitatively the applicability of current theoretical models in XAFS analysis. Vaarkamp (1) et al., 1993 has carried out a systematic investigation to compare the phase shifts and backscattering amplitudes obtained from high quality EXAFS data of reference compounds with those calculated from first principles. The theoretical standards were based on different prescriptions for scattering potentials and self-energies or exchange potentials. The most straightforward way of comparing calculated and experimental XAFS is by looking at a single absorber-backscatterer pair in a single scattering process. Only mono-atomic metals were used to avoid the difficulties of charged atoms in calculations. From each of the first three rows of the transition metals a readily available compound was selected (Cu, Rh, and Pt foil respectively).

The separation of the single scattering first shell contribution from the complete EXAFS spectrum was carried out by Fourier filtering. The Fourier filtering suffers from truncation errors due to the limited data range used in both the forward and the inverse Fourier transform. To verify that Fourier filtering introduces errors which are negligible compared to the differences in the theoretical standards Vaarkamp (1) et al., 1993 carried out a model study to quantify the Fourier filtering errors, which is summarized in section X.2.5.1. From the statistical errors in the data and the differences between experimental data and fit (based on theoretically obtained phase shifts and backscattering amplitudes), goodness of fit values and errors in the determined parameters are calculated. The goodness of fit values and the number of free parameters are used to test whether the differences in fit quality between the different theoretical references are statistically significant.

X.2.5.1.   Fourier Filtering Truncation Errors

To quantify the errors made by Fourier filtering a test single (first) shell copper foil EXAFS spectrum was calculated with the phase shift and backscattering

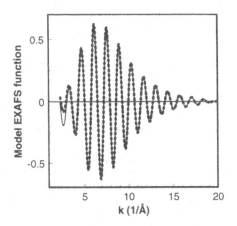

Figure X.4. — Model EXAFS function (solid line) calculated from theory (FEFF) with N=12, R=2.56 Å., σ^2= 0.0033. Dotted line is obtained by Fourier filtering of model EXAFS function; k^3, Δk: 2.3 - 19.4 Å^{-1}, ΔR: 1.3 - 3.1 Å.

amplitude obtained with FEFF. This test single shell EXAFS spectrum was Fourier filtered over a number of forward and inverse ranges. By analyzing the Fourier filtered EXAFS spectrum with the phase shift and backscattering amplitude used to calculate the original test first shell EXAFS spectrum, filtering errors were quantified. The agreement between the original test first shell EXAFS and the Fourier filtered spectrum is expressed as the k^n variance:

$$k^n \text{ variance} = 100 \frac{\int \left\{ k^n \left(\chi_{\text{model}}(k) - \chi_{\text{exp}}(k) \right) \right\}^2}{\int \left\{ k^n \chi_{\text{exp}}(k) \right\}^2} \qquad (X.10)$$

As Fourier truncation errors in the Fourier filtered EXAFS spectrum are largest at the start and end of the forward Fourier transform interval the range and weighing of analysis were varied. To see the effect of window functions, forward transforms ending and starting at nodes or maxima and minima were compared. Analysis of a Fourier filtered function obtained with a forward range of Δk: 2.3 - 19.4 Å^{-1} and an inverse range of Δr: 0-8 Å resulted in a variance of zero, implying that in this case Fourier filtering errors do not occur. Analysis of a Fourier filtered EXAFS function obtained with a forward range of Δk: 2.3 - 19.4 Å^{-1} and an inverse range of Δr: 1.3 - 3.1 Å resulted in a non-zero variance. Figure X.4 shows the original calculated first shell test spectrum (solid line) and the Fourier filtered EXAFS function (dotted line) using Δk: 2.3 - 19.4 Å^{-1} and Δr: 1.3 - 3.1 Å. Figure X.5a shows the decrease of the variance upon shrinking the k-interval for analyses. By not using 1.0 Å^{-1} at the start and 0.8 Å^{-1} at the end excellent agreement is obtained between original and Fourier filtered data (k^1 variance 0.07). Using this range for fitting the Fourier filtered data leads to coordinations parameters which are nearly identical to the original values. Maximum deviations are: coordination number ±2%, Debye-Waller factor ±0.0002 Å^2, coordination distance ±0.003 Å, and E_0 ±0.5 eV.

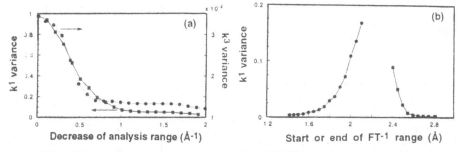

Figure X.5. — Variance of the analysis of a Fourier filtered EXAFS spectrum using its original phase shift and backscattering amplitude. a) Dependence of the variance on the analysis range in k-space. Squares: changing the start of the analysis range while keeping the end fixed at 18.5 Å^{-1}. Dots: changing the end of the analysis range while keeping the start fixed at 3.3 Å^{-1}. b) Dependence of variance for k^1 weighed analysis (from 3.5 - 18.5 Å^{-1}) on the inverse FT range. Dots: changing the start of the inverse FT range (keeping the end at 2.8 Å). Squares: changing the end of the inverse FT range (keeping the start at 1.3 Å).

While keeping the forward Fourier transform range at 2.3 - 19.4 Å^{-1}, the end and start of the inverse Fourier transform range were varied separately (see Fig. X.5b). As long as the complete main peak is included in the inverse Fourier transform excellent agreement over the complete analysis range can be obtained. Omitting only a small part of the main peak results in relatively large errors over the complete Fourier transform range. Changing the start and/or end of the forward Fourier transform range from nodes to maxima and minima does not result in a significant increase of the variance. From this it is clear that the use of special window functions is not necessary as long as the complete main peak is included in the inverse Fourier transform. Moreover, special window functions will alter the inverse Fourier transform.

X.2.5.2   Comparison of Theoretical Models

The comparisons were carried out with k^1 weighed EXAFS spectra to emphasize the differences at low k. The choice for k^1 weighed comparisons was based on the importance of the low k-region in EXAFS data analysis. Virtually all applications of EXAFS consist of measuring the XAFS of relatively heavy atoms in a matrix of low Z atoms. To determine the contribution of these low Z scatterers analysis of the heavy scatterers in the low k-region of the spectrum has to be accurate. Errors in phase or backscattering amplitude of heavy scatterers at low k will perturb the EXAFS signal at low k and thus'results in unreliable results for low Z scatterers. Vaarkamp (1) et al., 1993 also carried out k^3 weighed comparisons,

Table X.1. — Numerical results of the k^1 weighed fits of copper foil over a k-range 3.5 - 18.0 Å^{-1}.

Standard	N	R(Å)	$\Delta\sigma^2$ (x 10^{-3} Å^2)	ΔE_0 (eV)	k^1 variance
FEFF	11.01	2.542	1.66	-3.63	1.2
EXCURVE90	10.27	2.534	3.95	13.90	2.3
MUFPOT	10.00	2.527	4.16	9.27	2.6
McKale et al.	5.57	2.547	0.68	-9.38	6.2
Teo and Lee	7.73	2.499	2.20	-7.37	40.7
Input parameters for standard	12.00	2.560	σ^2 (x 10^{-3} Å^2) 3.28		

Table X.2. — Numerical results of the k^1 weighed fits of rhodium foil over a k-range 3.5 - 18.0 Å^{-1}.

Standard	N	R(Å)	$\Delta\sigma^2$ (x 10^{-3} Å^2)	ΔE_0 (eV)	k^1 variance
FEFF	9.19	2.698	0.17	-1.88	2.1
EXCURVE90	11.15	2.682	1.86	14.36	2.5
MUFPOT	10.57	2.671	1.84	12.36	3.2
McKale et al.	3.87	2.721	-0.84	1.71	3.5
Teo and Lee	3.59	2.691	-1.83	3.73	9.2
Input parameters for standard	12.00	2.680	σ^2 (x 10^{-3} Å^2) 1.40		

differences in such comparisons are much smaller and the obtained parameters more accurate. The reader is further referred to Vaarkamp (1) et al., 1993 for the details of the comparisons between the experimental first shell EXAFS functions and the model functions calculated with the different theoretical approaches. Figure X.6 shows the differences between the experimental data and the fits obtained for the different theoretical standards over the important low k data range. Numerical results are listed in Tables X.1 to X.3 for copper, rhodium, and platinum foil, respectively.

Table X.3. — Numerical results of the k^1 weighed fits of platinum foil over a k-range 3.5 - 18.0 Å^{-1}.

Standard	N	R(Å)	$\Delta\sigma^2$ (x 10^{-3} Å^2)	ΔE_0 (eV)	k^1 variance
FEFF	9.37	2.764	-0.72	-8.09	1.2
EXCURVE90	9.88	2.749	1.06	13.23	6.3
MUFPOT	10.17	2.739	1.30	6.00	6.1
McKale et al.	3.49	2.769	-0.76	-8.19	5.4
Teo and Lee	3.07	2.772	-2.90	-4.80	9.2
Input parameters for standard	12.00	2.774	σ^2 (x 10^{-3} Å^2) 2.61		

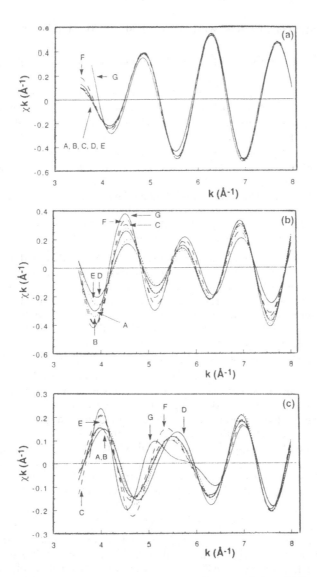

Figure X.6. — Results of analysis: a) Copper, b) Rhodium, c) Platinum. Curves: A) Experimental data, B) Experimental reference (see Vaarkamp (2), 1993), C) FEFF, D) EXCURVE90, E) MUFPOT, F) McKale et al., G) Teo and Lee.

Comparing Fig. X.6c with Fig. X.6b and X.6a leads to the conclusion that the differences between the various references are bigger for platinum than for copper or rhodium. For platinum differences in both phase and amplitude are evident below 12 Å^{-1}. It is important to note that for platinum only references obtained from calculations with a Hedin-Lundqvist exchange potential give acceptable fits. The coordination numbers obtained with McKale et al. and Teo and Lee are unacceptable in all cases.

Debye-Waller factors are heavily dependent on the type of phase shift and backscattering amplitude used. Calculations with $X\alpha$ exchange potentials always yield positive values. Calculations with a Hedin-Lundqvist exchange potential (FEFF), the tables of McKale et al. and the tables of Teo and Lee result in either positive or negative values with both tabulations giving large deviations in either positive or negative direction. The values for E_0 are also very dependent on the method of calculation. References calculated with a Hedin-Lundqvist exchange potential yields negative values, the $X\alpha$ exchange potential gives rather large positive values, the tables of McKale et al. and the tables of Teo and Lee show inconsistent behaviour. The differences between the various codes are ascribed to the different definition of the energy zero among the theoretical standards. However, changes with element as in the tables of McKale et al. and the tables of Teo and Lee indicate an improper definition of the energy zero.

The higher variance values for heavier elements indicate that theoretical calculation of EXAFS spectra of heavy elements is not as accurate as for light elements. Analysis of EXAFS spectra with theoretically obtained phase shifts and backscattering amplitudes will result in accurate distance determinations (± 0.03 Å). Other parameters vary with the type of calculation used. To obtain accurate coordination numbers a scaling factor, whose value will depend on the description of the exchange potential and the treatment of the inelastic losses, is needed. The value of this scaling factor is not transferable from one element to another. Theoretically it is expected that a scaling factor S_0^2 is required to correct the calculations for the decrease in the overlap of the passive electrons between the initial and final states of the absorbing atom. The required scaling factors to correct the coordination numbers are within the expected spread for the *ab initio* calculations. It appears that the choice of an energy dependent self-energy, as used in FEFF is the most important consideration, and that the method of potential construction is secondary, at least in monoatomic metals. The use of ground state, $X\alpha$ or energy independent exchange, as in the McKale et al. tables or the codes EXCURV90 and MUFPOT is found to be inadequate and leads to large phase and amplitude errors.

The variance values found in the quantification of Fourier filtering errors (Fig. X.5) is about an order of magnitude less than the errors found in the comparison of the theoretical standards. Coordination parameters are virtually unaffected by Fourier filtering. Fourier filtering errors can be minimized by using a smaller part of the filtered EXAFS spectrum than the original. Typically the first 1.0 Å^{-1} and the last 0.8 Å^{-1} of the filtered EXAFS spectrum should not be used.

The reliability of the EXAFS analysis depends upon the transferability of the reference phase shift and amplitude functions. Great care must be taken in selecting the reference compounds for providing the reference amplitude and phase functions. It is of crucial importance to measure the EXAFS spectra of reference compounds at liquid nitrogen temperatures. In most cases the dynamic part of the Debye Waller factor increases strongly in going from liquid nitrogen to room temperature. The amplitude of the EXAFS spectra measured at liquid nitrogen is higher with increasing k-values (compare in Fig. X.7a solid with dotted line) which gives the opportunity to apply a forward Fourier transform over a larger range in k-space.

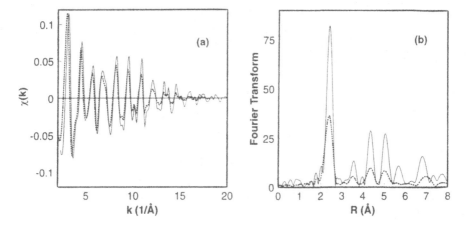

Figure X.7. — a) EXAFS spectra of rhodium foil (20 μ thickness) measured at SSRL (solid line: liquid nitrogen temperature, dotted line: room temperature). b) Fourier transform of spectra shown in a. (solid line, Δk: 2.8 - 19.4 Å⁻¹; dotted line, Δk: 2.6 - 15.8 Å⁻¹).

This leads to peaks in a Fourier transform which have a smaller width and higher amplitude (compare also in Figure X.7b solid line with dotted line). If theoretically calculated functions are used, they should be optimized by testing them on reference compounds (van Dijk et al., 1990). The accuracy of the structural parameters obtained from EXAFS analysis depends upon many factors such as the quality of the data, the length of the data range, the amount of disorder, the choice of the reference compound and the complexity of the system being studied. Typical accuracies are 1% for the determination of coordination distances, 10 to 20% for the coordination numbers and 10 to 20% for Debye-Waller factors.

X.2.6.   Phase- and Amplitude Corrected Fourier Transforms.

Use of Imaginary Part

As pointed out in section X.2 and shown in Fig. X.1 the phase factor $\varphi(k)$ is a non-linear function of k and the backscatterings amplitude $F(k)$ is a function of k. This implies that the Fourier transformation of an EXAFS function does not lead to an optical transform (is transform of function with constant phase and amplitude). Generally, the peaks of the Fourier transform of an EXAFS function are asymmetric. For high Z elements the k-dependence of the phase and the backscattering amplitude may even lead to the appearance of multiple peaks in the Fourier transform of an EXAFS function describing a single absorber-backscatterer pair. This have led in the past to a lot of confusion in analyzing EXAFS data containing contributions of high Z elements. A normal Fourier transform can be converted to an optical transform by removing the phase function and the backscattering amplitude. A phase- and

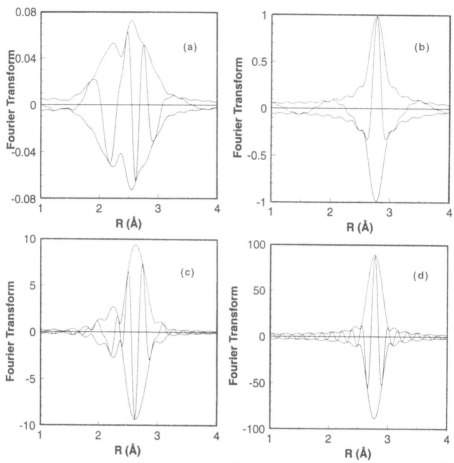

Figure X.8. — Fourier transforms with Δk: 3.1 - 18.6 Å^{-1}. a) k^1; b) k^1, Pt-Pt phase and ampl. corr.; c) k^3; d) k^3, Pt-Pt phase and ampl. corr.

amplitude corrected (optical) Fourier transform can be obtained by transformation of the following function (van Zon et al., 1985; Lytle et al., 1985):

$$\chi(k) \frac{e^{-i\varphi(k)}}{F(k)}. \qquad (X.11)$$

Figure X.8a and b shows a normal k^1 and k^3 weighted Fourier transform of a single Pt-Pt EXAFS function calculated with $N = 5$, $R = 2.77$, $\Delta\sigma^2 = 0.003$ and $\Delta E_0 = 0$. The phase shift and backscattering function were obtained from EXAFS measurements on Pt foil (thickness 4μ) measured at liquid nitrogen temperature. Since the Fourier transformation as defined in eq. X.4 is complex it is possible to calculate both the magnitude (see Fig. X.8 envelop) and the imaginary part (oscillations) of the transform. It can be seen in Fig. X.8 that the magnitude of the k^1 weighted Fourier transform of a single Pt-Pt contribution is splitted in three peaks, whereas the imaginary part shows a complex behavior as a function of r. The applied weight factor k^n influences the k-dependence of the amplitude of the $k^n\chi(k)$ function. The magnitude of the k^3 weighted Fourier transform appears now as one asymmetric peak. Application of a phase- and amplitude corrected Fourier transform leads in all cases to a single peak with a symmetrical imaginary part, having its maximum in the top of the magnitude at the right coordination distance (see Figs. X.8c and d).

The use of optical Fourier transforms can be of great help (Kampers, 1988) in the identification of different types of neighbours by application of different phase- and/or amplitude corrections. A Fourier transform, phase- and amplitude corrected for an X-Y absorber-backscatterer pair must have a positive imaginary part peaking at the maximum of its magnitude if the EXAFS function indeed originates form an X-Y pair. Fourier transform peaks which are not symmetrical are a superposition of more than one contribution.

X.2.7. Detection of Low Z Scatterers

It is tempting to apply a k^2 or k^3 weighting to the EXAFS spectrum during the fitting in k-space to compensate for the decay in amplitude of the spectrum with k. Also Fourier transformation of a function which has an equalized amplitude results in less broadened peaks, which are easier to filter for inverse transformation. However, applying a k^2 or k^3 emphasizes the high Z contributions to the spectrum since high Z elements have more scattering power at high values of k than low Z elements. As pointed out in section X.2.1 low Z elements have the highest amplitude at low values of k. Therefore, the use of a k^2 or k^3 weighted EXAFS spectrum or Fourier transform makes the analysis much less sensitive for the presence of low Z contributions in the EXAFS data. This is demonstrated in Fig. X.9. Pt-Pt and Pt-O EXAFS model functions have been calculated with the coordination parameters as given in Table X.4. These parameters are in the range of typical parameters found in the EXAFS analyses of data collected on small metal particles dispersed on high surface area supports. Pt-Pt and Pt-O phase shift functions and Pt and O backscattering amplitudes have been obtained from EXAFS data collected from Pt-foil and $Na_2Pt(OH)_6$, which were measured at liquid nitrogen temperature. Figure X.9a shows the individual Pt-Pt (solid line) and Pt-O (dotted line) EXAFS spectra calculated with the parameters of Table X.4. The sum of the Pt-Pt and Pt-O EXAFS functions is also plotted (dashed line) and this function mimics the experimental data. The difference between the EXAFS spectrum with only a platinum contribution and the EXAFS spectrum with a platinum and an oxygen contribution is most pronounced below 6 Å^{-1}. In the k^3 weighted Fourier transform of the spectra (shown in Fig. X.9c) the difference between the spectrum with only a platinum contribution and the simulated experimental EXAFS spectrum (with a platinum and an oxygen contribution) is hardly detectable. The k^1 weighted Fourier transform (Fig. X.9b)

Table X.4. — Parameters used in the calculation of the model spectra.

Coordination	N	R(Å)	$\Delta\sigma^2$ (Å^2)	ΔE_0 (eV)
Pt-Pt	5.0	2.77	0.003	0
Pt-O	1.9	2.65	0.006	0

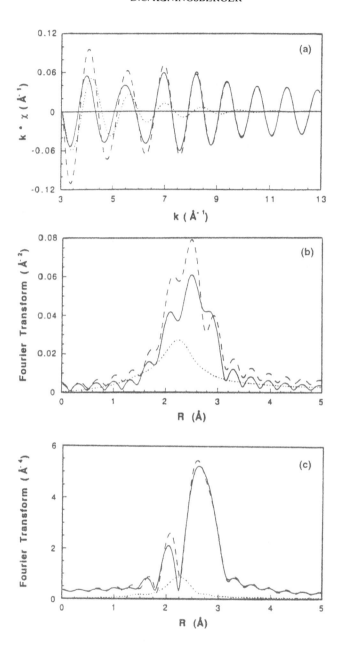

Figure X.9. — Simulated EXAFS spectra of Pt-Pt (solid line), Pt-O (dotted line) and Pt-Pt + Pt-O (dashed line). a) EXAFS functions, b) k^1 weighed Fourier transform (Δk: 3.5 - 13.0 Å^{-1}), c) k^3 weighed Fourier transform (Δk: 3.5 - 13.0 Å^{-1}).

shows much larger differences. From this model study it is obvious that for a proper analysis of low Z contributions present in EXAFS spectra k^1 weighted fits and/or Fourier transforms should be applied.

X.2.8. Correlation Between N and $\Delta\sigma^2$. Use of Both k^1 and k^3 Weighted Fourier Transforms

The determination of a unique set coordination parameters for a particular contribution to an EXAFS spectrum is often difficult due to the correlation between the value of the coordination number N and the Debye-Waller factor $\Delta\sigma^2$. Kampers, 1988 showed that by simultaneous minimization of the difference between fit and data in a k^1 and k^3 weighted Fourier transform, a unique set of parameters can be found. This can be rationalized by examining eq. X.2 and noting that the amplitude of the EXAFS spectrum depends on both N and $\Delta\sigma^2$. More particularly, different combinations of N and $\Delta\sigma^2$ will lead to the same peak amplitude of the Fourier transform of an EXAFS spectrum. However, this set of combinations depends on the weight factor, which has been used for the Fourier transform. The combination of N and $\Delta\sigma^2$ that gives a good fit both in a k^1 and k^3 weighted Fourier transform offers a unique set of parameters. An example of this approach is given in Fig. X.10. Figure X.11b displays the k^3 weighed Pt-Pt phase- and amplitude corrected Fourier transforms of three EXAFS spectra calculated with the combinations of N and $\Delta\sigma^2$, which compose the k^3 curve of Fig. X.10. At the onset of the main peak some differences are present, but the amplitude of the main peak of the transform of these spectra is equal. However, the k^1 weighed Pt-Pt phase- and amplitude corrected Fourier transforms of the same three EXAFS spectra (Fig. X.11a) are very different. Only if Fourier transforms with both k^1 and k^3 weighting show good agreement between model and experimental data a "good" combination of coordination number and Debye-Waller factor has been selected. It is essential to use phase- and

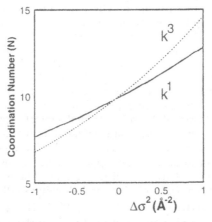

Figure X.10. — Combinations of coordination number (N) and Debye-Waller factor ($\Delta\sigma^2$) giving good fits in k^1 or k^3 Fourier transform.

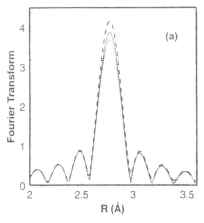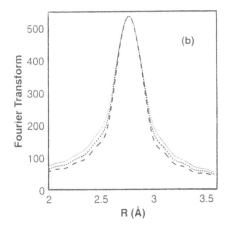

Figure X.11. — Fourier transforms (Δk: 3.1 - 18.6; Pt-Pt phase and amplitude corr.) of EXAFS model functions. Solid line: N=8.2, $\Delta\sigma^2$= -0.0005 Å^2; Dotted line: N=10, $\Delta\sigma^2$= 0 Å^2; Dashed line: N=12.3, $\Delta\sigma^2$= 0.0005 Å^2. a) k^1 weighed, b) k^3 weighed.

amplitude corrected Fourier transforms when applying this method, because otherwise the asymmetry of the peaks will obscure the results.

X.3.   Stereo Chemistry: Metal-Support Interaction of Small Iridium Clusters Dispersed on MgO (An Example of the Uniqueness of EXAFS Spectroscopy)

X.3.1.   Introduction

Most industrial metal catalysts are dispersed as small particles on high-area metal oxide supports to allow access of reactants to most of the metal atoms. The dispersion is stabilized by metal-support interactions that hinder particle migration and coalescence that lead to loss of catalytic activity by loss of metal surface area. Metal-support interactions also affect the activities of catalysts, but little is understood about how (Stevenson et al., 1987; Haller et al., 1989). The understanding is limited by the lack of precise methods for structural characterization of microscopic metal particles and metal-support interfaces. Van Zon et al., 1993 have shown that small, uniform supported metal clusters offer an excellent opportunity for precise characterization of the structure of the metal clusters and the metal-support interface by EXAFS spectroscopy; the smaller the metal cluster, the larger the fraction of the EXAFS signal that arises from the metal-support interface and the greater the opportunity for accurate characterization of the interface structure. The [Ir$_4$(CO)$_{12}$] metal carbonyl cluster was chosen to be the precursor, and partially hydroxylated MgO was chosen to be the support, as the chemistry of Ir carbonyls on the basic MgO surface is relatively well understood (Maloney et al., 1990) being similar to that occurring in basic solutions and allowing synthesis in high yields of [HIr$_4$(CO)$_{11}$]$^-$ on the surface (Maloney et al., 1991). The goal was to form and decarbonylate this cluster anion on the support without changing its nuclearity. The MgO support offers several advantages for characterization of the samples by X-ray

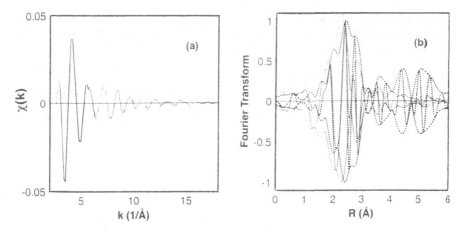

Figure X.12. — a) Raw EXAFS data of Ir_4/MgO. b) Normalized Fourier transforms (k^1, Δk: 2.6 - 13.5 Å^{-1}) of Ir_4/MgO (solid line) and Ir powder (dotted line).

absorption spectroscopy. In contrast to the structurally more complex γ-Al_2O_3 and TiO_2, MgO (which has the rock salt structure) exhibits predominantly (100) faces, even in the powder form (Henrich, 1985). Thus the surface of MgO is relatively simple and well defined and a good choice for characterization of the metal-support interface. Furthermore, as both the O^{2-} and Mg^{2+} ions are present in the same surface layers, both the Ir-O and Ir-Mg interactions are expected to be characterized by the EXAFS data. EXAFS evidence of interactions between supported metals and cations of the support has been reported only rarely (Koningsberger et al., 1992).

Table X.5. — Coordination parameters obtained from fitting the Ir_4/MgO EXAFS spectrum.

Parameters	N	R(Å)	$\Delta\sigma^2 \times 10^{-3}$ (Å^2)	ΔE_0 (eV)
Coordination:				
Ir-Ir	2.6 ± 0.3	2.713 ± 0.004	0.6 ± 0.7	-5.2 ± 0.7
Ir-O	3.4 ± 0.1	2.63 ± 0.01	3.8 ± 0.7	-0.8 ± 0.6
Ir-Mg	0.3 ± 0.1	1.60 ± 0.02	13.6 ± 4.3	-2 ± 4

X.3.2.   Results

The sample designated Ir_4/MgO represents that prepared by decarbonylation of $[HIr_4(CO)_{11}]^-$ supported on MgO. The raw EXAFS data characterizing Ir_4/MgO is shown in Figure X.12a. The data quality is high. A preliminary indication of the Ir cluster size is given by the Fourier transforms of Figure X.12b, which is a comparison of a k^1-weighted Fourier transform of data characterizing Ir powder and data characterizing the Ir_4/MgO sample. The Fourier transform of the EXAFS data characterizing the Ir powder in this figure is scaled to the main peak of the Fourier transform of the EXAFS data characterizing Ir_4/MgO. Figure X.12b shows both the magnitude (the envelope) and the imaginary part (the oscillations) of the k^1-weighted Fourier transform. It can be seen that the second, third, and fourth Ir-Ir shells are not detectable in the Ir_4/MgO sample. The presence of low-Z neighbors of Ir is also demonstrated in Figure X.12b. The peak at the position of the first Ir-Ir shell in Ir powder seems to be split into three peaks; this splitting is caused by the k dependence of the phase shift and backscattering amplitude of the Ir-Ir absorber-backscatterer pair (see also section X.2.6). The contrast between the data in the first-shell region for the two samples (Fig. X.12b), both the magnitudes and the imaginary parts of the Fourier transforms) demonstrates the presence of low-Z scatterers in the immediate neighborhood of Ir in the Ir_4/MgO sample; these are inferred to be present in the metal-support interface.

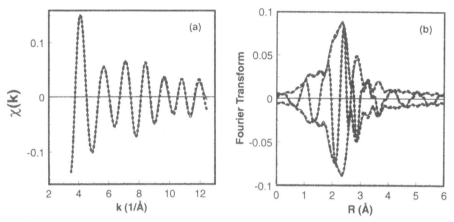

Figure X.13. — Results of multiple shell fitting (dotted line) of the EXAFS data characterizing Ir_4/MgO (solid line). a) Fit in k-space (k^1, Δk: 3.5 - 12.5 $Å^{-1}$) and b) fit in R-space (FT: k^1, Δk: 3.5 - 12.5 $Å^{-1}$).

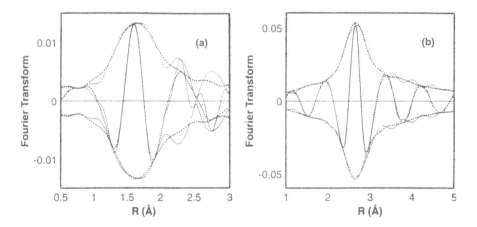

Figure X.14. — a) Fourier transform (k^3, Δk: 3.5 - 12.5 $Å^{-1}$, Ir-Mg phase corrected) of difference file (isolated EXAFS minus calculated Ir-Ir + Ir-O) (solid line) and calculated Ir-Mg EXAFS function (dotted line). b) Fourier transform (k^3, Δk: 3.5 - 12.5 $Å^{-1}$, Ir-O phase corrected) of difference file (isolated EXAFS minus calculated Ir-Ir + Ir-Mg) (solid line) and calculated Ir-O EXAFS function (dotted line).

The final results of the EXAFS data analysis is presented in Figure X.13. Fits obtained by a nonlinear least squares multiple-shell fitting routine are shown in k space and in r space with k^1 weighting. The coordination parameters and their standard deviations (calculated from the covariance matrix including the statistical multiple-shell fitting routine), are given in Table X.5. The Ir-O and Ir-Mg errors of the experimental points, obtained from the nonlinear least squares contributions are evidence of the metal-support interactions. These contributions are shown for the Ir_4/MgO sample in Figure X.14a and b. The Ir-O EXAFS function was determined by subtraction of the calculated sum of the (Ir-Ir) + (Ir-Mg) contributions from the primary EXAFS data by using the coordination parameters obtained from the best fit. Similarly, the Ir-Mg EXAFS function was obtained by subtraction of the calculated (Ir-Ir) + (Ir-O) contribution from the primary data. The Ir-O phase-corrected Fourier transform peaks at about 2.6 Å, and the Ir-Mg phase-corrected Fourier transform peaks at about 1.6 Å. More details of this analysis are given by van Zon et al., 1993. In particular, the authors discuss in detail the statistical significance of the Ir-O and Ir-Mg contributions.

X.3.3.   Discussion

A unique advantage of the organometallic precursors is the opportunity they offer for preparation of extremely small supported metal clusters that are almost optimally suited to precise characterization with EXAFS spectroscopy. The results of the work of van Zon et al., 1993 give evidence that the tetrairidium cluster can be decarbonylated on the support with little change in nuclearity (Table X.5). Since the clusters are so small, the EXAFS signal is determined in large measure by the metal-support interactions and not just the metal-metal interactions. Consequently, the EXAFS data offer the prospect of detailed characterization of the structure of the metal-support interface.

The Ir-Ir coordination number of 2.6 is consistent with the inference that the cluster nuclearity was maintained. However, within the experimental error

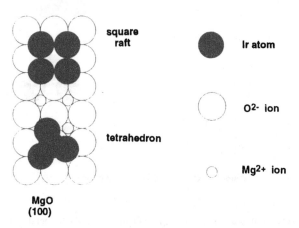

Figure X.15. — Structural model of Ir_4/MgO based on the EXAFS data. The MgO (100) surface is assumed.

(Table X.5), this value is less than 3.0, the value characteristic of the tetrahedron of the metal frame of the precursor $[Ir_4(CO)_{12}]$ or $[HIr_4(CO)_{11}]^-$, which implies that the tetrahedral frame of the precursor was not retained in all the clusters. The average Ir-Ir coordination number of 2.6 cannot be reconciled with raft-like structures alone either, as the first-shell Ir-Ir coordination number in a square Ir_4 raft is 2, and much larger rafts must be assumed to account for N = 2.6 (but these imply sizeable third and fourth Ir-Ir shells). A small fraction of clusters with nuclearities greater than four may have been present; however, as the third and fourth Ir-Ir shells are almost absent, these must have been rare. There is no conclusive evidence that raft-like structures would be exclusively square; various other arrangements can not be ruled out. The EXAFS data have been modeled as simple as possible; they have been found to be consistent with a mixture of Ir_4 tetrahedra and square rafts (see Fig. X.15). Taking the accuracy of their analysis results, the authors infer that 40-50% of the Ir clusters are tetrahedra and the remainder square rafts.

The Ir-O and Ir-Mg contributions are pronounced in the EXAFS data. These contributions are attributed to the metal-support interface. The actual interfacial iridium-oxygen coordination number for the Ir_4/MgO sample can be calculated directly from the coordination number determined for the Ir-O shell (3.4) using the earlier mentioned estimated fraction of iridium clusters present as a tetrahedron on the MgO (100) surface (0.5). In the tetrahedron 75%, and in the raft 100% of the iridium atoms are present in the metal-support interface. This leads to the following equation: $(0.5 * 0.75 * N(i))_{tet} + (0.5 * 1 * N(i))_{raft} = 3.4$, resulting in $N(i) = 3.9$. Taking into account the experimental errors in the EXAFS data (Table X.5), we might

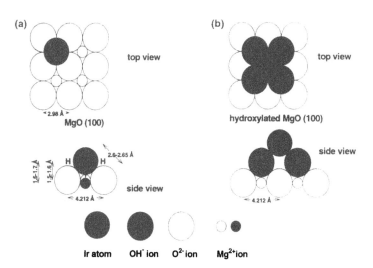

Figure X.16. — Structural models of the Ir_4/MgO interface. A) Ir metal atom on a partially hydroxylated MgO (100) surface with hydrogen in the metal-oxygen interface. B) Ir metal atom on a fully hydroxylated MgO (100) surface.

therefore suggest that the Ir-MgO interface could be modeled with an Ir-O coordination number of four; a fourfold Ir-O coordination can be visualized on the (100) face of MgO. This is the predominant face (Henrich, 1985).

The EXAFS observations raise fundamental questions about the nature of the interactions at metal-support interfaces. Van Zon et al., 1993 discuss extensively the suggestions that the long metal-oxygen distances (2.6-2.8 Å) observed in this work and for numerous other samples (Koningsberger et al., 1992) may indicate the presence of hydrogen in the metal-support interface or, alternatively, interactions of metal atoms with hydroxyl groups of the surface (Koningsberger et al., 1992). As outlined above, the Ir at the interface is inferred to be coordinated principally to four oxygens. The oxygens on the surface of partially hydroxylated MgO include O^{2-} as well as oxygen of OH^- groups of more than one kind, including protonated O^{2-} ions terminating the MgO lattice and OH groups bonded to Mg^{2+} ions terminating the lattice (Lamb et al., 1988). This results in two limiting cases for the Ir atoms at the MgO surface (Fig. X.16).

The two simple models can be described as follows: (1) If the support surface has no hydroxyl groups in contact with the clusters, individual Ir atoms on the MgO (100) face are positioned between the oxygens and directly above the Mg^{2+} ions (Fig. X.16a). From the MgO crystallographic distance (Henrich, 1985) and an Ir-O coordination distance of 2.6 Å (Table X.5), an Ir-Mg distance of 1.5-1.6 Å is expected. (2) If the MgO (100) face is fully hydroxylated, then the Mg^{2+} ions are covered with OH groups. Ir atoms on this surface that have an Ir-O coordination number of four would have to be placed between four OH groups, each positioned on top of a Mg^{2+} surface ion (Fig. X.16b). Using the same coordination distances as in the former model and a Mg-OH distance of 2.106 Å (equal to the Mg-O distance in MgO), an Ir-Mg distance of 4.1-4.2 Å is calculated. The Ir-Mg distances indicated by EXAFS (1.6-1.7 Å, Table X.5) are in agreement with the first but not the second of these limiting-case models.

X.4.   Determination of The Electronic Structure: White Line Intensities of L_{II} and L_{III} Edges as Probe for the d-Band Density of States of Transition Metals (XANES)

X.4.1.   Introduction

In the following sections a very fundamental issue in catalysis by metals has been addressed. The intrinsic metal properties leading to very important catalytic applications can be changed by interaction with the support or by adding so-called promotor ions. Catalytic and modern spectroscopy studies strongly suggest that catalytic properties like activity and selectivity can be modified by charge transfer or polarization at the interface between very small metal particles and the support or the promoting ions (Boudart et al., 1984).

Information about the d-band density of states in transition metal clusters can be obtained from the white line intensities of the L_{II} (transition from $2p_{1/2}$ to $5d_{3/2}$) and L_{III} (transition from $2p_{3/2}$ to $5d_{5/2}$ and $5d_{3/2}$) X-ray absorption edges. A basic theory of white lines has been given by Mott, 1949; Brown et al., 1977 and Mattheiss et al., 1980. The theoretical calculations of the unoccupied Pt d-states show that the J=5/2 final state is predominant. Brown et al. showed that the J=5/2 states contribute about 14 times more to the final d-states than the J=3/2 states. Mattheis et al. calculated that the ratio of the unoccupied states ($h_{5/2}/h_{3/2}$)ranges from 3.5 within 0.5 eV of the Fermi level to 2.9 over the entire unoccupied conduction band. These calculations explain why the intensity of the white line of the L_{III} X-ray absorption edge is much higher than the intensity of the L_{II} edge in bulk platinum.

Several authors have used the L_{III} X-ray absorption edge spectra to characterize the chemical state of the absorbing atom in transition metal compounds. Lytle et al., 1976 and 1979 have shown for iridium, platinum, and gold that the intensity of the white line of the L_{III} X-ray absorption edge is proportional to the d-electron vacancies. Samant et al., 1991 reported a decrease in the intensity of the L_{III} X-ray absorption edge of platinum upon removal of hydrogen from the surface for measurements carried out at low temperature, whereas Lytle (2) et al., 1985 reported a slight intensity increase of the L_{III} edge upon hydrogen removal. However, a decrease in intensity of the L_{II} edge upon hydrogen removal was shown. The important consequence of this observation is that both the L_{III} and the L_{II} edge have to be examined before any statement about a change in electron density can be made.

From the factors influencing the white line intensity the particle size and the support are of course of most interest in catalysis. The relation between LDOS and particle size is of interest as this might explain the particle size sensitivity of some reactions. The establishment of the effect of the support on the LDOS is expected to give a physical key to the changes in reactivity of small metal clusters on different supports or in the presence of different cations.

Gallezot et al., 1977 and 1979 reported that the white line intensity of the L_{III} edge of 1-nm Pt clusters in Pt/NaY is larger than 3-nm clusters. Mansour et al., 1984 carried out a more quantitative study including the white line of the L_{II} absorption edge for Pt/SiO_2 with an average particle size of 15 Å (72 atoms) and determined that there were 14% more d-holes than in bulk platinum metal. An increase in the d-band density of states of surface atoms due to the decreased degree of delocalization has been discussed by Gordon et al., 1977 and Saillard et al., 1984. A recent theoretical study, using local-density methods on neutral and charged Ir_4 and Ir_{10} clusters (Ravenek et al., 1989) indicates a higher number of d-electrons for the Ir_4 cluster in comparison to the Ir_{10} cluster. However, introduction of a core hole, to mimic the X-ray absorption process, into the clusters results in a lower number of d-electrons for the Ir_4^+ compared to the Ir_{10}^+ cluster. The higher electron deficiency for the smaller cluster is due to a less efficient screening of the core hole.

Changes in the intensity of the L_{III} X-ray absorption edge of a Pt/SiO_2 catalyst have been related to alterations in the d-band density of states resulting from a change in metal-support interaction after reduction at high temperatures (Lytle (2) et al., 1985. Gallezot et al., 1977 and 1979 reported that platinum clusters in Ce-promoted NaY zeolite have a larger number of holes in the d-band than platinum in NaY zeolite. Recently, the number of unfilled d-states of platinum particles supported on silica were reported to become more electron deficient loading the sample with increasing amounts of Na (Yoshitaka et al., 1991).

The recent results obtained by Vaarkamp (2), 1993 will be discussed in the following. He has shown that chemisorbed hydrogen is a first order effect on the d-band density of states of small metal particles. The important outcome of his study is that X-ray absorption data have to be taken under vacuum before any conclusion can be drawn about changes in the d-band density of states due to support or promotor effects.

X.4.2.   Experimental

A 1.0 wt% Pt/γ-Al$_2$O$_3$ catalyst was prepared by impregnation of Ketjen CK-300 (200 m^2/g, 0.6 cm^3/g) with an aqueous solution of hydrochloric platinic acid. The catalyst was dried in air overnight at 120°C before it was reduced at 450°C for 4 hours. After reduction the catalyst was passivated at room temperature (further details of the preparation procedure are given by Vaarkamp (2) et al., 1993).

The X-ray absorption data were obtained at the Synchrotron Radiation Source in Daresbury, U. K., Wiggler Station 9.2 in the transmission mode at liquid nitrogen temperature. The estimated resolution at the Pt L$_{III}$ (11564 eV) and L$_{II}$ (13273 eV) edge is 3 eV. The Si(220) monochromator was detuned to 50% intensity to avoid the effects of higher harmonics. In order to obtain an absolute energy calibration of the data a third ion chamber was used with a platinum metal foil (thickness 4 μ) placed between the second and the third ion chamber. Each spectrum was separately calibrated at the L$_{III}$ and L$_{II}$ edge. The spectra were normalized by the edge jump at 50 eV above the edge. Self-supporting wafers ($\mu x = 2.5$) were reduced in flowing H$_2$ in an *in-situ* cell (Kampers et al., 1989). The Pt/γ-Al$_2$O$_3$ sample was rereduced at 300 (LTR) and 450°C (HTR). After collecting the data on the reduced catalyst the cell was evacuated to 10^{-5} Torr at RT and heated to the evacuation temperature. Maintaining the vacuum, the cell was held at the evacuation temperature for an additional hour and cooled to RT. After collecting the data on the evacuated catalyst, hydrogen was admitted into the cell at RT. Subsequently the sample was heated to 200°C under flowing hydrogen and held there for one hour. Data on the reduced samples were obtained in the presence of H$_2$, data on the evacuated samples were obtained under dynamic vacuum.

X.4.3.   Results

X.4.3.1.   Structural characterization

Characterizations of the catalyst by H$_2$ TPD, hydrogen chemisorption, and EXAFS are reported by Miller et al., 1993 and Vaarkamp (3) et al., 1993. The structure of the metal-support interface changes with reduction temperature. After LTR a Pt-O contribution at a distance of 2.7 Å is present in the EXAFS spectrum. Based on hydrogen TPD this long Pt-O distance has been attributed to a structure were atomic hydrogen is present between the platinum and the oxygen atoms of the support. Treatment in either inert or hydrogen atmosphere at high temperature removes this interfacial hydrogen. Hence, the Pt-O distance, as determined by EXAFS, is shortened to 2.2 Å. The EXAFS first-shell coordination numbers, which are important for an evaluation of the white line results are given in the Table X.6.

X.4.3.2.   White line intensity determination

The measured Pt L$_{II}$ and L$_{III}$ X-ray absorption edges are a combination of a smooth function representing the transition of electrons from core-levels into the continuum and a function representing the transition of core-level electrons into unfilled states in the valence band. Hence, to evaluate the number of unfilled states these two functions have to be separated. Horsley, 1982 showed that the Pt L_{III} edge of platinum foil can be deconvoluted in a smooth arctangent function and a Lorentz function representing atoms transferred to the Pt 5d level (Fig. X.17a). However, this deconvolution of the edge is unable to cope with the asymmetric form of the X-ray absorption edge of the supported platinum catalysts when hydrogen is chemisorbed on the surface. A representative example of this deconvolution for a Pt/K-LTL sample reduced at 300°C is shown in Figure X.17b. It is clear that the fitted Lorentz function is not representative of the area of the white line of Pt/K-LTL, hence an alternative approach is needed. Vaarkamp (2), 1993 checked whether the white line consists of more than one Lorentz function by calculating the second derivative of the spectrum, thereby diminishing the contribution of the smooth background (Lytle et al., 1990). Only one negative peak was present indicating that the white line consists of a single transition. To quantify the differences in white line intensity between the catalysts and platinum foil the approach described by Mansour et al., 1984 was used.

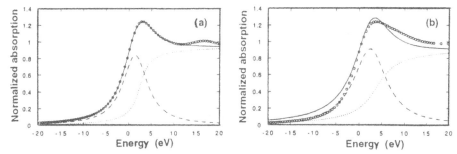

Figure X.17. — The Pt L_{III} edge (circles) of a) platinum foil and b) Pt/K-LTL reduced at 300°C with an arctangent (dotted line) and Lorentz (dashed line) obtained by non linear fitting, solid line: sum of arctangent and Lorentz.

The normalized XAFS spectra of the Pt L_{II} and L_{III} edge of the platinum foil and the catalyst were aligned at there inflection points. After subtraction of the platinum foil data from the data of the catalyst, the resulting curves were numerically

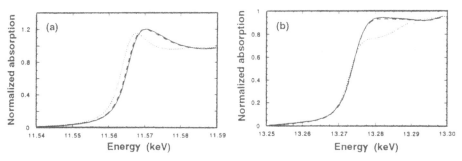

Figure X.18. — X-ray absorption edge of Pt/γ-Al$_2$O$_3$ after reduction at 300°C (solid line), subsequent evacuation at 300°C (dotted line) and subsequent admission of hydrogen at 200°C (dashed line). a) L_{III} edge and b) L_{II} edge.

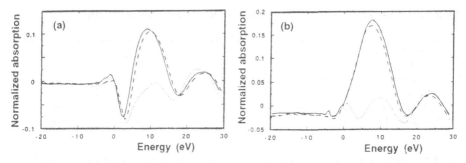

Figure X.19. — Difference between the X-ray absorption edges of Pt/γ-Al$_2$O$_3$ and the corresponding edge of platinum foil after reduction at 300°C (solid line), subsequent evacuation at 300°C (dotted line) and subsequent readmission of hydrogen at 200°C (dashed line). a) L$_{III}$ and b) L$_{II}$ edge.

integrated between -2 and +17 eV for both the L$_{III}$ (ΔA$_3$) and the L$_{II}$ (ΔA$_2$) edge. The resulting areas were combined to give the difference in the total number of unfilled states in the d-band (Δh$_T$) (Mansour et al., 1984) in comparison to bulk platinum:

$$\Delta h_T = 2.25 \, (\Delta A_3 + 1.11 \, \Delta A_2) \tag{X.11}$$

A change in h$_T$ of 7 (which is approximately the largest change observed in this study) corresponds to 0.08 electron per atom. The alignment of the X-ray absorption edge of platinum foil and the catalyst has a dramatic effect on the difference spectrum of the platinum foil and the catalyst (Mansour et al., 1984). To circumvent the problem of the changing edge position by an increase in white line intensity one should align the smooth background function in the edges of interest. The position of the smooth background function can be deduced from the position of the EXAFS wiggles. However, in the samples which were studied by Vaarkamp (2), 1993 the EXAFS wiggles are subject to change. Consequently a systematic alignment procedure based on the EXAFS wiggles is not possible. To ensure a systematic approach for the determination A$_3$ and A$_2$ Vaarkamp (2), 1993 aligned the inflection points of the platinum foil and the sample.

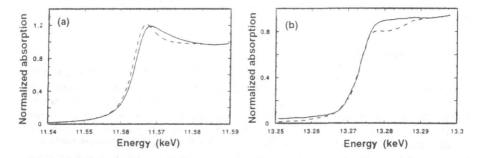

Figure X.20. — X-ray absorption edge of Pt/γ-Al$_2$O$_3$ after reduction at 450°C (solid line) and after subsequent evacuation at 450°C (dashed line). a) L$_{III}$ and b) L$_{II}$ edge.

Table X.6. — EXAFS Pt-Pt coordination numbers and whiteline intensities of Pt/γ-Al$_2$O$_3$.

Teatment	N	ΔA_3	ΔA_2	Δh_T
Reduction 300°C	5.0	0.57	1.58	5.23
Evacuation 300°C	3.8	-0.58	-0.15	-1.70
Hydrogen 200°C	4.9	0.48	1.43	4.66
Reduction 450°C	5.0	0.35	1.14	3.64
Evacuation 450°C	4.1	-0.09	0.26	0.44

X.4.3.3.   Influence of chemisorbed hydrogen on the white line intensity

The influence of chemisorbed hydrogen on the position and intensity of the white line has been investigated for the Pt/γ-Al$_2$O$_3$ sample reduced at 300 and 450°C. The Pt L$_{III}$ and L$_{II}$ X-ray absorption edges of after reduction at 300°C, evacuation at 300°C, heating in hydrogen to 200°C are shown in Figure X.18. Removal of chemisorbed hydrogen causes a dramatic decrease in the intensity of both the Pt L$_{III}$ and L$_{II}$ X-ray absorption edge. The white line intensity is completely restored by readmission of hydrogen at RT and subsequent heating to 200°C. The L$_{III}$ edge of the evacuated sample is shifted to lower energies by -1.5 eV. Readmission of hydrogen causes an upward shift of 1.4 eV, almost completely restoring the original edge position. The L$_{II}$ edge of the evacuated sample does not show the downward shift. In addition to the position change of the L$_{III}$ edge, also a change in shape (sharpening) is evident upon removal of the chemisorbed hydrogen. The difference between these spectra and the corresponding X-ray absorption edge of platinum foil are plotted in Figure X.19, results of the numerical integration of these curves between -2 and 17 eV are listed in Table X.6.

The Pt L$_{III}$ and L$_{II}$ X-ray absorption edges after reduction at 450°C and subsequent evacuation at 450°C are shown in Figure X.20. Evacuation causes a downward shift of the position of the Pt L$_{III}$ edge of -0.7 eV, which is approximately half the shift obtained for the sample reduced at 300°C. Again no shift of the L$_{II}$ edge was observed. The results of the integration are listed in Table X.6. The decrease in white line intensity by removal of chemisorbed hydrogen is not as large as after reduction and evacuation at 300°C. Furthermore, the white line intensity in the presence of chemisorbed hydrogen is lower after reduction at 450°C than after reduction at 300°C.

X.4.4.   Discussion

The removal of chemisorbed hydrogen leads to a decrease in white line intensity (i.e. a decrease in the number of unfilled d-states). This is in agreement with literature data (Lytle (2) et al., 1985; Samant et al., 1991; McHugh et al., 1990) where the whiteline is reported to be affected by the presence of either hydrogen or helium during the measurement. The larger decrease in white line intensity in our measurement compared to literature data originates from the smaller particles in this study. In smaller particles a larger fraction of the atoms is exposed to adsorbed gases and thus a larger effect of adsorbed gases on the white line intensity is to be expected.

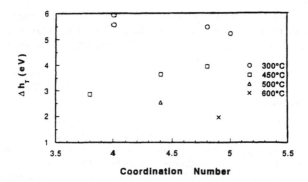

Figure X.21. — Correlation between first shell Pt-Pt coordination number and total number of unfilled d-states after reduction at 300 (squares), 450 (circels), 500 (triangle) and 600 °C (cross) for different supported platinum catalysts.

After reduction and evacuation at 300 °C the total number of unfilled d-states is less than the total number of unfilled d-states in the platinum foil (Δh_T = -1.7), indicating that the platinum particles have become electron rich in comparison to bulk platinum. In contrast, reduction and evacuation at 450 °C results in a total number of unfilled d-states that is larger than the total number of unfilled d-states of platinum foil (Δh_T = 0.44), viz. the platinum particles are electron deficient in comparison to bulk platinum. The change in the number of unfilled d-states as a function of reduction temperature can not be explained by a change in particle size. The platinum particle size (= coordination number) did not change as a function of the reduction temperature. However, Vaarkamp (3) et al., 1993 showed that the structure of the metal-support interface changes when the reduction temperature is increased from 300 to 450 °C. After reduction at 300 °C hydrogen is present in the metal-support interface. In the absence of interfacial hydrogen, e.g. after reduction at 450 °C, the interfacial platinum atoms are in direct contact with the support. The effect of the support on the number of unfilled d-states of supported platinum particles is not *a priori* known.

The results obtained by Vaarkamp (2), 1993 open for the first time the possibility to estimate the lower limit of the magnitude of the electron withdrawing or donating properties of γ-Al_2O_3 from or to small platinum particles. Intuitively, the removal of interfacial hydrogen is expected to have the same effect on the white line intensity as the removal of chemisorbed hydrogen, i.e. it is expected to decrease the white line intensity. Therefore it is logical to assume that removal of the interfacial hydrogen, (experimentally not possible after evacuation at 300 °C !!) still present after evacuation at 300 °C would lead to a further decrease of the white line intensity. After evacuation at 300 °C Δh_T = -1.7. Removal of the interfacial hydrogen will then result in a further decrease of the white line intensity to a value lower than -1.7. An estimation can be made about a lower limit of this further decrease. The total decrease in the value of Δh_T due to the removal of chemisorbed hydrogen amounts about 7. For a platinum cluster of about 12 atoms at least 30% of the total amount of atoms will be present in the metal-support interface. Therefore, removal of interfacial hydrogen would lead to a total value for Δh_T of at least -4. The white line intensity of the sample after evacuation at 450°C is found to be 0.44. This is caused by the influence of the support, since the interfacial platinum atoms are now in direct contact with the support. This counteracts the effect of the removal of interfacial hydrogen. The lower limit of the magnitude of the influence of the support can now be estimated to be at least 4.4, which is comparable with 7, the influence of chemisorbed hydrogen.

Vaarkamp (2), 1993 studied a number of different types of supported platinum catalysts. The changes of the number of unfilled d-states with the first shell Pt-Pt coordination number e.g. particle size, after reduction are shown in Figure X.21. It is evident that the changes in the number of unfilled d-states are mainly determined by the reduction temperature. The number of unfilled d-states for a particular reduction temperature is constant in the range of extremely small particles used in this study. A particle size effect is expected to be present when the range of particle sizes is extended to much larger particle (Yoshitaka et al., 1991). The decreases of the white line intensity with increasing reduction temperature can now be explained straightforward. Increasing the reduction temperature changes the structure of the metal-support interface and decreases the hydrogen chemisorption capacity (Vaarkamp (2) et al., 1993). The decrease in the amount of chemisorbed hydrogen decreases the white line intensity. This effect is apparently in the same order of magnitude as the effect of the change in the structure of the metal-support interface, e.g. increase in intensity due to the removal of interfacial hydrogen and an unknown effect of the support. Hence, to establish the effect of the support on the electronic structure of small metal particles, measurements in the absence of both chemisorbed and interfacial hydrogen should be performed. This has to be the subject of future research.

REFERENCES

ASHLEY C.A., DONIACH S., 1975 - Phys. Rev. B11, 1279.

BINSTED N., COOK S.L., EVANS J., GREAVES G.N., PRICE R.J., 1987 - J. Am. Chem. Soc. 109, 3669.

BOUDART M, DJEGA-MARIADASSOU, 1984 - Kinetics of Heterogenous Catalytic Reactions - Princeton University Press, Princeton N.J.

BROWN M., PEIERLES R.E., STERN E.A., 1977 - Phys. Rev. B15, 738.

BUNKER B.A., STERN E.A., 1983 - Phys. Rev. B27, 1017.

CITRIN P.H., EISENBERGER P., KINCAID B.M., 1976 - Phys. Rev. Lett. 22, 3551.

COOK Jr. J.W., SAYERS D.E., 1981 - J. Appl. Phys. 52, 5024.

van DIJK M.P., van VEEN J.A.R., BOUWENS S.M.A.M., van ZON F.B.M., KONINGSBERGER D.C., 1990 - Proc. 2nd European Conf. on Progress in X-Ray Synchrotron Radiation Research - SIF (Bologna) 139-142.

FONTAINE A., 1993 - Neutron and Synchrotron Radiation for Condensed Matter Studies, Vol I, Theory, Instruments and Methods - Springer-Verlag (Berlin) and Les Editions de Physique (Les Ulis, France).

GALLEZOT P., DATKA J., MASSARDIER J., PRIMET M., IMELIK B., 1977 - Proc. 6th Int. Congr. Catal., London 1976, - Chem. Society (London) 696.

GALLEZOT P., WEBER R., DALLA BETTA R.A., BOUDART M., 1979 - Z. Naturforsch A34, 40.

GORDON M.B., CYROT-LACKMANN F., DESJONQUERES M.C., 1977 - Surf. Sci. 68, 359.

HALLER G.L., RESASCO D.E., 1989 - Adv. Catal. 36, 173.

HENRICH V.C., 1985 - Rep. Progr. Phys. 48, 1481.

HORSLEY J.A., 1982 - J. Chem. Phys. 76, 1451.

McHUGH B.J., LARSEN G., HALLER G.L., 1990 - J. Phys. Chem. 94, 8621.

McKALE A.G., KNAPP G.S., CHAN S.-K., 1986 - Phys. Rev. B33, 841.

McKALE A.G., VEAL B.W., PAULIKAS A.P., CHAN S.-K., KNAPP G.S., 1988 - J. Am. Chem. Soc. 110, 3763.

KAMPERS F.W.H., 1989 - Exafs in Catalysis; Instrumentation and Applications - PhD Thesis, Eindhoven University of Technology.

KAMPERS F.W.H., MAAS T.M.J., van GRONDELLE J., BRINKGREVE P., KONINGSBERGER D.C., 1989 - Rev. Sci. Instr. 60, 2635.

KIP B.J., DUIVENVOORDEN F.B.M., KONINGSBERGER D.C., PRINS R., 1986 - J. Am. Chem. Soc. 108, 5633.

KIP B.J., DUIVENVOORDEN F.B.M., KONINGSBERGER D.C., PRINS R., 1987 - J. Catal. 105, 26.

KONINGSBERGER D.C., GATES B.C., 1992 - Catal. Lett. 14, 271.

LAMB H.H., GATES B.C., KNOEZINGER H., 1988 - Angew. Chem. Int. Ed. Engl. 27, 1127.

LEE P.A., PENDRY J.B., 1975 - Phys. Rev. B27, 95.

LENGELER B., 1986 - J. Phys. (Paris) 47, 75.

LYTLE F.W., 1976 - J. Catal. 43, 376.

LYTLE F.W., WEI P.S.P., GREGOR R.B., VIA G.H., SINFELT J.H., 1979 - J. Chem. Phys. 70, 4849.

LYTLE F.W.(1), GREEGOR R.B., MARQUES E.C., SANDSTROM D.R., VIA G.H., SINFELT J.H., 1985 - J. Catal. 95, 546.

LYTLE F.W.(2), GREEGOR R.B., MARQUES E.C., BIEBESHEIMER V.A., SANDSTROM D.R., HORSLEY J.A., VIA G.H., SINFELT J.H., 1985 - ACS Symp. Ser. 288, 280.

LYTLE F.W., SAYERS D.E., STERN E.A., 1988 - Physica B158, 701.

LYTLE F.W., GREEGOR R.B., 1990 - Appl. Phys. Lett. 56, 192.

MALONEY S.D., van ZON F.B.M., KELLEY M.J., KONINGSBERGER D.C., GATES B.C., 1990 - Catal. Lett. 5, 161.

MALONEY S.D., KELLEY M.J., KONINGSBERGER D.C., GATES B.C., 1991 - J. Phys. Chem. 95, 9406.

MANSOUR A.N., COOK Jr. J.W., SAYERS D.E., 1984 - J. Phys. Chem 88, 2330.

MARTENS J.H.A., PRINS R., ZANDBERGEN H., KONINGSBERGER D.C., 1988 - J. Phys. Chem. 92, 1903.

MATTHEISS L.F., DIETZ R.E., 1980 - Phys. Rev. B22, 1663.

MILLER J.T., MEYERS B.L., MODICA F.S., LANE F.S., VAARKAMP M., KONINGSBERGER D.C., 1993 - J. Catalysis, in press.

MOTT N.F., 1949 - Proc. Phys. Soc. London 62, 416.

MUSTRE DE LEON J., REHR J.J., ZABINSKY S.I., ALBERS R.C., 1991 - Phys. Rev. B44, 4146.

PENDRY J.B., 1974 - Low Energy Electron Diffraction - Academic Press (London).

RAVENEK W., JANSEN A.P.J., van SANTEN R.A., 1989 - J. Phys. Chem. 93, 6445.

SAILLARD J.Y., HOFFMANN R., 1984 - J. Am. Chem. Soc. 106, 2006.

SAMANT M.G., BOUDART M., 1991 - J. Phys. Chem. 95, 4070.

SAYERS D.E., 1987 - X-Ray Absorption: Principles, Applications, Techniques of Exafs, Sexafs and Xanes - John Wiley & Sons (New York), 211-253.

STERN E.A., SAYERS D.E., LYTLE F.W., 1975 - Phys. Rev. B11, 4836.

STEVENSON S.A., DUMESTIC J.A., RUCKENSTEIN E., Editors, 1987 - Metal-Support Interactions in Catalysis, Sintering and Redispersion - Van Nostrand Reinhold (New York).

TEO B.K., LEE P.A., 1979 - J. Am. Chem. Soc. 101, 2815.

VAARKAMP M.(1), DRING I., OLDMAN R.J., STERN E.A., KONINGSBERGER D.C., 1993 - Phys. Rev. B., in press.

VAARKAMP M. (2), 1993 - The Structure and Catalytic Properties of Supported Platinum Catalysts - PhD Thesis, Eindhoven University of Technology.

VAARKAMP M.(3), MODICA F.S., MILLER J.T., KONINGSBERGER, 1993- J. Catal., in press.

YOSHITAKE H., IWASAWA Y., 1991 - J. Phys. Chem. 95, 7368.

van ZON J.B.A.D., KONINGSBERGER D.C., van 't BLIK H.F.J., SAYERS D.E., 1985 - J. Chem. Phys. 82, 5742.

van ZON F.B.M., MALONEY S.D., GATES B.C., KONINGSBERGER D.C., 1993 - J. Amer. Chem. Soc., in press.

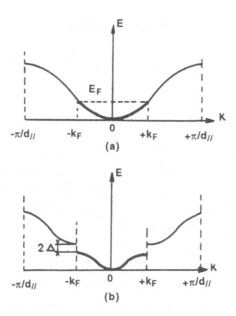

Figure XI.1. — Band structure of a 1D electron gas above (a) and below (b) the Peierls transition.

The first 1D conductor ever studied was the Krogmann salt $K_2Pt(CN)_4O.3Br-xH_2O(KCP)$ formed of well separated ($d_\perp$ = 10 Å) chains of $Pt(CN)_4$ planar complexes ($d_{//}$ = 2.9 Å). The conduction band, built with the d_{z^2} orbitals of this complex, is emptied by the transfer to the Br^- of 0.3 electron per formula unit. With a chain periodicity of 2 formula units this leads to $2k_F$ = 0.3 c^* (c^* = G/2). At room temperature the anisotropy of conductivity of KCP is $\sigma_{//}/\sigma_\perp \sim 5\ 10^4$, with $\sigma_{//} \sim$ 200 S/cm. Its 1D metallic behavior is probably best illustrated by reflectivity measurements which exhibit a plasma edge only when the electric field of the polarized light is applied in chain direction. Inorganic 1D conductors of present interest are the molybdenum bronze $K_{0.3}MoO_3$, where chains are composed of clusters of MoO_6 octehadra, and the niobium triselenide $NbSe_3$, where chains are formed of $NbSe_6$ antiprisms. For such complex chains several conduction bands cut the Fermi level at incommensurate values of k_F. Another important class of 1D conductors consists of the organic salts like TTF-TCNQ whose structure is composed of segregated stacks of donors (the tetrathiafulvalene molecule TTF) and of acceptors (the tetracyanoquinodimethane molecule TCNQ), and where the metallic properties are due to a partial charge transfer from the donors to the acceptors. In TTF-TCNQ the charge transfer is incommensurate and $2k_F$ = 0.295 b^* for both the bands of donor and acceptor chains.

XI.2.   The Peierls Instability[1,2]

Peierls demonstrated in the 1950s that at T = 0°K a 1D metal is unstable against a periodic lattice distortion (PLD) of the chain with a wave vector $2k_F$ allowing a gap

CHAPTER XI

LOW DIMENSIONAL CONDUCTORS
J.P. POUGET

Low dimensional electronic systems have been the object of intensive studies from both the theoretical and experimental points of view. The main reason is that interactions such as the electron-phonon and the electron-electron couplings as well as disorders display enhanced effects when the number of electronic dimensions is restricted. In particular, this is the case of the electron-phonon coupling in one dimensional (1D) conductors which drives the charge density wave-Peierls instability. Its basic physical aspects are now well understood especially thanks to X-ray and neutron scattering studies performed these last 20 years. A brief summary of such investigations is the purpose of the present chapter.

XI.1. Definition and examples of 1D conductors

1D metallic properties are found among anisotropic materials where the crystallographic repeat distance in one (chain) direction, $d_{//}$, is substantially smaller than in perpendicular (interchain) directions, $d_\perp$. In addition, the wave function of the conduction band of most 1D conductors is built with orbitals, such as d atomic orbitals (in inorganic materials containing transition metal elements) or p_π molecular orbitals (in organic charge transfer salts) that exhibit a much larger overlap in chain direction ($t_{//}$) than between chains ($t_\perp$). If the electron mean free path, ℓ, is such that the electronic movement is coherent in chain direction ($\ell_{//} > d_{//}$) and incoherent (or diffusive) between chains ($\ell_\perp < d_\perp$), only the intrachain wave vector components, $k_{//}$, are required to label the electronic wave function. In this case, the dispersion in energy of the conduction band is given, in the tight binding approximation, by:

$$\varepsilon(k_{//}) = 2\, t_{//} \cos(k_{//}\, d_{//}) \tag{XI.1}$$

(see Fig. XI.1a). With an average of ρ conduction electrons per repeat unit cell and two electrons of opposite spin occupying each $\Delta k = \dfrac{2\pi}{Nd_{//}}$ state [N is the number of unit cells (of sites, below) per chain], the 1D conduction band is thus filled from $-k_F$ to $+k_F$ with:

$$2 \times 2k_F = N\rho \times \frac{2\pi}{Nd_{//}} \tag{XI.2}$$

In most 1D conductors the $2k_F$ wave vector, separating occupied states from empty ones at the Fermi energy (E_F), is generally incommensurate with the reciprocal lattice periodicity in chain direction, $G = \dfrac{2\pi}{d_{//}}$.

2Δ to open at $\pm$ k_F in the band structure (Fig. XI.1b). In the presence of this gap, the 1D electronic dispersion becomes:

$$E\left(k_{//}\right) = E_F \pm \sqrt{\varepsilon^2\left(k_{//}\right) + \Delta^2} \qquad \text{(XI.3)}$$

with $\varepsilon(k_{//})$ given by (XI.1). The stabilization of an insulating ground state leads to a reduction of electronic energy over all occupied states below E_F of:

$$E_{e\ell} = -N(E_F)\,\Delta^2 \ln \frac{2E_F}{\Delta} \qquad \text{(XI.4)}$$

per chain site. In (XI.4), $N(E_F) = \dfrac{2d_{//}}{\hbar\pi v_F}$ is the density of states at E_F, including both spin directions, with $\hbar v_F = \dfrac{\partial\varepsilon\left(k_{//}\right)}{\partial k_{//}}$ The establishment of a $2k_F$ long range PLD, where the site n is displaced by:

$$u(n) = u_{2k_F}\,\sin\,(2k_F\,n\,d_{//} + \varphi) \qquad \text{(XI.5)}$$

costs an elastic energy:

$$E_L = \frac{1}{2}\,K\,u^2_{2k_F} \qquad \text{(XI.6)}$$

where K is the elastic modulus of the chain. For a modulation of small amplitude Δ is proportional to u_{2k_F}; i.e. $\Delta = g\,u_{2k_F}$, where g is a coupling constant between the electrons and the lattice degrees of freedom (the phonons). Thus $E_{e\ell}$, proportional to $\Delta^2\,\ell n\,\Delta$ always overcomes E_L proportional to Δ^2. More precisely the total energy $E_{e\ell} + E_L$ exhibits a minimum for:

$$\Delta_0 = 2\,E_F\,\exp\,(-1/\lambda) \qquad \text{(XI.7)}$$

with $\lambda = g^2\,N(E_F)/K$. The Peierls ground state has been observed in KCP, $K_{0.3}MoO_3$, $NbSe_3$ and TTF-TCNQ, with $\Delta_0 \sim 50$ meV.

The PLD given by (XI.5) modulates the intersite distances and thus the intrachain electronic density. The $2k_F$ modulation of the electronic density leads to the formation of a charge density wave (CDW). More precisely the change of the 1st neighbour distance from $d_{//}$ to $d = d_{//} + u(n+1) - u(n)$ (assuming a longitudinal displacement) leads to a variation of the intrachain transfer integral, $t_{//}$, which for small displacements and in a continuous notation $(nd_{//} \rightarrow x)$ amounts to:

$$t_{//}(d) \approx t_{//}(d_{//}) + \frac{\partial t_{//}}{\partial d}\,\frac{\partial u}{\partial x}\,d_{//} \qquad \text{(XI.8)}$$

This induces a modulation of the electronic density whose spatial variations are given by:

$$\rho(x) = \rho + \rho_{2k_F}\,\cos\,(2k_F\,x + \varphi), \qquad \text{(XI.9)}$$

with ρ_{2k_F} proportional to u_{2k_F} and $\partial t_{//}/\partial d$ (or g). Because of the presence of the gradient term $\dfrac{\partial u}{\partial x}$ in (XI.8), there is a phase shift of $\pi/2$ between the CDW and PLD.

The PLD is detected by diffraction techniques. The amplitude diffracted by a chain whose site positions are longitudinally displaced by u(n), given by (XI.5), is:

$$A(Q) = \sum_n F(Q) \exp i \left(Q\left[nd_{//} + u_{2k_F} \sin\left(2k_F nd_{//} + \varphi \right) \right] \right) \qquad (XI.10)$$

where F(Q) is the form factor of the repeat unit, assumed to be rigidly displaced. Using the expansion of the exponential in Bessel functions:

$$\exp(i\, z\, \sin\theta) = \sum_{\nu=-\infty}^{+\infty} J_\nu(z) \exp\left(i\, \nu\theta \right)$$

(XI.10) becomes:

$$A(Q) = \sum_{n,\nu} F(Q) J_\nu \left(Qu_{2k_F} \right) \exp i \left[(Q + \nu\, 2k_F)\, nd_{//} + \varphi \right] \qquad (XI.11)$$

When φ is constant, the diffracted intensity I(Q) is independent of the phase of the PLD. In the case of an incommensurate modulation one gets:

$$I(Q) = |A(Q)|^2 = \sum_\nu \sum_{n,m} \left| F(Q) J_\nu \left(Qu_{2k_F} \right) \right|^2 \exp i \left[(Q + \nu 2k_F)(m-n)\, d_{//} \right] \qquad (XI.12)$$

$$= N \sum_\nu \sum_{\ell=m-n} \left| F(Q) J_\nu \left(Qu_{2k_F} \right) \right|^2 \exp i \left[(Q + \nu 2k_F)\, \ell d_{//} \right] \qquad (XI.13)$$

and, in the limit of large N (p and $\nu \in Z$):

$$I(Q) \cong N \sum_\nu \sum_p \left| F(Q) J_\nu \left(Qu_{2k_F} \right) \right|^2 \delta \left(Q + \nu\, 2k_F - pG \right) \qquad (XI.14)$$

The diffracted spectrum is thus composed of sharp main Bragg reflections, located at $Q_0 = pG$ and of intensity $I_0(Q) = |F(Q) J_0 (Qu_{2k_F})|^2$, surrounded by sharp satellite peaks, located at $Q = p G - \nu\, 2k_F$ and of intensity $I_\nu(Q) = | F(Q) J_\nu (Qu_{2k_F})|^2$. Their separation gives directly the $2k_F$ modulation wave vector. With typically $u_{2k_F} \sim 0.05$ Å and $d_{//} \sim 4$ Å, Qu_{2k_F} is smaller than 1, so that:

$$J_\nu \left(Qu_{2k_F} \right) \cong \frac{1}{\nu!} \left[\frac{Qu_{2k_F}}{2} \right]^\nu \qquad (XI.15)$$

Thus generally only 1st order satellite reflections (of intensity $I_1 \sim 10^{-2}\, I_0$) and sometimes 2nd order satellite reflections (of intensity $I_2 \sim 10^{-4}\, I_0$) are detected, as in $K_{0.3}MoO_3$. If the modulation is not sinusoidal, as in Nb Se_3, the intensity of higher order ($\nu \geq 2$) satellite reflections will be enhanced by the contribution of the $\nu 2k_F$ Fourier components of the PLD.

In the limit of small displacements the intensity of 1st order satellite reflections is proportional to the square of the amplitude of the distortion, u_{2k_F}. The temperature dependence of $I_1(Q)$ allows the variation of the PLD to be measured This is illustrated by Fig. XI.2 showing the continuous vanishing of u_{2k_F} and the 2nd order nature of the Peierls transition of $K_{0.3}MoO_3$ occuring at $T_p \sim 180$ K. The decrease of u_{2k_F} upon heating, and thus of Δ, is caused by the reduction of the electronic gain, $E_{e\ell}$, via the thermal excitation of the electrons through the Peierls gap. When fluctuations are neglected, the mean field theory predicts a critical temperature T_p^{MF} of:

$$k_B T_p^{MF} = 0.57 \, \Delta_0, \qquad (XI.16)$$

with Δ_0 given by (XI.7). However, for reasons related to the 1D nature of the driving force of the Peierls instability (see next section) T_p^{MF} given by (XI.16) is generally much larger than the observed T_p.

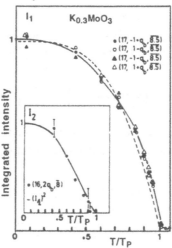

Figure XI.2. — Temperature dependence of the intensity of first order (I_1) and second order (I_2, insert) satellite reflections in the $2k_F$ Peierls modulated state of $K_{0.3}MoO_3$[3].

XI.3.   Fluctuations and Dynamics of the Peierls Chain[2]

The Peierls transition stabilizes a PLD through a 2nd order phase transition. In the weak electron-phonon coupling limit ($\lambda < 1$) considered in section XI.2, the lattice instability is announced in the 1D metallic regime, above T_p, by the softening around the critical wave vector $2k_F$ of a particular phonon branch. The mechanism of this lattice instability is illustrated by Fig. XI.3. Let us consider a periodic chain of ions of mass M, whose displacements around the equilibrium position are described by the normal coordinate u_q, solution of the equation of motion:

$$M \, \ddot{u}_q + K \, u_q = 0 \qquad (XI.17)$$

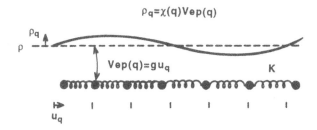

Figure XI.3. — CDW response of an electron gas in the presence of a periodic potential $V_{ep}(q)$ caused by a periodic displacement u_q of the ions.

The eigenvalues of (XI.17) are the phonon pulsations $\Omega_0 = \sqrt{\dfrac{K}{M}}$. The ionic displacement u_q causes a potential $V_{ep}(q) = g\, u_q$ acting on the 1D electron gas. The electrons respond to it by setting a CDW, whose q Fourier component is given, in the framework of the linear response theory, by:

$$\rho_q = \chi(q,T)\, V_{ep}(q) \tag{XI.18}$$

In (XI.18), $\chi(q,T)$ is the electron-hole response of the electron gas. Its thermal and wave vector (for q close to $2k_F$) dependences are given for the case of 1D non interacting electrons by:

$$\chi(q,T) = N(E_F)\left[\text{Log}\,\frac{1.13\,E_F}{k_B T} - \xi_0^2\,|q - 2k_F|^2 \right] \tag{XI.19}$$

where ξ_0 is the length associated with the thermal broadening of the Fermi surface $\left(\hbar\, v_F\, \xi_0^{-1} \approx \pi k_B T \right)$, and are illustrated by Fig. XI.4. The interaction between the CDW and the ions:

$$H_{ep} = \sum_q \rho_{-q}\, V_{ep}(q)$$

provides an external force:

$$F_q^{ext} = \frac{\partial H_{ep}}{\partial u_{-q}} \tag{XI.20}$$

on the right side of the equation of movement (XI.17), which thus becomes:

$$\ddot{u}_q + \Omega_0^2\left[1 - \frac{g^2}{K}\,\chi(q,T) \right] u_q = 0 \tag{XI.21}$$

and presents the eigenvalues:

$$\omega^2(q) = \Omega_0^2\left[1 - \frac{g^2}{K}\,\chi(q) \right] \tag{XI.22}$$

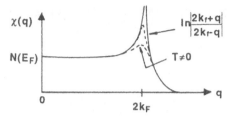

Figure XI.4. — Electron-hole response function of a 1D non interacting electron gas. The solid line is for T = 0°K and the dashed lines are for finite T.

In (XI.22), the renormalization of the phonon pulsations is due to the screening of the Coulomb interactions between the ionic cores by the CDW. As shown Fig. XI.4, $\chi(q,T)$ exhibits a well defined maximum for $q \sim 2k_F$. This leads to a sizeable softening of $\omega(q)$, forming a so-called Kohn anomaly in the phonon spectrum. Figure XI.5 shows the giant Kohn anomaly measured by inelastic neutron scattering at room temperature around the $2k_F$ wave vector in the acoustic branch polarized in chain direction of KCP[4].

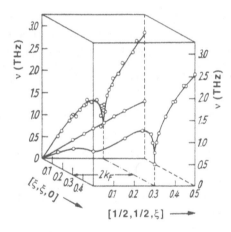

Figure XI.5. — Room temperature phonon dispersion surface of $K_2Pt(CN)_4$ 0.3 Br - xH_2O showing the $2k_F$ giant Kohn anomaly is the acoustic branch polarized along the chain direction[4].

Upon cooling the $2k_F$ phonon frequency ($v_K = \omega(2k_F)/2\pi$ in Fig. XI.6a) exhibits a critical softening. (XI.19) included into (XI.22) leads to the mean field thermal dependence:

$$\omega^2 (2k_F) = \Omega_0^2 \, \mathrm{Log} \, \frac{T}{T_P^{MF}} \qquad (XI.23)$$

where ω^2 (2k$_F$) vanishes at T_P^{MF} given by (XI.16). Such a softening is

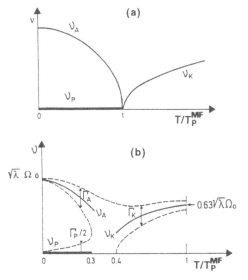

Figure XI.6. — (a) Mean field behavior of the frequency of the Kohn anomaly (ν_K), the amplitude mode (ν_A) and the phase mode (ν_P) at 2k$_F$. (b) Schematic representation of the temperature dependence of the maxima in frequency (ν_i) and the fullwidth at half maximum (Υ_i) of the 2k$_F$ dynamical response function of the Peierls chain.

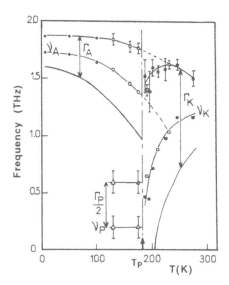

Figure XI.7. — Temperature dependence of the quasiharmonic frequency (ν_i) and the damping (Υ_i) for the Kohn anomaly (i = K), the amplitude mode (i=A) and the phase mode (i = P) of K$_{0.3}$MoO$_3$ at 2k$_F$[5].

observed in $K_{0.3}MoO_3$ (Fig. XI.7)[5]. In addition to the softening, a giant damping (Υ_K) of the $2k_F$ critical phonon mode is observed upon cooling. It is due to the decay of $2k_F$ phonons into electron-hole pairs and to anharmonicity arising from lattice fluctuations of large amplitude near T_p.

Together with the growth of a sharp Kohn anomaly in a phonon branch, short range $2k_F$ CDW develop in chain direction, with a correlation length of:

$$\xi_{//} = \xi_0 \frac{\Omega_0}{\omega(2k_F)} \tag{XI.24}$$

The diffraction from these local modulations, uncorrelated from chain to chain, gives rise in reciprocal space to diffuse scattering (see chapter IX Vol. I) consisting of sheets perpendicular to the chain direction. These diffuse sheets, of width $2\xi_{//}^{-1}$ in chain direction, are located at $\pm 2k_F$ from the layers of main Bragg reflections. Their intersection with the Ewald sphere gives rise to "$2k_F$" diffuse lines on the diffraction pattern, as illustrated in Fig. XI.8 for the case of KCP[6].

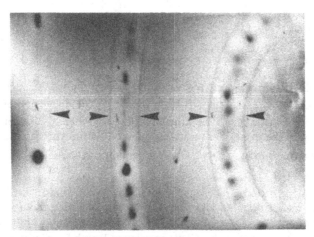

Figure XI.8. — X-ray diffuse scattering pattern from $K_2Pt(CN)_40.3$ Br -xH_2O at room temperature. The $2k_F$ diffuse lines are shown by black arrows. The chain axis c is horizontal[6].

Below T_p the collective excitations of the incommensurate PLD (or CDW) are those associated with the spatio-temporal variations of the amplitude, $\delta u_{2k_F}(n,t)$, and of the phase, $\delta\varphi(n,t)$, of the modulation wave (XI.5), as schematically illustrated in Fig. XI.9. Accordingly there are two phonon branches of dispersion:

$$v_A^2(q) = v_A^2(2k_F) + \frac{1}{3} v_\varphi^2 |q - 2k_F|^2 \tag{XI.25}$$

$$v_P(q) = v_\varphi |q-2k_F| \tag{XI.26}$$

called amplitudon and phason. Their frequency dependence has been measured in $K_{0.3}MoO_3$ (Fig. XI.10)[5,7]. At $T = 0°K$, $v_A^0(2k_F) = \sqrt{\lambda}\Omega_0 / 2\pi$ amounts to 1.7 THz.

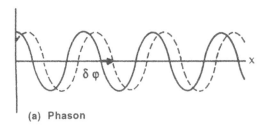

(a) Phason

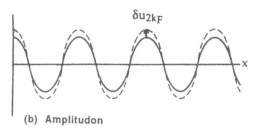

(b) Amplitudon

Figure XI.9. — Collective excitations of the phase φ (a) and of the amplitude u_{2k_F} (b) of a CDW-PLD.

The mean field theory predicts the vanishing of $v_A(2k_F)$ at T_p^{MF} (Fig. XI.6a), a behavior not observed experimentally (see Fig. XI.7). The dispersion of the phason branch is of the acoustic type. The vanishing of $v_p(q)$ when $q\rightarrow 2k_F$ corresponds to the free sliding of the incommensurate modulation with respect to the underlying lattice. The phason velocity $v_\varphi \approx v_A^o\left(2k_F\right)\xi_0$ amounts to about $2 \times 10^4 ms^{-1}$ ($\sim v_F/10$) near T_p in $K_{0.3}MoO_3$.

So far we have compared experimental results with mean field predictions which neglect lattice fluctuations. Thermal fluctuations play a crucial role in 1D systems because they suppress any kind of phase transition at finite temperature (Landau theorem). As a result the CDW correlation length $\xi_{//}$ diverges only at 0°K. Numerical simulations of the Peierls chain[8] still give a softening of the Kohn anomaly upon cooling, but its damping increases so much that the soft mode becomes overdamped at about $0.4\ T_p^{MF}$ (Fig. XI.6b). However below about $0.3\ T_p^{MF}$, $\xi_{//}$ becomes long enough so that the fluctuations of the amplitude and of the phase of the PLD are partially decoupled.

In real materials a modest interchain coupling $V_\perp$ is able to drive the 3D ordering of the 1D local CDW at finite temperature when $\xi_{//}$ becomes long enough upon cooling. More precisely a 3D Peierls transition occurs when:

$$\frac{\xi_{//}\left(T_p\right)}{d_{//}}\ V_\perp \cong k_B T_p \qquad (XI.27)$$

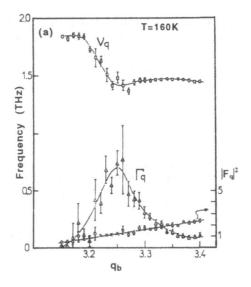

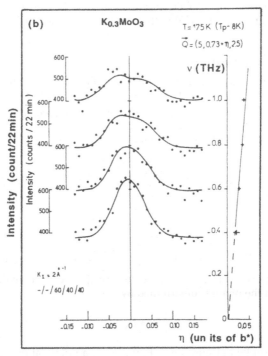

Figure XI.10. — (a) Wave vector dependence of the quasiharmonic frequency (v_q), the damping (Υ_q) and the square of the structure factor ($|F_q|^2$) of the amplitudon branch, and (b) constant frequency neutron scans through the phason branches of $K_{0.3}MoO_3$[5].

Its critical temperature Tp is significantly lower than T_p^{MF}. However even in the presence of a 3D Peierls transition the $2k_F$ dynamics observed in $K_{0.3}MoO_3$ (Fig. XI.7) resembles that predicted (Fig. XI.6b) for the Peierls chain, in particular for the amplitude mode which emerges on the high frequency side of the overdamped Kohn anomaly.

XI.4.   CDW Phase deformation

Fröhlich proposed in the 1950s that an incommensurate CDW, whose coupling energy with a perfect lattice is independent of φ, could slide freely under the action of an electric field, E. In real materials where defects pin the phase of the CDW such a collective sliding of the CDW is observed below T_p for E greater than a threshold field E_T related to the pinning energy. The understanding of CDW collective transport phenomena thus requires a description of the CDW pattern perturbated by the applied electric field and the defects. The most likely deformation is via the phase of the CDW whose collective excitations have the lowest energy [Note that because of the pinning effects the phason frequency v_p remains finite at q = $2k_F$ (see Fig. XI.7)]. The effects of the electric field on a sliding CDW (for E > E_T) and of the defects on the texture of the 3D CDW pattern will now be illustrated by recent high resolution investigations performed with the synchrotron radiation.

XI.4.1.   Effect of the electric field

The acoustic like dispersion of the phason mode allows a CDW phase elasticity to be constructed[9]. To a local displacement u_x of the CDW in chain direction corresponds a phase shift φ = $2k_F u_x$. With the relative change of the CDW wave length $\dfrac{\delta\lambda_{CDW}}{\lambda_{CDW}}$, one can associate a strain:

$$e_{xx} = \frac{\partial u_x}{\partial x} = \frac{1}{2k_F}\frac{\partial \varphi}{\partial x} \qquad (XI.28)$$

related to the stress:

$$\sigma_{xx} = (2k_F)^2\, K_\varphi\, e_{xx} \qquad (XI.29)$$

by the intrachain CDW elastic constant $K_\varphi = \dfrac{\hbar v_F}{2\pi s}$, where s is the chain cross section area. In equilibrium with an external force density applied in chain direction, F_x, the spatial variation of the phase is determined by:

$$\frac{\partial \sigma_{xx}}{\partial x} + F_x = 0 \qquad (XI.30)$$

In the case of an applied electric field, E, inducing a polarization $P_x = \rho e\, u_x$ of the CDW, (XI.30) becomes with $F_x = \rho\, eE$:

$$\frac{\partial^2 \varphi}{\partial x^2} = -\frac{\rho eE}{2k_F\, K_\varphi}. \qquad (XI.31)$$

(XI.31) predicts a linear variation of $2k_F$ between the electric contacts producing the electric field E (Fig. XI.11a):

$$\Delta 2k_F = \frac{\partial \varphi}{\partial x} = -\frac{\rho e E}{2k_F K_\varphi}(x - x_0) \qquad (XI.32)$$

Such a $2k_F$ variation has been recently observed[10] by an accurate measurement of the satellite positions in function of the impact point of the X-ray beam on a single crystal of NbSe$_3$ for $E > E_T$ (Fig. XI.11b).

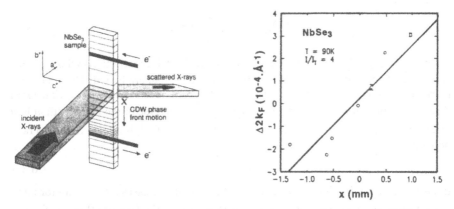

Figure XI.11. — (a) Schematic representation of the elastic deformation of the phase of a CDW moving under the action of an electric field. (b) Shift of the $2k_F$ CDW wave vector as a function of the position x of the X-ray beam in the NbSe$_3$ chain direction and in the experimental configuration of (a)[10].

XI.4.2. Effect of defects

In the presence of defects the phase of the CDW is no longer spatially uniform. Thus, for wave vectors in the vicinity of 1st order satellite reflection positions, (XI.12) must be replaced by:

$$I(Q) = I_1(Q) \sum_n \sum_\ell \exp i\left[(Q \pm 2k_F)\ell d_{//} + \varphi_\ell\right] \qquad (XI.33)$$

where φ_ℓ is the phase difference between the sites m and n: $\varphi_m - \varphi_n$. The spatial variations of the phase lead to a broadening of the satellite reflections which has been recently studied by X-ray high resolution experiments[11,12]. In principle, φ_ℓ averaged over the volume of the sample irradiated by the X-ray beam can be obtained from the profile analysis of I(Q). In NbSe$_3$ doped with Ta impurities (Fig. XI.12) and in K$_{0.3}$MoO$_3$ perturbated by irradiation defects intrinsic Lorentzian squared (LS) profiles are found. Such a profile can be explained either by the breaking up of the 3D CDW into domains of random size or by the perturbation of the CDW by random fields induced by the defects.

Let us consider the first explanation which is the simplest to describe. If in presence of defects the texture of the CDW is such that there are regions, where the phase is nearly constant, separated by sharp domain walls (nucleated by defects) where the phase change so abruptly that the CDW phase-phase correlation is lost over a few lattice spacings, (XI.33) becomes:

$$I(Q) = I_1(Q) \sum_n \sum_\ell \sigma(n, n+\ell) \exp i \left[(Q \pm 2k_F) \ell d_{//} \right] \tag{XI.34}$$

where $\sigma = 1$ if n and n+ℓ belong to the same domain and $\sigma = 0$ otherwise. If the probability that two sites separated by ℓ belong to the same domain is defined by:

$$G(\ell) = \frac{1}{N} \sum_n \sigma(n, n+\ell), \tag{XI.35}$$

(XI.34) becomes:

$$I(Q) = N \, I_1(Q) \sum_P G(\delta Q) \tag{XI.36}$$

where $\delta Q = Q - pG \pm 2k_F$. (XI.36) leads to broadened satellite reflections whose profile is given by $G(\delta Q)$, the Fourier transform (FT) of $G(\ell)$ (generally a 3D FT is required because 3D CDW domains are formed). If there is a narrow (i.e. Gaussian) distribution of domain sizes $G(\delta Q)$ is Gaussian. The inverse of the half width at half maximum of the satellite reflection in the ith direction, ΔQ_i^{-1}, is related to the average domain size in this direction by:

$$L_i \cong \pi / \Delta Q_i \tag{XI.37}$$

If there is a broad distribution of domain sizes $G(\delta Q)$ decreases less rapidly for large δQ. For a random distribution of domain sizes with κ the (constant) probability to cross a domain wall per unit length, $G(\ell)$ is given by the differential equation:

$$G(\ell + d\ell) = G(\ell) [1 - \kappa d\ell] \tag{XI.38}$$

Its solution is $G(\ell) = \exp(-\kappa \ell)$, whose 3D FT is a L.S.

$$G(\delta Q) = \frac{1}{\left(\kappa^2 + \delta Q^2 \right)^2} \tag{XI.39}$$

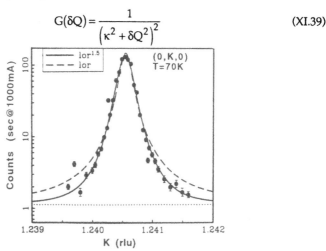

Figure X.12. — High resolution profile in the chain direction of a $2k_F$ satellite reflection in NbSe$_3$ doped with 0.25% Ta impurities. The solid line (Lor$^{1.5}$) is the profile expected for an intrinsic LS line shape integrated over the out of scattering plane direction by the vertical divergence of the X-ray beam[11b].

Further investigations are however necessary in order to discrimate between the various descriptions of the CDW pattern textured by the presence of defects and to study the interaction of the CDW with the defects on a microscopic scale.

XI.5.   Final remarks

In this chapter electron-electron interactions have been neglected. $2k_F$ CDW survive in the presence of important electron-electron interactions, as observed in the organic charge transfer salts of the TTF-TCNQ family. In the limit of very strong electron-electron interactions 1D charge localization phenomena occur with an average separation between charges of $d_{///}\rho$. The 1st Fourier component of the associated CDW, which corresponds to the $4k_F$ wave vector, has been observed in several organic charge transfer salts. The pretransitional dynamics of such a CDW instability is however not known.

In presence of electron-electron interactions the 1D electron gas can also exhibit instabilities towards a $2k_F$ spin density wave (SDW) ground state. A $2k_F$ SDW can be viewed as the difference between two CDW shifted by π, each possessing a given spin. A SDW is stabilized by the exchange potential whose $2k_F$ Fourier component opens a gap at the Fermi level as does the $2k_F$ lattice potential in the case of the Peierls transition. Such a SDW ground state occurs below T_{SDW} = 12 K in (TMTSF)$_2$PF$_6$ (TMTSF is the tetramethyltetraselenafulvalene molecule). Until now neutron scattering experiments have failed to reveal its magnetic periodicity, probably because the amplitude of the SDW is very small ($\sim 1/10\ \mu_B$). Its finding , as well as the study of the cascade of field induced SDW phase transitions that occurs when (TMTSF)$_2$ ClO$_4$ is submitted to a transverse magnetic field are challenges for future research.

REFERENCES

[1]   BERLINSKY A.J., 1976 - Contemp. Physics 17 331.

[2]   TOMBS G.A., 1978 - Physics Report C40 181.

[3]   POUGET J.P., NOGUERA C., MOUDDEN A.H. and MORET R., 1985 - J. Physique (France) 46, 1731 (Erratum 47, 145 (1986)).

[4]   COMES R., RENKER B., PINTSCHOVIUS L., CURRAT R., GLASER W. and SCHEIBER G., 1975 - Phys. Status Solidi B71, 171.

[5]   POUGET J.P., HENNION B., ESCRIBE-FILIPPINI C. and SATO M., 1991 - Phys. Rev. B43, 8421.

[6]   COMES R., LAMBERT M., LAUNOIS P. and ZELLER H.R., 1973 - Phys. Rev. B8, 571.

[7]   HENNION B., POUGET J.P. and SATO M., 1992 - Phys. Rev. Lett. 68, 2374.

[8]   TUTIS E. and BARISIC S., 1991 - Phys. Rev. B43, 8431.

[9]   FEINBERG D. and FRIEDEL J., 1988 - J. Physique (France) 49, 485.

[10]   DI CARLO D., SWEETLAND E., SUTTON M., BROCK J.D. and THORNE R.E., 1993 - Phys. Rev. Lett. 70, 845.

[11] a) SWEETLAND E., TSAI C.Y., WINTNER B.A., BROCK J.D. and THORNE R.E., 1990 - Phys. Rev. Lett. 65, 3165.
b) DI CARLO D., SWEETLAND E., BROCK J.D. and THORNE R.E., manuscript in preparation

[12] DELAND S.M., MOZURKEWICH G. and CHAPMAN L.D., 1991 - Phys. Rev. Lett. 66 2026.

CHAPTER XII

LAYERS - MULTILAYERS - SUPERLATTICES
S. FERRER AND J.L. MARTINEZ

XII.1. Introduction

Multilayers are artificial one-dimensional periodic structures consisting in alternate layers of two materials. The thickness of the layers is normally on the order of the nm.

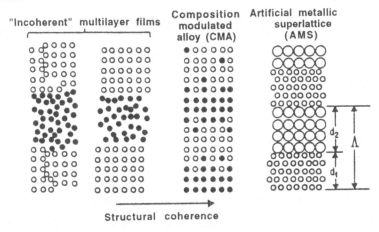

Figure XII.1. — Schematic classification of multilayer structures.

Multilayers are usually grown from the vapour phase on the top of substrates as flat as possible. Note that a 0.1 degree miscut of a crystal face results in flat areas on the order of 1000 Å. Advances in Molecular Beam Epitaxy Techniques have allowed to monitor the fabrication process, atomic layer by atomic layer. In this way, perfection can be achieved at atomic scale. In addition to the substrate quality, the morphological perfection of a multilayer is the result of many physical processes such as, among others, homogeneous and heterogeneous surface nucleation, surface diffusion and bulk interdiffusion and also of a variety of states such as surface roughness, elastic strain within the layers, pseudomorphic metastable structures and misfit dislocations among others.

There are many applications of multilayers. As X-ray optical devices, advantage is taken for the fabrication of multilayers with a preselected period for X-ray monochromatization via Bragg law. As magnetic storage systems, magnetic thin films alternate with non-magnetic ones exhibiting extremely interesting magnetic behaviours. Furthermore, the electronic states localized at the interfaces of the multilayers are model cases for two dimensional electronic systems where very interesting quantum effects can be studied.

This lecture gives an introduction to some basic aspects of the multilayers. We will discuss their stability from the thermodynamic point of view and also their basic diffraction characteristics and the influence on the diffraction of very simple defects and interdiffusion.

XII.2.   Stability of multilayers

XII.2.1.   Thermodynamics of binary solid solutions

The equilibrium state of a binary system at a well defined temperature, can be determined by minimizing the free energy $F = U-TS$ of the system. For a system consisting in atoms A and B distributed in N lattice sites , let c be the concentration of atoms A and cN = total number of A atoms. The energy of a random mixture of A and B can be expressed as

$$U = \frac{1}{2} NZ \left[c^2 V_{AA} + (1 - c)^2 V_{BB} + 2c(1-c) V_{AB} \right]$$

(XII.1)

Z = number of nearest neighbours, Vij energy of a bond (<0).

The energy of the pure and separate components is

$$U_{pure} = \frac{1}{2} NcZV_{AA} + \frac{1}{2} N(1-c)Z V_{BB}$$

(XII.2)

The energy of mixing is defined as $\Delta U = U - U_{pure}$:

$$\Delta U = NZc(1-c)\,\varepsilon; \qquad \varepsilon = V_{AB} - 1/2\,(V_{AA} + V_{BB}): \text{interaction parameter.}$$

For $\varepsilon > 0$, $1/2\,(V_{AA} + V_{BB}) < V_{AB}$ and the like atoms prefer to stay together (phase separation) . For $\varepsilon < 0$ homogeneization is energetically preferred.

Note that $\Delta U = \Delta U(c)$ is a parabola that has a curvature that depends on the sign of ε. $\Delta U'' > 0$ if $\varepsilon < 0$.

The difference in entropy between the binary system and that of the pure and separate components is, in the simplest approximation, the well known entropy of mixing:

$$\Delta S = - k_B N\,(c \log c + (1 - c) \log (1 - c))$$

(XII.3)

Note that $\Delta S'' < 0$ for c between 0 and 1.

The free energy of the mixture $F = \Delta U - T\Delta S$ may show different behaviours depending on the sign of ε and on the temperature.

If $\varepsilon < 0$, $F'' > 0$ for all c. This is the case of a mixing solution:

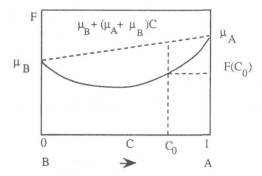

Figure XII.2. — Free energy versus concentration for a mixing solution.

At $c = c_0$, $F(c_0)$ is smaller than the value that one would obtain for an inhomogeneous system.

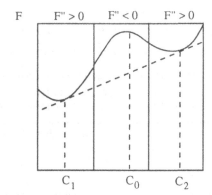

Figure XII.3. — Free energy versus concentration for a solution with mixing and phase separation possibilities.

If $\varepsilon > 0$, then, depending on the temperature and concentration, one may have phase separation. In the $F'' > 0$ regions there is homogenization. For $F'' < 0$ there is phase separation. At $c = c_0$ the system splits into two phases of concentrations c_1 and c_2, since F is lower than for a single phase. Upon increasing T the regions $F'' < 0$ tend to disappear since the entropy term $-T\Delta S$ tends to dominate and $(-T\Delta S)'' > 0$.

These are the general results for binary solutions of bulk systems where surface and interface effects have a negligible contribution to the free energy. This is not the case of multilayers, especially when the period is only of a few atomic layers.

XII.2.2.   Stability of inhomogeneous systems with strong concentration variations

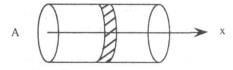

Figure XII.4. — Schematics of system exhibiting concentration variations along x.

If A is the lateral dimension of the system and there is an inhomogeneity c(x) along the x direction, then one may write $F = A \int f(c)\, dx$, f(c) free energy p.u. volume.

If c(x) varies strongly with the position as in the low period multilayers, then f(c) depends not only on the value of c at x, but also on the values of c at neighbouring positions. Within an approach similar to the Ginzburg-Landau theory of phase transitions where the free energy is expanded in a power series, one may modify the last expression by inserting a term that depends on the magnitude of the gradient of concentration:

$$F = A \int \left(f(c) + \kappa \left(\frac{dc}{dx} \right)^2 \right) dx$$

(XII.4)

κ is the gradient energy coefficient. In a nearest neighbour interaction model one finds:

κ = Nydε where N = number of atoms p.u. area, y = number of bonds

between interface atoms and d = distance between planes.

Note that for ε > 0 (phase separation energetically preferred for bulk systems), then κ > 0 and if dc/dx diminishes then κ (dc/dx)^2 diminishes and also F. This means that the interfacial contribution of the free energy of the system (at the interfaces dc/dx is most important) is lowered if the system mixes. This is therefore an opposite contribution to the bulk term that energetically tends to non-mix.

XII.2.3.   Interdiffusion

Due to the strong concentration variations accross the interfaces of the multilayers, the classical diffusion equations need to be modified in order to predict correctly the mass transport accross the interfaces when the temperature changes.

In a multicomponent system the mass transport along a direction x vanishes when the chemical potential of every species is constant along x:

$\mu_i (x)$ = constant   i = A, B,....   or μ_A = dF/dc = constant

If there is a gradient in chemical potential, then there is mass transport given by:

J = -M grad m.   M: mobility (> 0),   J: flux

These expressions need to be modified.

In the last paragraph we have written instead of f(c), f(c) + κ(dc/dx)^2. This causes that, instead of df/dc = constant as stability condition, one has to use

$$\alpha = \frac{df}{dc} - 2\kappa \frac{d^2c}{dx^2} = \text{constant}$$

(XII.5)

The gradients of α are the driving forces for mass transport:

J = -M Ω dα/dx   Ω: atomic volume

and the diffusion equation reads:

$$\frac{dc}{dt} = -\operatorname{div} J = M\Omega \left[f'' \frac{d^2 c}{dx^2} - 2\kappa \frac{d^4 c}{dx^4} \right]$$

(XII.6)

$D = M\Omega f''$ diffusion coefficient (bulk).

For multilayers one may look for solutions of the form:

$c(x) = c_0 + A(t) \cos \beta x \quad \beta = 2\pi / \Lambda. \quad \Lambda = $ multilayer period.

By inserting this in the diffusion equation, one finds that

$$A(t) = \exp\left[-D\beta^2 (1 + \frac{2\kappa\beta^2}{f''}) t \right]$$

(XII.7)

and that

$$R = \frac{d \log A}{dt} = D_\Lambda \beta^2 \quad ; D_\Lambda = -D(1 + \frac{2\kappa\beta^2}{f''})$$

(XII.8)

Therefore the rate of change of the composition amplitude depends on β^2 (i.e. the higher Λ is, the smaller is the decay rate, as one could guess by intuition) and on a coefficient that depends on κ, f'' D and Λ.

$D_\Lambda > 0$ corresponds to the usual downhill diffusion i.e. particles go where they are missing. On the contrary $D_\Lambda < 0$ means uphill diffusion: A particles tend to accumulate in regions where there are already A particles.

The above result gives rise to peculiar behaviours of the interdiffusion accross the multilayers interfaces that may be summarized as follows:

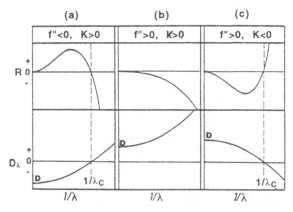

Figure XII.5. — Dependence of the amplification factor, R, and the interdiffusivity. D_λ, on the modulation repeat length for a phase separating system inside the spinodal (a), and outside the spinodal (b), and for an ordering system (c).

• Case (a): $\varepsilon > 0$ (phase separation) and hence $\kappa > 0$ and $f'' < 0$.

For long period multilayers, $D_\Lambda < 0$ and temporal evolution (or a moderate temperature increase) cause the components of the system to separate. This holds up to a critical modulation length Λc where the behaviour is reversed and the compositional modulation diminishes upon time evolution.

$$\Lambda_c = 2\pi\sqrt{\frac{2\kappa}{|f''|}}$$

(XII.9)

The intermixing is due to the increased role of the interfacial energy at low periods. As $\kappa > 0$, mixing (decrease of dc/dx) causes a decrease in the free energy.

• Case (b): $f'' > 0$, $\kappa > 0$.

$D_\Lambda > 0$ for all Λ. Therefore there is downhill diffusion i.e. mixing. In this case the interfacial energy and the bulk terms tend to mix ($f'' > 0$) and as there is no competition, there is no critical period.

• Case (c): $f'' > 0$, $\kappa < 0$.

Here the bulk terms tend to mix ($f'' > 0$) and the interface terms tend to separate, therefore there is a critical Λ.

Note also that at Λc there is a balance and the temporal stability of the composition profile is the optimum.

XII.3. Diffraction from Multilayers

The figure shows the most perfect type of layered structure designed as phase locked superlattice.

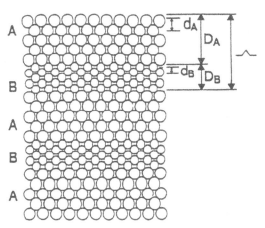

Figure XII.6. — Ideal phase-locked superlattice.

The period is $\Lambda = n_A d_A + n_B d_B = D_A + D_B$ repeated N times.

The general problem of N identical objects evenly spaced at distances Λ can be formulated in terms of a function

$$f(x) = \sum_{-(N-1)/2}^{(N-1)/2} \delta(x - n\Lambda) * g(x)$$

(XII.10)

$g(x)$ is the "pattern" function defined within one period. The diffracted amplitude is then

$$F(q) = G(q) \frac{\sin(\pi N \Lambda q)}{\sin(\pi \Lambda q)}$$

(XII.11)

The second term plots

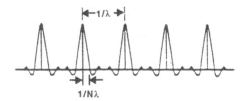

Figure XII.7. — Diffracted amplitude for N units with periodicity Λ.

Between two main maxima of the diffracted intensity, $|F|^2$, there are N-2 smaller maxima.

To simplify and see the essential, assume that B is transparent compared to A i.e. $Z_B << Z_A$ then

$$g(x) = f_A \sum_{-(n_A-1)/2}^{(n_A-1)/2} \delta(x - m d_A)$$

(XII.12)

$$G(q) = f_A \frac{\sin(\pi n_A d_A q)}{\sin(\pi d_A q)}$$

and

(XII.13)

i. e. the same interference function as before. In practice $N >> n_A$ for example $N = 100$ stacks and $n_A = 5$ layers, the widths of the peaks of the two interference functions are very different and the diffracted amplitude is the product of two functions as

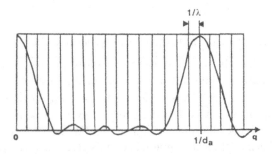

Figure XII.8. — Schematic representation of the amplitudes for a multilayer system with a large number of slacks.

In general one obtains peaks at $n1/d_A$ with intensities that depend on how well the condition $m\,1/\Lambda = n\,1/d_A$ is fulfilled.

XII.3.1.   Effect of imperfections in the period

In order to illustrate one of the effects of the imperfections in the diffracted intensity, let us assume again that B is transparent and that D_B fluctuates and therefore that Λ also fluctuates around a mean value Λ_0 with a standard deviation σ of a normal distribution function.

To calculate the diffracted intensity we will average on Λ. However, one may average the intensity or the amplitude i.e. one may calculate $<|F|^2>$ or $|<F>|^2$. The adequate way to proceed will depend on how the coherence lengths of the X-rays compare with the characteristic lengths of the "domain" in the sample. To give a simple example let us evaluate $<|F|^2>$.

$$\langle F(q)\,F^*(q)\rangle = |G(q)|^2 \left\langle \frac{\sin^2(\pi N\Lambda q)}{\sin^2(\pi\Lambda q)} \right\rangle \;; \qquad P(\Lambda) = \frac{1}{\sqrt{\pi}\sigma}\,\exp\left(-\frac{(\Lambda-\Lambda_0)^2}{2\sigma^2}\right) \qquad \text{(XII.14)}$$

we should evaluate

$$\int \frac{\sin^2(\pi N\Lambda q)}{\sin^2(\pi\Lambda q)}\,P(\Lambda)d\Lambda$$

Let us examine qualitatively the result that we may expect

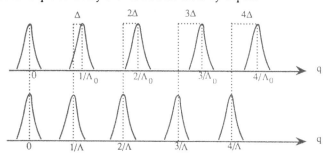

Figure XII.9. — Schematic illustration of the effect of the imperfections in the period.

For $q = 0$ one has always the same intensity irrespective of the value of Λ; if $\Lambda > \Lambda_0$ as in the figure, the intensity at $1/\Lambda_0$ will be diminished with respect to the perfect case since the peak at $1/\Lambda$ is shifted Δ. At $q = 2/\Lambda_0$, the intensity will still be smaller since the peak at $2/\Lambda$ is shifted 2Δ respect to the peak at $1/\Lambda_0$. This means that upon adding terms, the intensities at $q = n/\Lambda_0$ will decrease rapidly upon increasing n. In this model for the imperfection this gives an envelope that shows an attenuation with q as a Debye-Waller factor:

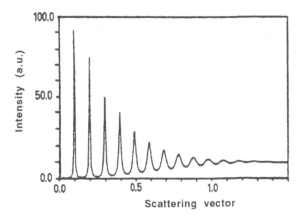

Figure XII.10. — Debye-Waller decay of the intensity for increasing values of the scattering vector.

XII.3.2.   Effect of the interdiffusion

Consider a perfect multilayer that has an electronic density perpendicular to the packing direction that has the form of a rectangular function:

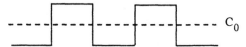

Figure XII.11. — Rectangular density profile.

It may be expressed as a Fourier series

$$c(z) = c_o + \sum_p C_p \cos\left(\frac{2\pi \, pz}{\Lambda}\right) \qquad\qquad C_p = \frac{2}{\pi p} \sin\left(\frac{\pi p \, D_A}{\Lambda}\right) \qquad\qquad (XII.15)$$

As we may see the amplitude of the harmonics decreases as $1/p$ when the harmonic number increases.

If interdiffusion occurs then $c(z)$ is not so rectangular but has rounded edges and resembles more a sinusoidal function. In this case the amplitude of the Fourier cofficients of high harmonics is decreased since the high harmonics are the ones that cause the sharpness of $c(z)$.

It is a completely general result of diffraction theory the fact that the diffracted amplitude of a periodic system with periodicity Λ is represented by a set of delta functions equally spaced with separation $1/\Lambda$ in q, each delta funtion having a weight equal to the Fourier coefficient of the periodic function describing the system.

Then, an ideal multilayer with abrupt interfaces will have a diffraction pattern that will show peaks with intensities decreasing as $1/p^2$ where p is the harmonic number and p/Λ is the corresponding scattering vector. If interdiffusion occurs the decrease in the intensities will be faster and only the lowest order harmonics will appear in the diffraction pattern.

Another interesting and simple result is that one may extract in some cases more detailed information from the simple analysis that we are considering.

If pD_A/Λ = integer then the peak at p/Λ is missing. If $p + 1$ is the following non-missing peak then $D_A = \Lambda/p$. The following example illustrates this result.

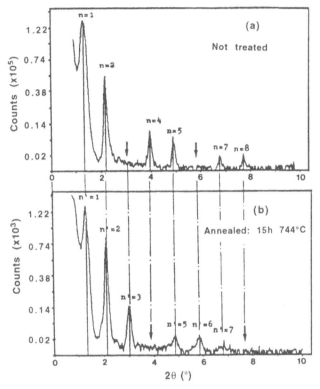

Figure XII.12. — Intensity for a carbon-tungsten multilayer showing the decrease of the carbon layer thickness after annealing.

There is a large number of peaks indicating sharp interfaces. Before annealing the third and sixth peaks are missing meaning that as Λ = 107Å, $\Lambda/3$ = 35.6 Å is the thickness of the C layer. After annealing the $p = 4$ peak is missing showing that the C layer thickness has decreased to 107/4 = 26.7 Å due to interdiffusion. Also, the $p = 7$ peak decreases markedly after annealing due to interface smoothing.

XII.4.   Examples of multilayered systems

There are so many examples of multilayered or closely related systems that we will restrict ourselves to some specific cases that are interesting from the point of view of neutron or X-ray scattering.

We will focus on magnetic transport and structural properties of metallic layers, multilayers and superlattices and we will not discuss several important examples such as intercalated compounds in graphite, natural layered minerals (clay), the almost perfect bi-dimensional systems such as high-temperature superconductors, and the artificial semiconductor structures.

Multilayers and Superlattices are systems of Chemically Modulated Films. We have to distinguish between Chemical and Structural order.

The chemical order will give blocks of amorphous materials like Fig. XII.13(a). If the growing facility is reliable (shutters....) the result will be a modulated film of sharp and flat interfaces, but with structural disorder. The chemical modulation will be a perfect square wave.

If both materials have close lattice parameters and belong to the same space group, there is a big chance that both materials will be soluble. As a consequence, the growth will be single crystalline with a complete interdiffusion. Such a case will have structural order with a sinusoidal chemical modulation (see Fig. XII.13(c) and XII.14(b)).

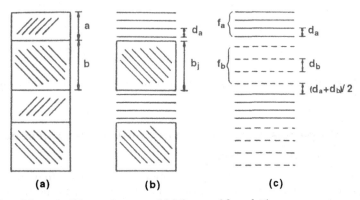

(a)   (b)   (c)

Figure XII.13. — Schematic difference between a Multilayer and Superlattice.

After the chemical order is obtained, the classification is made with respect to the structural order. The case of Multilayers (ML) corresponds to amorphous blocks or alternative regions of amorphous and textured polycrystalline regions, see Fig. XII.13(a) and (b). The case of "perfect" three dimensional order is a Superlattice (SL) (Fig. XII.13(c)).

In the case of superlattices usually there is a small difference between lattice parameters of both materials. Then the layers could be constrained to match at the interface by introducing a coherent strain. There could be a balance between the strain energy and the energy to produce misfit dislocations. Depending on this competition, the latter case could produce a complete incoherent structure. (See the example in Fig. XII.14(a) and (b)). In Figure XII.14(a) one single crystal block is under uniaxial stress and the other under uniaxial strain. In Figure XII.14(b) this energy is released by an incoherent interface, which relieves the strain and produce dislocations.

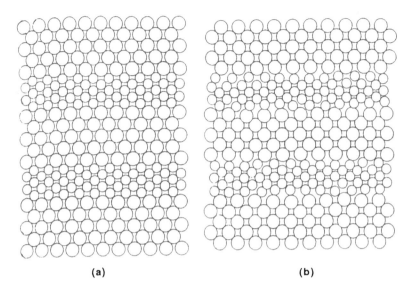

(a)   (b)

Figure XII.14. (a) — A *perfect* superlattice with coherent strains, (b) Superlattice with incoherent interfaces.

XII.5.   Properties of metallic layers, multilayers and superlattices

XII.5.1.   Experimental techniques

There is a great variety of experimental techniques that give valuable information about the magnetic, transport and structural properties of metallic Layers, Multilayers and Superlattices. We may enumerate some of them related to magnetic properties as:

- Surface Science Techniques like:
> Spin Polarized Photoemission
> Polarized Auger Spectroscopy
> Spin Polarized Low Energy Electron Diffraction...

All of them are Ultra-High-Vacuum techniques. Others are more structural in character, for example:

- Electron Microscopy Techniques, including:
> Scanning Electron Microscopy
> Scanning Tunneling Microscope
> Magnetic Force Microscope

These techniques are useful to characterize the growing process and obtain information about the texture of multilayers.

- Magnetometry, including:
> Mössbauer Spectroscopy
> Magnetic Resonance
> Magnetization and Magneto-Optical techniques

The last group of techniques are specific to obtain information about the magnetic properties. The magnetic resonance gives information about the anisotropy (in-plane and out-of-plane) of the magnetic moments. It is very simple and gives valuable information especially from the point of view of possible applications. For instance the change of the magnetic anisotropy with the thermal treatment in Co/Au superlattices (den Broeder *et al.*) is shown in Fig. XII.15.

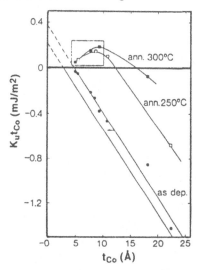

Figure XII.15. — Dependence of the magnetic anisotropy on the Co thickness and the thermal treatments for Co-Au superlattices.

Classical magnetization measurements (SQUID, VSM...) provide important information about the bulk magnetic properties. The high sensitivity of these measurements allows to measure a monolayer of Fe on $1 cm^2$ substrate, although some problems usually arise from the strong signal of the substrate (usually diamagnetic).

The most interesting magneto-optical techniques are SMOKE and Brillouin scattering. SMOKE (Surface Magneto Optic Kerr Effect) is devoted to the study of the polarization plane of the light reflected from a magnetized surface. It is a simple experimental set-up which works in UHV conditions and it is not very expensive, (small laser (mW), polarizer and photodetector). By applying an external magnetic field (± 1000 Oe) the magnetization (proportional to the Kerr signal) can be reversed and a hysteresis loop is obtained, which explores the magnetic character of the top layers.

Brillouin scattering is a more sophisticated technique, which involves the analysis of the energy (wavelength) of the scattered light from the multilayer or superlattice (see Fig. XII.16).

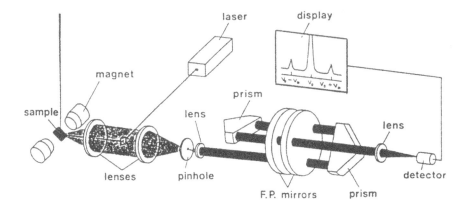

Figure XII.16. — Schematic drawing of the Brillouin scattering set-up for metallic samples.

Brillouin scattering is useful to study in certain cases the magnetic collective excitations in a metallic superlattice.

- Neutron Scattering Techniques

Neutron Scattering is a powerful method to study structural and magnetic properties in Layers, Multilayers and Superlattices. One may obtain information at the microscopic level, like magnetization profile through the layer and interface (magnetic modulation: square or sinusoidal wave), interface roughness, magnetic moments at the interface between magnetic and diamagnetic layers (magnetic dead monolayers).

The case of individual layers can be also studied by neutron reflectometry. This technique allows to probe with polarized neutrons a few Monolayers (ML) of magnetic materials over a substrate and to measure the magnetic moment. The principle of the technique is as follows. The reflection coefficient of a neutron of wavelength (λ) arriving at a glacing angle θ, at the boundary between two media of constant reflective index n_i n_j is:

$$r_{ij} = \frac{(q_i - q_j)}{(q_i + q_j)}$$

(XII.16)

with

$$q_k = \frac{2\pi}{\lambda} \sqrt{n_k^2 - \cos^2\theta}$$

(XII.17)

where q_k is the component of the neutron wavevector within the kth medium. The neutron interacts with the material as a wavefield with a continuous medium of refractive index n_k, which for a ferromagnetic layer of atomic density r_k, Fermi length b_k and a magnetic induction B_k in the layer with an interaction potential V_k is

$$n_k^2 = 1 - 2\frac{M_n \lambda^2}{h^2} V_k$$

(XII.18)

$$V_k = \frac{h^2}{2\pi M_n} \rho_k b_k - \mu_m B_k$$

(XII.19)

The reflectivity depends on the spin orientation of the neutron (+ parallel, - antiparallel) with respect to the magnetic induction in the medium (B_k). For the case of a three medium system (i.e. vacuum-magnetic layer-substrate)

$$r_{jkl} = \frac{r_{jk} + r_{kl}\, e^{2iq_k t_k}}{1 - r_{jk}\, r_{kl}\, e^{2iq_k t_k}}$$

(XII.20)

t_k is the thickness of the kth film (see Fig. XII.17).

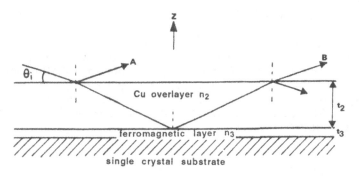

Figure XII.17. — Schematic set-up to study magnetic monolayers by neutron techniques.

By the estimation of B_k (for a saturated ferromagnetic film $B_k \approx \mu_0 M_k$, M_k magnetization of the layer) it is possible to estimate μ_k magnetic moment per atom.

In the case of Multilayers and Superlattices, neutron diffraction is a powerful technique specially with the help of polarization analysis. A schema of the instrument is presented in Fig. XII.18.

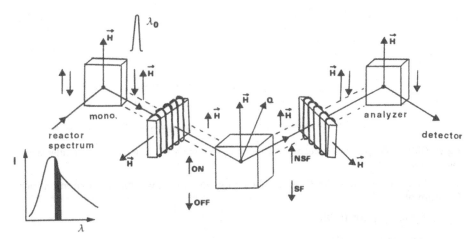

Figure XII.18. — Typical neutron diffraction experiment with polarization analysis to study superlattices.

- X-Ray Diffraction

It is a widely used technique. We will only emphasize the use of anomalous X-ray scattering to study Multilayers (ML) and Superlattices (SL). Because of the dependence of the X-ray scattering factor on the number of electrons, ML and SL with close atomic numbers will give very poor "*contrast*". Anomalous diffraction in ML and SL is a normal diffraction technique, which is performed at an incident energy close to the absorption edge of one of the elements of the ML or SL. In that case the scattering factor has a strong variation, and ML or SL satellites peaks can be visible. This technique needs a continuous variation of the photon energy (i.e. Synchrotron Radiation).

XII.5.2. Experimental results

XII.5.2.1. Layers

The first example we would like to present is the determination of the magnetic moment and transition temperature of one or few Monolayers (ML) of Co and Fe, over a substrate of Cu(100). For instance, from Spin-polarized critical angle neutron scattering measurements on a 18 Å layer of Co on Cu(100) substrate, with a 130 Å coating of Cu. The coating has a strong effect as an enhancement of the signal as it is shown in Fig. XII.19 (Bland *et al.*)

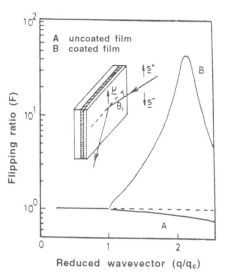

Figure XII.19. — Flipping ratio (proportional to Co magnetic moment) *vs.* wavevector for different Co thicknesses.

The results of the reflectivity can be analyzed and show a magnetic moment per Co atom of

1.8 ± 0.25 μ_B for 10 ML

2.3 ± 0.3 μ_B for 4 ML

Therefore, Co films in the range 2-10 ML thickness exhibit essentially bulk like behaviour, with a magnetic moment close to the value of bulk hcp-Co of 1.7 μ_B. Also, there is not much dependence between 4-400 K. This is an important result because there are some calculations, which predict a strong enhancement of the magnetic moment for reduced dimensionality.

Another interesting example is obtained from SMOKE technique on Co layers and sandwiches on Cu(100). There is the first possibility of studying the dependence of the ordering temperature on the Co coverage. In this case, the strong decrease in T_C from T_{CBulk} depending on coverage (T_{CBulk} = 1400 K) was measured, see Fig. XII.20 (Schneider *et al.*)

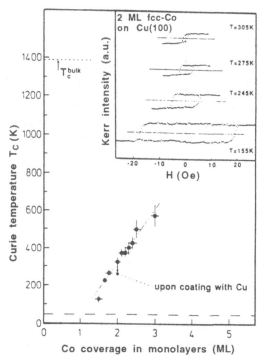

Figure XII.20. — Dependence of the ordering temperature T_C on the coverage of Co monolayers.

Another possibility is to investigate the coupling in Co/Cu/Co sandwiches depending on the Cu thickness. This study has been made also with SMOKE as shown in Fig. XII.21 and XII.22 (Cebollada *et al.*).

The definition of M_r is I_{Kerr} at H = 0, and H_c is H for I_{Kerr} = 0. The first case in Fig. XII.21 is for Cu(001) substrate-3MLCo/4MLCu/nMLCo, that shows a ferromagnetic coupling through the Cu interlayer.

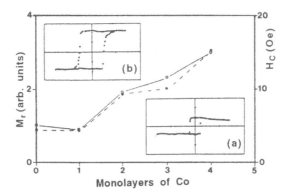

Figure XII.21. — Remanent magnetization (from SMOKE data) *vs.* Co coverage for substrate-3MLCo/4MLCu/nMLCo system.

The second case (Fig. XII.22) is the substrate 3MLCo/7MLCu/nMLCo. Now the coupling is Antiferromagnetic and for 3MLCo, the system is compensated and $M_r = 0$. For $n \neq 3$ ML of Co the difference comes from the non-compensation. This oscillatory behaviour depending on the thickness of the intermediate Cu layer seems to indicate a RKKY coupling mechanism against a dipolar one.

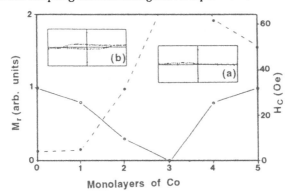

Figure XII.22. — Remanent magnetization (from SMOKE data) *vs.* Co coverage for substrate-3MLCo/7MLCu/nMLCo system.

XII.5.2.2 Superlattices

Rare Earth superlattices are the best quality system for metallic SL. There is a great number of systems already very well characterized like: Gd/Y, Dy/Y, Er/Y, Gd/Dy...

Usually the SL is grown over a sapphire substrate, with a Nb buffer, later an Y seed layer is deposited and the SL on-top, finishing with an Y protection layer to avoid contamination (see Fig. XII.23).

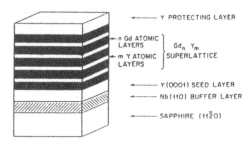

Figure XII.23. — Schematic drawing of a typical rare earth superlattice.

The quality and coherence can be observed in the number of satellites at high q-values (due to the single-crystal character). Around each Bragg peak of the average structure ($d_{ave} = [(N_A d_A + N_B d_B)/(N_A + N_B)]$). There are structural satellites corresponding to the chemical modulation wavelength ($\pm n \, (2\pi/L)$). In addition, around each Bragg peak there could also be magnetic satellites corresponding to the magnetic modulation (Ferro, Antiferro, Helical..). In the simple case of AF they appear at ($\pm m \, (2\pi/L_{Mag})$, where $L_{Mag} = 2L$).

The first results we will discuss on is the magnetization measurements in Gd/Dy SL shown in Fig. XII.24(a) and (b) (Camley *et al.*).

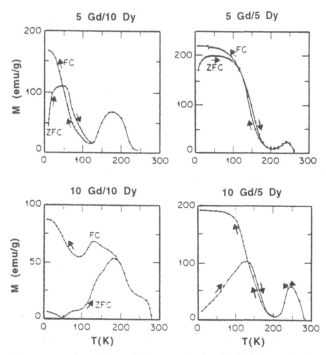

Figure XII.24. — Temperature dependence of the magnetization for Gd-Dy superlattices.

Figure XII.24(a) shows the experimental data for different Gd/Dy SL measured in a SQUID magnetometer. In (b) the theoretical results for the magnetization as a function of temperature are shown .

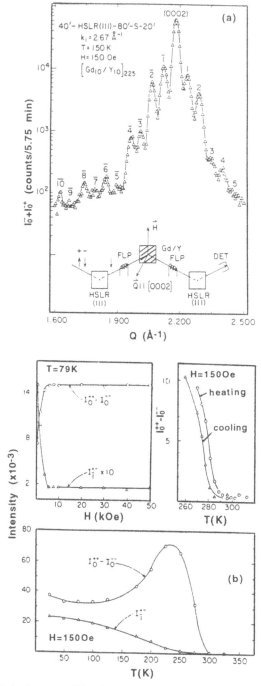

Figure XII.25. — Polarized neutron diffraction data for Gd-Y superlattices.

For neutron scattering Rare Earth SL are ideal systems because of the high values of the magnetic moments per atom. One of the classical results for Gd/Y SL is presented in Fig. XII.25(a) and (b) (Majkrzak *et al.*).

The even numbers m = ± 2, 4... correspond to the chemical modulation L. The odd numbers n = ± 1, 3, 5... correspond to a periodicity 2L. Polarization analysis shows (Fig. XII.24(b)) the temperature dependence of the magnetic component. Different couplings than AF are also observed like helical ordering in Gd/Dy SL.

The magnetic coupling mechanism through the SL is another problem to be studied in detail. In Gd/Y there is an oscillatory behaviour depending on the Y thickness (see Fig. XII.26) (Yafet *et al.*).

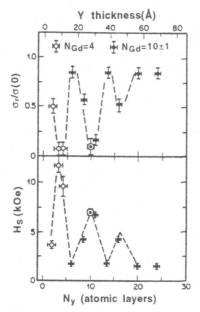

Figure XII.26. — Oscillatory dependence of the magnetic coupling as function of Y thickness in Gd-Y superlattices.

A mechanism of the RKKY-type through the diamagnetic Y layer is likely to occur.

3d metals superlattices are also an interesting and much less studied example. In this case the contrast in X-Ray is usually small. The more interesting cases are: Co/Cu, Fe/Cr, Fe/Ru and Ru/Mn.

In the case of Fe/Cr the magnetization measurements indicate that Fe layers are ferromagnetically ordered and the coupling between subsequent Fe layers, through Cr is antiferromagnetic, for Cr thickness smaller than 30 Å (see Fig. XII.27) (Baibich *et al.*)

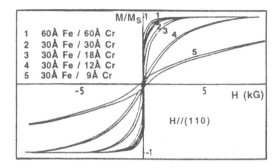

Figure XII.27. — Hysteresis loops on Fe-Cr superlattices for different Cr thicknesses.

The case of Co/Cu on Cu(100) substrate is also interesting. Polarized neutron diffraction indicates the good SL quality and that the Co layers are ferromagnetically ordered (1.7 μ_B per Co atom) and the coupling between subsequent Co layers is

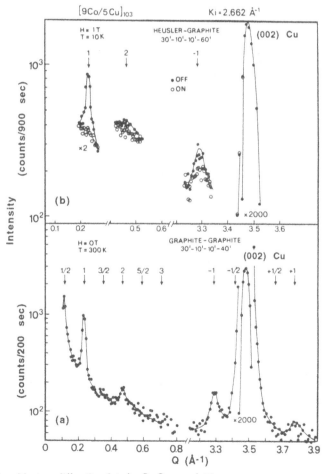

Figure XII.28. — Neutron diffraction data for Co-Cu superlattices.

antiferromagnetic (see Fig. XII.28) (Cebollada *et al.*). The data show the new superlattice statellites of magnetic origin (n/z index). The use of polarized neutrons (upper part) with the analysis of the spin state of the outcoming neutrons (flipper on and off) allows to evaluate the magnetic moment per Co atom.

This result is in agreement with the magnetic coupling observed by SMOKE in Co/Cu/Co sandwiches. As it has been pointed out before, due to the oscillatory behaviour depending on the thickness of the Cu layer, the RKKY coupling mechanism is expected.

The system Co/Cu is not very recommended for X-Ray due to the fact that there is only a difference of two electrons between both atoms. However anomalous diffraction near the absorption edge of Co, increase the X-Ray contrast i.e. the intensity of the SL satellites (see Fig. XII.29) (A. de Andrés).

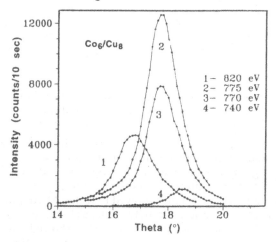

Figure XII.29. — Anomalous X-ray diffraction for Co-Cu superlattices.

The case of Fe/Ru is also very illustrative of the advantages of the use of complementary techniques. Very good quality SL of Fe/Ru have been grown on a sapphire substrate with a Ru buffer. The quality of the SL is evident from the SL satellite at small and high q-values observed by X-Ray diffraction. Especially clear is the anomalous diffraction near the absorption edge of Fe. The intensity of the SL satellites increases very much depending on the photon energy (see Fig. XII.30, A. de Andrés *et al.*) around the energy of this edge.

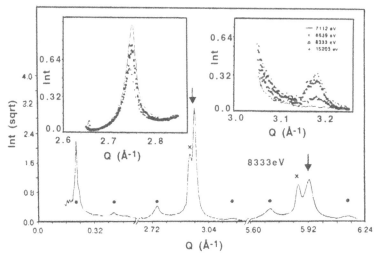

Figure XII.30. — Anomalous X-ray diffraction for Fe-Ru superlattices.

Complete analysis of the data allows to calculate the modulation wavelength and interface roughness. In Figure XII.30 the arrows indicate the SL average Bragg peaks, the dots the structural satellites (crosses x the Ru buffer).

A very interesting and rather new results is the complementary use on polarized and unpolarized neutron diffraction and resonant magnetic X-ray diffraction in Ni/Ag multilayers. The initial work by B. Rodmacq *et al.* showed by neutron diffraction the antiferromagnetic coupling between Ni layers in Ni(7Å)/Ag(11Å) multilayers. Also they showed how this AF coupling is reversed by application of an external magnetic field ($\approx$ 0.25T). More recent experiments (Tonnerre *et al.*) of X-ray diffraction near the L_{III} of Ni showed a resonant magnetic signal with an strong intensity corresponding to the AF satellites. Moreover this intensity vary as a function of the orientation of the external magnetic field. This is shown if Fig. XII.31, where the left indicate the variation of the AF satellite intensity

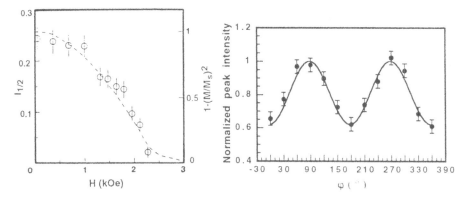

Figure XII.31. — Dependence of the satellite intensity *vs.* (a) external applied magnetic field (by neutron diffraction) and (b) external applied field orientation (by X-ray resonant diffraction) in Ni/Ag multilayers.

as a function of the external magnetic field. The right part shows the variation of the X-ray diffraction intensity of the AF satellite as a function of the relative orientation with an external applied magnetic field.

In contrast, the situation with neutron diffraction is less clear (the samples are the same). See Fig. XII.32 (Piecuch *et al.*). The top part of the figure indicate the low angle region with the corresponding weak satellites. The bottom portion indicates the high-q region showing peaks from the Al_2O_3 substrate and Al plate, as well as the Bragg peaks from the Fe/Ru SL with the corresponding satellites (+1, -1, -2), which are very weak in intensity.

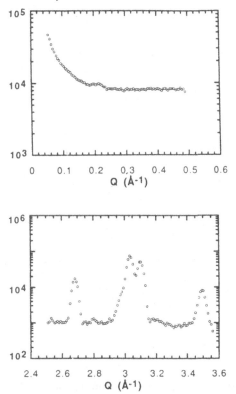

Figure XII.32. — Neutron diffraction for Fe-Ru superlattices.

The explanation resides in the unfortunate small difference in the Fermi lengths for Fe and Ru (b_{Ru} = 7.21 fm, b_{Fe} = 9.54 fm (24%)), in comparison with Co/Cu (70%). This problem is specially important for the small q satellites. The Bragg peaks of the average SL indicate a good quality of the epitaxial growth.

The last case to be presented will be Ru/Mn SL. Preliminary results indicate the existence of SL satellites at small q-values corresponding to an AF coupling between Mn layers (b_{Mn} = -3.73 fm). See Fig. XII.32. These magnetic peaks are suppresed by the application of an external magnetic field of 2T (at 15 K).

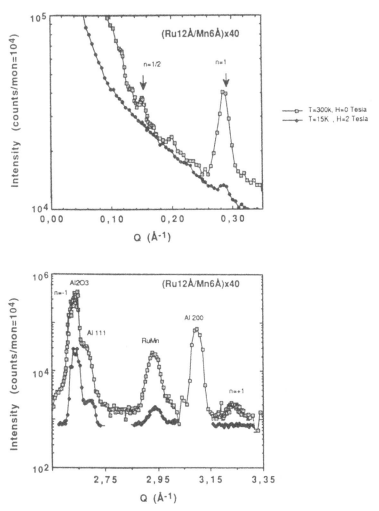

Figure XII.33. — Neutron diffraction data for Ru-Mn superlattices.

Until now most of the work presented in metallic SL have been in the area of characterization (magnetic and structure). When the quality of metallic SL will achieve the case of semiconductor SL, some new aspect in basic research will be explored like phonon folding, magnetic collective excitations, competing interactions FE/AF, AF/Superconductor....

GENERAL READINGS

On thermodynamics of binary systems:

KITTEL C. and KROEMER H. ,1980 - "Thermal Physics", Freeman and Co. San Francisco. or any other textbook on thermodynamics and statistical physics.

On stability, interdiffusion etc.:

DHEZ P. and WEISBUCH C., editors, 1988 - "Physics, Fabrication, and Applications of Multilayers Structures", Nato ASI Series B: Physics Vol. 182. Plenum Press, New York, paper by F. Spaepen.

On diffraction:

CHAMBEROD A. and HILLARIET J., editors ,1990 - "Metallic Multilayers", Materials Science Forum, Volumes 59 and 60, Trans Tech Publications, Switzerland, paper by M. Piecuch and L. Nevot.

OTHER GENERAL REFERENCES

CHANG L.L and GIESSEN B.C. ,1985 - "Synthetic Modulated Structures". Academic Press.

JIN B.Y. and KETTERSON J.B., 1989 - "Artificial Metallic Superlattices". Adv. in Phys, vol. **38**, 189.

FALICOV L.M. *et al.*, 1990 - "Surface, Interface and Thin-Film Magnetism". J. Mat. Res. vol. **5**, 1299.

FALICOV L.M. *et al.*, 1990 - "Magnetic Properties of Low-Dimensional Systems II". Ed. Springer Proc. Physics. Vol. **50**.

MAJKRZAK C.F., KWO J., HONG M., YAFET Y., GIBBS D., CHIEN C.L. and BOHR J., 1991 - "Magnetic Rare Earth Superlattices". Adv. in Phys. vol. **40**, 99.

SPECIFIC REFERENCES

de ANDRES A. - (Private communication).

de ANDRES A. *et al.*, 1990 - Proc. X-ray studies by Synchrotron Radiation II. Editors Balerna et al. SIF ed. Bologna. Rome.

BAIBICH M.N. *et al.*, 1988 - Phys. Rev. Lett. **61**, 2472.

BLAND J.A.C. *et al.*, 1987 - Physica Scripta **T19**, 413.

den BROEDER F.J.A. *et al.*, 1988 - Phys. Rev. Lett. **60**, 2769.

CAMLEY R.E. *et al.*, 1990 - Phys. Rev. Lett. **64**, 2703.

CEBOLLADA A. *et al.*, 1989 - Phys. Rev. **B39**, 9726.

CEBOLLADA A. *et al.*, 1991 - J. Mag. Mag. Mat. **102**, 25.

MAJKRZAK C.F. *et al.*, 1986 - Phys. Rev. Lett. **56**, 2700.

PIECUCH M. *et al.*, (Private communication).

RODMACQ B. *et al.*, 1991 - Europhys. Lett. **5**, 503.

SCHNEIDER C.M. *et al.*, 1990 - Phys. Rev. Lett. **64**, 1059.

TONERRE J.M. *et al.*, 1993 - J. Mag. Mag. Mat. **121**, 230.

YAFET Y. *et al.*, 1987 - Phys. Rev. **B36**, 3948.

shall restrict ourselves only to the description of the spin dynamics; we shall omit the lattice dynamics, while it is also an important topics, because it appears to be not so crucial for revealing the unusual properties of high-T_C superconductors.

XIII.2.   Phase diagram of high-T_C superconductors

High-T_C superconductors are all layered cuprate materials containing CuO_2-sheets and charge reservoir layers. The T_C-value is strongly correlated with the amount of holes, n_h, in oxygen 2p-orbitals, which is transferred into CuO_2-layers. From the large amount of experimental works using various techniques, a schematic phase diagram (Fig. XIII.1) can be drawn and six typical regimes can be defined. Five of them are present in the $YBa_2Cu_3O_{6+x}$ phase diagram (Fig. XIII.2). $YBa_2Cu_3O_{6+x}$ is a quite interesting and flexible system because by changing just the amount of oxygen content in Cu(1) planes (CuO_x-layers), from $x = 0$ to $x = 1$, the different typical regimes can be successively investigated. However the relationship between x and n_h is not trivial (see Uimin et al.).

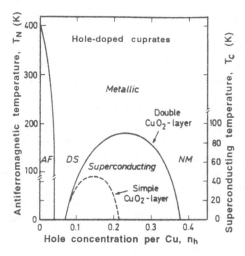

Figure XIII.1. — Schematic phase diagram of high-T_C cuprate materials as a function of hole doping in CuO_2-layers. AF, DS, NM identify the antiferromagnetic state, the disordered spin state and the normal metallic state, respectively.

These typical regimes can be defined for the following ranges of x:

- the AF-state for $0 < x < 0.2$, where no hole are transferred into CuO_2-layers.

- the doped AF-state for $0.2 < x < 0.4$, a few holes are transferred which strongly affect the AF-order. The long range AF-state is destroyed by a critical hole content as small as $n_h^c \approx 0.02$ hole/Cu.

- the weakly-doped metallic state for $0.4 < x < 0.6$ where an insulator-metal transition occurs and a superconducting ground state takes place above a hole

CHAPITER XIII

MAGNETIC EXCITATIONS IN HIGH-Tc SUPERCONDUCTORS

J. ROSSAT-MIGNOD

XIII.1. Introduction

Since the discovery of superconductivity in $La_{2-x}Ba_xCuO_4$ in 1986 by J.G. Bednorz and K.A. Muller a great deal of effort has been carried out to study the fundamental properties of copper oxide materials but there is not yet a well accepted understanding of their unusual properties in the normal state $(T > T_c)$ and the origin of the physical mechanism involved in the Cooper pair formation is still controversial.

More than two or three dozen of new cuprate superconductors have been successfully synthetized. They have in common chemical and crystal properties resulting from a regular stacking of CuO_2-sheets (single, double or triple CuO_2-sheets) and block layers. These block layers ($La_{1-x}Sr_x - O$, $Ba - O$ and $Cu - O$ chains in $YBa_2Cu_3O_{6+x}$, $Bi - O$, $T\ell - O$, etc...) are charge reservoir layers which supply CuO_2-sheets with charge carriers. So the essential physics in all the cuprate superconductors is that of CuO_2-sheets which can be either hole-doped or electron-doped. Actually, all experiments, especially X-rays absorption spectroscopy and inelastic neutron scattering ones, are pointing out that these doped CuO_2-sheets must be considered as strongly correlated electronic layers.

A crucial piece of information is the knowledge of the spin excitation spectrum and its behaviour as a function of temperature and doping of these CuO_2-sheets, in order to reveal the unusual features of the strongly correlated electronic CuO_2-layers and to built an appropriated theory. The inelastic neutron scattering (INS) technique is unique to provide both the wave-vector and the energy dependences of the spin excitation spectrum, i.e. the imaginary part of the dynamical spin susceptibility $\chi(\mathbf{q}, \hbar\omega)$. However, the power of the INS technique is moderated by the weakness of the scattering which requires large single-crystal samples. So far they are available only for the $(La - Sr)_2CuO_4$ and $YBa_2Cu_3O_{6+x}$ systems which have been extensively investigated In this lecture we shall concentrate, mainly, on the $YBa_2Cu_3O_{6+x}$ system because, actually, it is the only one which offers a wide range of hole doping up to $n_h \approx 0.25$ hole/Cu. Any doping can be achieved by changing continuously the oxygen content from x = 0 up to x = 1 and, moreover, using the same single-crystal sample. So, the $YBa_2Cu_3O_{6+x}$ system has provided us a unique possibility of investigating the five characteristic regimes of the phase diagram found in any high-Tc superconductor.

The aim of this lecture is to survey the most typical informations, characteristic of these different regimes, obtained by INS experiments in order to demonstrate the high potentiality and the uniqueness of the INS technique in solving solid state physics problems. Before presenting the results we got on the $YBa_2Cu_3O_{6+x}$ system, a brief description of the phase diagram will be given together with a short summary of informations provided by an INS experiment. Because of space limitations we content $n_h \approx 0.06 - 0.07$ hole/Cu. This regime extends up to a hole content $n_h \approx 0.12$ hole/Cu.

- the heavily-doped metallic regime for $0.6 < x < 0.94$, for which the hole doping ranges from $n_h \approx 0.12$ hole/Cu up to the optimum value $n_h \approx 0.25$ hole/Cu.

- the overdoped metallic regime for $0.94 < x < 1$ where T_C decreases and vanishes. Unfortunately for the $YBa_2Cu_3O_{6+x}$ system, this regime can only be investigated in a narrow concentration range.

- the normal metallic state, for which the doping is too large to accommodate any superconducting ground state cannot be studied up to now.

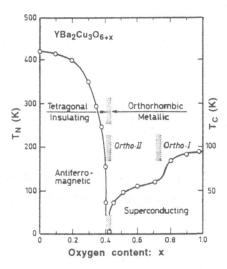

Figure XIII.2. — Phase diagram of the $YBa_2Cu_3O_{6+x}$ system as a function of the oxygen content, x, in Cu(1) planes.

XIII.3. The inelastic neutron scattering (INS) technique

Neutron scattering is an invaluable technique to investigate the structures of atomic and/or magnetic orders and their excitation spectra. The basic reasons are that the wavelength and energy of thermal neutrons are comparable with inter-atomic distances and energies of thermal fluctuations in solids.

The theory of the scattering of thermal neutrons by magnetic moments in crystals is described in many text books, we shall follow the notations used in Neutron Scattering (Sköld and Price). For an unpolarized neutron beam, the INS cross-section for magnetic atoms located in a single site is given by:

$$\frac{d^2\sigma(Q,\hbar\omega)}{d\Omega d\hbar\omega} = N\frac{k_1}{k_0}p^2g^2f^2(Q)e^{-2W(Q)}\mathcal{S}(Q,\hbar\omega) \qquad (XIII.1)$$

- $\mathbf{k}_o$ and $\mathbf{k}_1$ are the incident and scattered neutron wave-vectors, respectively, and $\mathbf{Q}$ is the scattering vector

- $\hbar\omega$ is the energy transfer given by $\hbar\omega = \dfrac{\hbar^2}{2m}(k_1^2 - k_o^2)$

- the magnetic moment μ can be accounted for by a pseudo-spin S such as $\mu = -g\mu_B S$ (for $Cu^{2+}, S = \dfrac{1}{2}$ and in copper oxides $g \approx 2.12$)

- $W(\bar{\mathbf{Q}})$ is the Debye-Waller factor

- $\mathscr{S}(\mathbf{Q},\hbar\omega)$ is the so-called dynamical scattering function which is related to the space and time Fourier transforms of the spin-spin correlation functions $\left\langle S^\alpha(\mathbf{R}_{\ell'}, t') S^\beta(\mathbf{R}_\ell, t) \right\rangle$:

$$\mathscr{S}(\mathbf{Q},\hbar\omega) = \sum_{\alpha,\beta=x,y,z} (\delta_{\alpha\beta} - Q_\alpha Q_\beta) \mathscr{S}^{\alpha\beta}(\mathbf{Q},\hbar\omega) \qquad \text{(XIII.2)}$$

where

$$\mathscr{S}^{\alpha\beta}(\mathbf{Q},\hbar\omega) = \frac{1}{2\pi\hbar} \int_{-\infty}^{+\infty} dt\, e^{-i\omega t} \left\langle S^\alpha_{-\mathbf{Q}}(0) S^\beta_{\mathbf{Q}}(t) \right\rangle \qquad \text{(XIII.3)}$$

and $S^\alpha_{\mathbf{Q}}$ is the spatial Fourier transform of $S^\alpha(\overline{\mathbf{R}}_\ell, t)$

According to the fluctuation-dissipation theorem, $\mathscr{S}^{\alpha\beta}(\mathbf{Q},\hbar\omega)$ is related to the imaginary part of the dynamical susceptibility, $\chi^{\alpha\beta}(\mathbf{Q},\hbar\omega)$, by the relation:

$$\mathscr{S}^{\alpha\beta}(\mathbf{Q},\hbar\omega) = \frac{1}{\pi} \frac{1}{1 - e^{-\hbar\omega/kT}} \operatorname{Im} \chi^{\alpha\beta}(\mathbf{Q},\hbar\omega) \qquad \text{(XIII.4)}$$

For the spin dynamics in CuO_2-planes only the diagonal terms $\chi^{xx}, \chi^{yy}, \chi^{zz}$ give a contribution, the values of which can be determined by varying the direction of the scattering vector $\mathbf{Q}$.

So, at low temperatures ($kT \ll \hbar\omega$), an INS experiment will give, after a proper normalization, a direct determination of $\operatorname{Im}\chi(\mathbf{Q},\hbar\omega)$. Let us consider the following two limiting cases.

i)   *The disordered state:*

In such a case spin excitations are purely of relaxation type and $\chi(\mathbf{Q},\hbar\omega)$ can be written as:

$$\chi(\mathbf{Q},\hbar\omega) = \chi_0(\mathbf{Q}) \frac{\Gamma_Q}{\Gamma_Q - i\hbar\omega} \qquad \text{(XIII.5)}$$

The spectrum (pole of χ) is purely imaginary, $\hbar\omega = -i\Gamma_Q$, and:

$$\mathrm{Im}\,\chi(\mathbf{Q},\hbar\omega) = \chi_0(\mathbf{Q})\hbar\omega\frac{\Gamma(\mathbf{Q})}{(\hbar\omega)^2 + \Gamma_Q^2} \qquad \text{(XIII.6)}$$

We can see in Fig. XIII.3a that $\mathrm{Im}\chi$ is maximum at $\hbar\omega = \Gamma$ and then decreases very smoothly as $\hbar\omega$ increases.

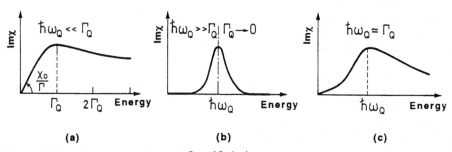

Figure XIII.3. — Energy dependence of $\mathrm{Im}\,\chi(\mathbf{Q},\hbar\omega)$ in three typical cases: a) for a spectrum of relaxation type ($\hbar\omega_Q \ll \Gamma_Q$), b) for propagative excitations ($\hbar\omega_Q \gg \Gamma_Q$), c) for overdamped excitations ($\hbar\omega_Q \approx \Gamma_Q$).

ii)   *The long-range ordered states:*

The spin dynamics is given by the transverse spin susceptibility $\chi^\perp(\mathbf{Q},\hbar\omega)$, the pole of which $\hbar\omega = \hbar\omega_Q - i\Gamma_\varrho$ has a real part, defining the excitation energy, and an imaginary part characterizing the damping of this excitation. Usually $\hbar\omega_Q \gg \Gamma_\varrho$ and well defined collective excitations are observed (see Fig. XIII.3b). We can write $\chi^\perp(\mathbf{Q},\hbar\omega)$ and $\mathrm{Im}\,\chi$ as:

$$\chi^\perp(\mathbf{Q},\hbar\omega) = -\frac{\chi_0^\perp(\mathbf{Q})}{2}\left(\frac{\hbar\omega_Q - i\Gamma_Q}{\hbar\omega - (\hbar\omega_Q - i\Gamma_Q)} - \frac{\hbar\omega_Q + i\Gamma_Q}{\hbar\omega + (\hbar\omega_Q + i\Gamma_Q)} \right) \qquad \text{(XIII.7)}$$

$$\mathrm{Im}\,\chi^\perp(\mathbf{Q},\hbar\omega) = -\frac{\chi_0^\perp(\mathbf{Q})}{2}\hbar\omega\left(\frac{\Gamma_Q}{(\hbar\omega - \hbar\omega_Q)^2 + \Gamma_Q^2} + \frac{\Gamma_Q}{(\hbar\omega + \hbar\omega_Q)^2 + \Gamma_Q^2} \right) \qquad \text{(XIII.8)}$$

When $\Gamma_\varrho \to 0$ $\mathrm{Im}\,\chi^\perp$ can be expressed in terms of δ-functions.

Between these two limiting cases we have the so-called critical regime with overdamped excitations ($\hbar\omega_\varrho \approx \Gamma_\varrho$) and $\mathrm{Im}\,\chi(\mathbf{Q},\hbar\omega)$ exhibits a pseudo-gap type of behaviour (see Fig. XIII.3c).

We shall see that, in the metallic state, the spin dynamics in high-T_c materials is not accounted for by any of the above cases which is a strong indication that it results from the quasi-particle dynamics.

XIII.4. The insulating AF-states

XIII.4.1. The undoped AF-state

The undoped high-T_c compounds are insulating materials with an energy gap of charge transfer type ranging from 1.5 to 2 eV. The Cu^{2+} spins, within CuO_2 –layers, exhibit a long-range AF-order which was first discovered in La_2CuO_4 (Vaknin et al., 1987) and then in $YBa_2Cu_3O_6$ (Rossat-Mignod et al., 1988 ; Tranquada et al., 1988). As an example, $YBa_2Cu_3O_6$ orders, below $T_N = 415 \pm 5K$, with an AF structure described by a wave-vector $\mathbf{k} = (\frac{1}{2}\frac{1}{2}0)$, i.e, corresponding to a magnetic unit cell $(a\sqrt{2}, a\sqrt{2}, c)$ (see Fig. XIII.4). The unit cell contains two Bravais sublattices and their moments are coupled antiferromagnetically. This AF-ordering is not affected by adding oxygen atoms in $Cu(1)$ planes up to $x = 0.20$, the low temperature ordered moment $m_0 = 0.64\mu_B$ and T_N remain unchanged. This result tell us that electronic transfers take place only within $Cu(1)$ planes: oxygen atoms transform neighbouring Cu^+ into Cu^{2+}.

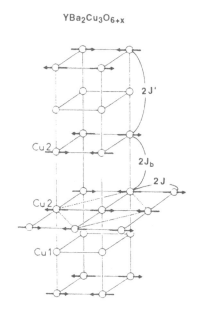

Figure XIII.4. — Antiferromagnetic structure of $YBa_2Cu_3O_{6+x}$. Only Cu ions are represented. The main exchange interactions are also indicated.

Due to the large in-plane superexchange interaction, strong AF-correlations persist above T_N. They are revealed by a strong magnetic scattering centered around the AF-rods like $(\frac{1}{2}\frac{1}{2}\ell)$. From the q-width of this scattering a magnetic correlation length has been deduced and has been found to diverge exponentially as the temperature is decreased. For La_2CuO_4 (Keimer et al.) the best fit is in agreement with quantum Monte-Carlo calculations:

$$\frac{\zeta}{a} = 0.276 \exp\frac{1.25J}{T}$$

In the AF-state the spin dynamics is well explained by the well known spin wave theory, provided a renormalization of the parameters, due to quantum fluctuations $(S = \frac{1}{2})$, is correctly taken into account. The spin wave spectrum can only be determined from INS experiments by performing energy or Q-scans through the Brillouin zone. A typical energy-scan, given in Fig. XIII.5, shows that well-defined peaks, as expected from Fig. 3b, are not observed. This mean that the Q-resolution is not good enough to resolve the spin wave modes of wave-vectors +q and −q (see Fig. XIII.6) because of very steep dispersion curves. However a deconvolution of the data allowed us to get the main features of the spin wave spectrum up to

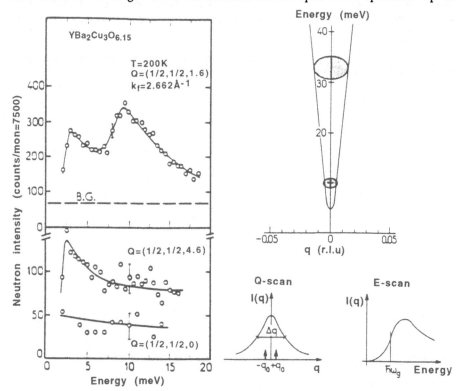

Figure XIII.5. — Energy scans for $YBa_2Cu_3O_{6.15}$ at T = 200 K, establishing the acoustic nature of the excitations of in-plane (low energy) and out-of-plane (high energy) spin components.

Figure XIII.6. — Sketch of the spin wave dispersion relation along $(qq0)$ showing the large spin wave velocity and the finite q-resolution. The deconvolution of q- and energy scans is schematically represented in the lower part.

$\hbar\omega \approx 90 meV$ and to determine the exchange parameters. In $YBa_2Cu_3O_{6+x}$, the main exchange interactions are the strong in-plane superexchange interaction, J, the direct coupling between the two CuO_2–sheets, J_b, which is responsible of the bilayer character, and the coupling between bilayers, J'. Moreover an anisotropic coupling between in-plane ($J^{xx} = J^{yy} = J^{\perp}$) and out-of-plane ($J^{zz} = J''$) spin components has to be taken into account.

The spin wave spectrum, shown in Fig. XIII.7, exhibits, along the $(q,q,0)$ direction, a single excitation energy, $\hbar\omega(q) = 4J \sin 2\pi q$, over almost the whole

Brillouin zone. However, near the zone center and the zone boundary the acoustic mode is well separated from the optical one. At the zone center the energy of the acoustic mode, $\hbar\omega(\mathbf{q}) = c_o|\mathbf{q}|$, yields a spin wave velocity $c_o = 2Ja\sqrt{2} = 1.0 \pm 0.05 eV\text{Å}$, i.e. a large in-plane exchange interaction $J_{Cu-Cu} = 2J / Z_c = 170 \pm 10 meV$ ($Z_c = 1.158$ is the quantum renormalization factor). At the zone boundary, the energy gap yields a weak planar anisotropy $(J^{\perp} - J'') / J = 2.10^{-14}$. The absence of an optical mode up to $\hbar\omega \approx 90 meV$ indicates a large coupling between the two CuO_2–layers, $J_b / J \geq 8.5.10^{-2}$.

So, all exchange parameters have been determined and in the AF-state the spin dynamics is well described by a spin wave theory.

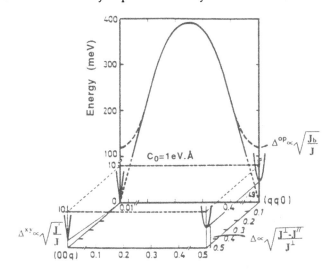

Figure XIII.7. — Schematic diagram of the spin wave excitation spectrum of $YBa_2Cu_3O_{6+x}$.

XIII.4.2. The doped-AF state

For $0.20 < x < 0.40$, T_N and m_o decrease (see Fig. XIII.8) smoothly up to $x = 0.35$ and more abruptly above, resulting in a sharp disappearance of the long-range AF-ordering at the critical oxygen content $x_c = 0.40$. Moreover, in addition to magnetic Bragg peaks, a diffuse scattering co-exists along the $(\frac{1}{2}\frac{1}{2}\ell)$ rod which has a finite q-width (Fig. XIII.9).

So, p-holes transferred in CuO_2–layers induce a disorder in the moment direction, within the basal plane, by creating a kind of magnetic polaron. From the q-width of the AF-rod, an estimation of the hole concentration can be obtained. Assuming that the amount of holes is given by $n_h \approx (\zeta / a)^2$, experiments on a sample with x=0.37 yields a hole concentration $n_h \approx 0.018$ hole/Cu. Therefore in both the $La_{2-x}Sr_xCuO_4$ and $YBa_2Cu_3O_{6+x}$ systems, a long-range AF-ordering cannot develop in CuO_2–layers above a critical hole content $n_h^c \approx 0.02$ hole/Cu.

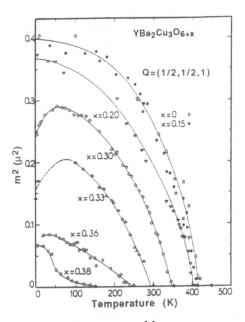

Figure XIII.8. — Intensity of the magnetic Bragg peak $(\frac{1}{2}\frac{1}{2}\ell)$, normalized to represent the square of Cu moment, as a function of temperature for $YBa_2Cu_3O_{6+x}$ at various oxygen contents.

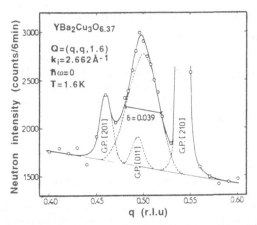

Figure XIII.9. — Elastic q-scan across the AF-rod at $Q = (\frac{1}{2},\frac{1}{2},1.6)$ for $YBa_2Cu_3O_{6.37}$ at $T = 1.6K$.

This small amount of holes strongly affect the spin dynamics at small energies near the zone center and the zone boundary, where a strong q-dependent renormalization of the spin wave energies has been observed (see Fig. XIII.10). In the sample $x = 0.37$ ($n_h \approx 0.18$ hole/Cu) the renormalization reaches about a factor two, the spin-wave velocity is reduced to $c = 450 \pm 50 meV.\mathring{A}$. So, the spin wave velocity softens when the hole content increases and seems to vanish at the critical value n_h^c, explaining the destruction of the 3d-long range AF-order.

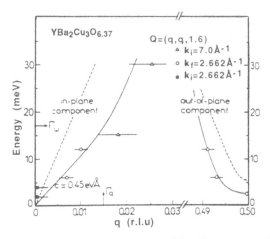

Figure XIII.10. — Dispersion of spin wave energies, around the AF-wave vector (contribution of out-of-plane spin components) and the Brillouin zone centre (contribution of in-plane spin components), along the $(qq0)$ direction for $YBa_2Cu_3O_{6.37}$. The results for the undoped AF-state is given in dotted line.

XIII.5. The metallic superconducting states

XIII.5.1. The weakly-doped metallic state

For an oxygen content larger than $x_c = 0.40$, an insulator-metal (I-M) transition takes place because a large amount of p-holes (10%) is suddenly transfer from Cu(1) planes into CuO_2 –layers. This sudden transfer occurs because of the tetragonal - orthorhombic phase transition which results from an oxygen atom ordering in Cu(1) planes to form a sequence of filled and empty CuO – chains.

We shall present recent measurements carried out on the three-axis spectrometer 2T at the Laboratoire Léon Brillouin with a sample prepared with an oxygen content $x = 0.51$. As can be seen in Fig. XIII.11, the magnetic scattering consists in a broad energy spectrum weakly temperature dependent, there are no more propagative excitations. Q-scans across and along the 2d-AF rods (see Fig. XIII.12 and XIII.13) clearly show that the scattering remains concentrated around these rods. The q-width measured along the $(q,q,0)$ direction is an important parameter ($\Delta q = 0.090 \pm 0.005 r.\ell.u.$) which appears to be directly related to the hole concentration and yields a correlation length $\xi / a \approx 2.5$. A quite important experimental fact is that this q-width is energy and temperature independent.

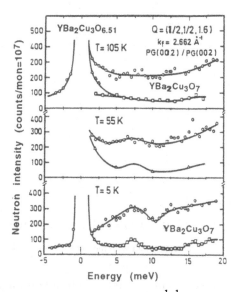

Figure XIII.11. — Low energy-scans performed at $Q = (\frac{1}{2},\frac{1}{2},1.6)$ as a function of temperature for $YBa_2Cu_3O_{6.51}$ (sample n°2). The background (Δ) is well represented by the scattering measured on the fully oxygenated sample $YBa_2Cu_3O_7$.

Q-scans along the AF-rod show a strong intensity modulation from which we can deduce that the AF-coupling between copper spins in the two adjacent CuO_2 –layers is not affected in the metallic state, even up to room temperature.

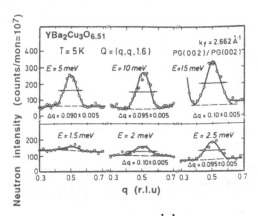

Figure XIII.12. — Q-scans across the AF-rod at $Q = (\frac{1}{2},\frac{1}{2},1.6)$ and increasing energy transfers for $YBa_2Cu_3O_{6.51}$ (sample n°2) in the superconducting state at $T = 5K$.

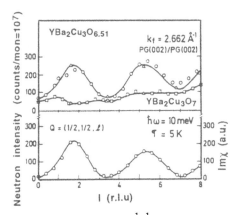

Figure XIII.13. — Q-scans along the AF-rod $\mathbf{Q} = (\frac{1}{2},\frac{1}{2},\ell)$ at $\hbar\omega = 10meV$ for $YBa_2Cu_3O_{6.51}$ (sample n°2). The background (Δ) value, together with measurements on $YBa_2Cu_3O_7$ yield an almost complete AF-coupling between the two CuO_2 −layers.

As was pointed out in section 3, the important physical quantity is the dynamical susceptibility, the imaginary part of which can be deduced from these measurements. The energy dependence of $\text{Im}\chi$ at the AF-wave-vector is given in Fig. XIII.14 for temperatures above and below $T_c \approx 40K$ (sample n°2).

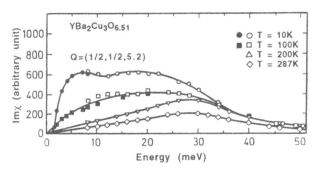

Figure XIII.14. — $\text{Im}\chi(\mathbf{Q},\hbar\omega)$ as a function of the energy at increasing temperature for $YBa_2Cu_3O_{6.51}$ (sample n°2).

At low temperatures, $\text{Im}\chi$ exhibits a broad maximum from 5 to 20 meV and falls off at high energy much faster than a sample Lorentzian function (see Fig. XIII.3a), above 50 meV the scattering is quite small. This is a characteristic feature of the metallic state.

By increasing the temperature, the low energy part of the spectrum is strongly depressed in a non trivial manner. It seems that the staggered susceptibility, $\chi(Q,\omega = 0)$, follows a kind of Curie-Weiss law. At high temperatures $\text{Im}\chi$ is maximum for a characteristic energy $\Gamma_\omega = 28meV$ which is temperature independent.

In the superconducting state, the low energy part of the spectrum (see Fig. XIII.15) clearly shows the opening of an energy gap in the spin fluctuation spectrum. Surprisingly, the gap has a quite small value ($E_G \approx kT_c$) which is the main characteristic of the weakly-doped metallic state.

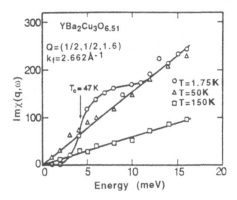

Figure XIII.15. — Low energy part of the spin fluctuation spectrum in terms of $\mathrm{Im}\,\chi(Q,\hbar\omega)$ for $YBa_2Cu_3O_{6.51}$ (sample n°1) below ($T = 1.75K$) and above ($T = 50, 150K$) the superconducting transition ($T_c = 47K$). An energy gap, $E_G = 4meV$, is clearly observed.

So, clearly the spin fluctuation spectrum in the metallic state has nothing to do with that of a disordered AF-system, actually it is typical of the quasi-particles dynamics within the strongly correlated electronic CuO_2 −layers.

XIII.5.2. The heavily-doped metallic state

The heavily-doped metallic state corresponds to hole-dopings which range from 0.12-0.15 hole/Cu up to optimal value of T_c, i.e. about $n_h^{max} = 0.25$ hole/Cu. We shall consider the example of the compound $x = 0.92$ which undergoes a superconducting transition below $T_c \approx 91K$.

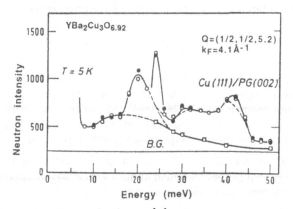

Figure XIII.16. — Energy-scan performed at $Q = (\frac{1}{2}, \frac{1}{2}, 5.2)$ for $YBa_2Cu_3O_{6.92}$ at $T = 5K$. Above the background (BG), the nuclear contribution is indicated (□).

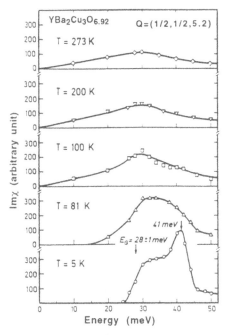

Figure XIII.17. — Spin fluctuation spectrum in terms of $\operatorname{Im}\chi(Q,\hbar\omega)$ at $Q=(\frac{1}{2},\frac{1}{2},5.2)$ for $YBa_2Cu_3O_{6.92}$ as a function of temperature below and above $T_c \approx 91K$. At low temperatures an energy gap, $E_G = 28meV$, is clearly seen together with a strong enhancement at $\hbar\omega = 41meV$.

Energy-scans, performed at increasing temperatures, exhibit rather complex lineshape. However, q-scans across and along the $(\frac{1}{2}\frac{1}{2}\ell)$ AF-rod, clearly, show evidence for a magnetic scattering at high energy transfers and allow us to define the various nuclear scattering contributions on top of a flat background.

So the obtained magnetic contribution, at different temperatures, is reported in Fig. XIII.17 in the form of $\operatorname{Im}\chi(Q,\hbar\omega)$. Several unusual features have to be underlined.

At $T = 5K$, in the superconducting state, there is no measurable magnetic scattering below an energy transfer $\hbar\omega = 25meV$, a sharp energy gap is observed in the spin excitation spectrum at $E_G = 28 \pm 1meV$. More surprising is the sharp drop of the spectrum also at high energies, above $45meV$, and the strong enhancement around $41meV$ which results from a q-narrowing of the spectrum.

At $T = 81K$, i.e. still below T_c, the magnetic scattering remains almost unchanged, except that the enhancement at $\hbar\omega = 41meV$ has disappeared which is a signature of its connection with the superconducting ground state.

Above T_c, the spectrum remains depressed at low energies and $\operatorname{Im}\chi$ recovers a linear $\hbar\omega$-dependence only above $T = 150K$. This shape of the spectrum is typical for the existence of a pseudo-gap (see Fig. XIII.3c). This pseudo-gap remains up to $T^* = 130K$, as shown by the temperature dependence of $\mathrm{Im}\,\chi$ at $\hbar\omega = 10meV$ given in Fig. XIII.18. The observation of a pseudo-gap, well above T_c, is the main characteristic of the heavily-doped state $(dT_c/dx > 0)$. In the overdoped regime $(dT_c/dx < 0)$, i.e. for $n_h > n_h^{max}$ which is realized in $YBa_2Cu_3O_7$, no pseudo-gap effect has been observed above T_c (see Fig. XIII.19).

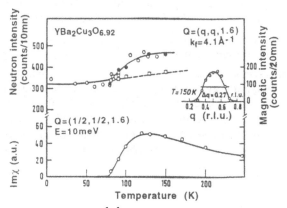

Figure XIII.18. — $\mathrm{Im}\,\chi(\mathbf{Q},\hbar\omega)$ at $\mathbf{Q} = (\frac{1}{2},\frac{1}{2},1.6)$ for $YBa_2Cu_3O_{6.92}$ measured at $\hbar\omega = 10meV$, as a function of temperature. Above $T_c \approx 91K$ a pseudo-gap effect is seen up $T^* = 130K$,

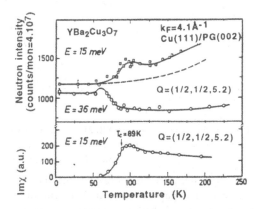

Figure XIII.19. — $\mathrm{Im}\,\chi(\mathbf{Q},\hbar\omega)$ at $\mathbf{Q} = (\frac{1}{2},\frac{1}{2},5.2)$ for $YBa_2Cu_3O_7$ at two typical energy transfers $\hbar\omega = 15meV$ and $36meV$, as a function of temperature.

Finally at high temperatures, $\mathrm{Im}\,\chi$ exhibits a maximum which defined a characteristic energy $\Gamma_\omega \approx 30meV$ which appears to be not dependent upon the hole-doping.

XIII.6. Conclusion

This rapid survey of INS experiments performed for the $YBa_2Cu_3O_{6+x}$ system, I hope, has revealed the powerfulness of the neutron scattering technique. Many informations have been obtained which show that the spin excitation spectrum of high-temperature superconductors exhibits quite unusual features compared with classical superconducting materials. The successful experiments carried out from the AF-state $(x = 0)$ up to the fully oxygenated state $(x = 1)$ have allowed us to define five typical regimes as a function of hole-doping in CuO_2 –layers.

In the insulating state, a long range AF-order always is present but the spin dynamics is strongly affected at small q-values.

In the metallic states only short-range dynamical AF-correlations remain. The q-width is, surprisingly, energy and temperature independent but increases almost proportionally to the doping in CuO_2 –layers (see Fig.XIII. 20) or to T_c. Then the two plateaux observed for T_c at $60K$ and $90K$, as a function of the oxygen content, have a different origin. The $60K$ plateau is due to the existence of the ortho-II phase which stops the charge transfer, whereas the $90K$ plateau is due to an overdoping.

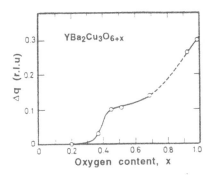

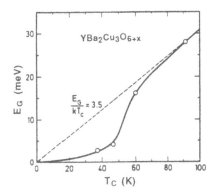

Figure XIII.20. — Q-widths of the spin excitation spectrum as function of the oxygen content in $YBa_2Cu_3O_{6+x}$

Figure XIII.21. — Energy gap found in the spin excitation spectrum of $YBa_2Cu_3O_{6+x}$ as a function of the value T_c of the superconducting transition.

For each superconducting samples, an energy gap has been found in the spin fluctuation spectrum below T_c. The gap value exhibits a quite unusual dependence as a function of T_c, as shown in Fig. XIII.21. Clearly a crossover regime occurs around $T_c \approx 55K$ $(x = 0.6)$ which leads us to define a weakly and a heavily-doped regimes. In the former regime the gap value is strongly reduced $(E_G = kT_c)$ whereas in the latter one a pseudo-gap remains above T_c.

A fundamental question which can be raised is the origin of the spin gap.

An important theoretical activity is developing and different explanations are growing up. However there is not yet a fully convincing model and moreover this theoretical activity is beyond the scope of this lecture.

Acknowledgments

This work has profited of the important collaboration with P. Bourges, P. Burlet, J.Y. Henry, L.P. Regnault and C. Vettier. The skilfulness of Mrs. C. Pomeau in the typing of this article has been appreciated.

REFERENCES

BEDNORZ J.G., MULLER K.A., 1986 - Z. Phys. B64, 189.

UIMIN G., ROSSAT-MIGNOD J., 1992 - Physica C 199, 251.

Neutron Scattering, vol. 23, 1991 - "Method of Experimental Physics, eds. K. Sköld and D.L. Price (Academic Press).

VAKNIN D., SINHA S.K., MONCTON D.E., JOHNSTON, NEWSAM J.M., SAFINYA C.R., KING H.E., 1987 - Phys. Rev. Lett. 58, 2802.

TRANQUADA J.M., COX D.E., KANNMENN W., MOUDDEN A.H., SHIRANE G., SUENAGA M., ZOLLIKER P., VAKNIN D., SINHA S.K., 1988 - Phys. Rev. Lett. 60, 156.

KEIMER B., BELK N., BIRGENEAU R.J., CASSANHO A., YEN C.Y., GREVEN M., KASTNER M.A., AHARONY A., ENDOH Y., ERWIN R.W., SHIRANE G. - preprint Submitted to Phys. Rev.

References for general informations on high-T$_c$ superconductors

See Proceedings of International Conferences on High Temperature Superconductors :Interlaken, Stanford, Kanazawa in Physica C, 153-155 (1988) 156, 162-164 (1989) 1269, 185-189 (1991) 86.

BATLOGG B., Summary of Kanazawa Conference, 1991 - Physica C185-189 and 1991 - Proceedings of Toshiba International School of Superconductivity, Kyoto, Eds. S. Maekawa and M. Sato, Springer.

Main references on our INS investigations

ROSSAT-MIGNOD J., BURLET P., JURGENS M.J., HENRY J.Y., VETTIER C., 1988 - Physica C 152, 19.

ROSSAT-MIGNOD J., BURLET P., JURGENS M.J., VETTIER C., REGNAULT L.P., HENRY J.Y., AYACHE C., FORRO L., NOEL H., POTEL M., GOUGEON P. LEVET J.C., 1988 - J. Physique 49, C8-2119.

ROSSAT-MIGNOD J., REGNAULT L.P., JURGENS M.J., BURLET P., HENRY J.Y., LAPERTOT G., 1991 - "Dynamics of magnetic fluctuations in high-temperature superconductors" G. Reiter, P. Horsch, G.C. Psaltakis, (Editors) NATO-ASI Series, Series B : Physics, vol. 246, 35.

ROSSAT-MIGNOD J., REGNAULT L.P., VETTIER C., BURLET P., HENRY J.Y., LAPERTOT G., 1991 - Physica B 169, 58.

ROSSAT-MIGNOD J., REGNAULT L.P., VETTIER C., BOURGES P., BURLET P., BOSSY J., HENRY J.Y., LAPERTOT G., 1991 - Physica C 185-189, 86.

ROSSAT-MIGNOD J., REGNAULT L.P., VETTIER C., BOURGES P., BURLET P., BOSSY J., HENRY J.Y., LAPERTOT G., 1992 - Physica B 180-181, 383.

ROSSAT-MIGNOD J., REGNAULT L.P., BOURGES P., VETTIER C., BURLET P., HENRY J.Y., 1992 - Proceedings of 12th EPS Conference, Prague, Physica Scripta T45, 74 and 1992 - the Kirchberg Winter School, (to be published by Springer).

A complete review can be found in:

ROSSAT-MIGNOD J., REGNAULT L.P., BOURGES P., BURLET P., VETTIER C., HENRY J.Y., 1993 in "Selected Topics in Superconductivity eds L.C. Gupta and M.S. Multani (World Scientific)", 265.

CHAPTER XIV

MEASUREMENT OF THE ENERGY GAP IN HIGH T_c SUPERCONDUCTORS BY ELECTRON SPECTROSCOPIES

Y. PETROFF

Abstract

Until the recent discovery of the high T_c superconductors, the value of the superconductor energy gap 2Δ was mostly obtained from infra-red or tunneling measurements. The purpose of this paper is to show that high resolution photoemission and energy loss spectrocopies can give very important informations on the value of the gap, its anisotropy, its temperature dependence and the symmetry of the superconducting order parameter.

XIV.1. High Resolution Photoemission

XIV.1.1 Introduction to angle resolved photoemission (ARPES)

If a crystal is illuminated with photons of sufficient energy (larger than the work function W), electrons are emitted into vacuum. The goal of angle resolved photoemission spectroscopy is to measure the energy and the wave vector of the electron emitted into the vacuum and to go back to the energy and wave vector of the same electron inside the crystal. This allows us to determine the wave vector dependence of the electronic energy levels of the valence band E_i ($\bar{K}$). From an experimental point of view, because electrons in vacuum are free electrons, the knowledge of the kinetic energy E_{kin} of these electrons, of the polar angle θ (see Fig. XIV.1)

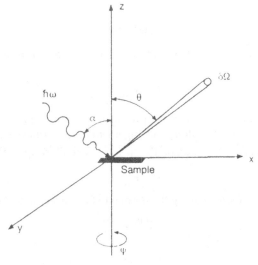

Figure XIV.1. — Principle of Angle resolved Photoemission Electron Spectroscopy (ARPES).

and the azimuthal angle φ allows to determine K_{II}^{out} and $K_{\perp}^{out}$ respectively the parallel and perpendicular components of their wave vector outside the crystal using the relationships:

$$K_{II}^{out} = \left[\frac{2m}{\hbar^2} \cdot E_{kin} \cdot \sin\theta \right]^{1/2} \qquad (XIV.1)$$

$$K_{\perp}^{out} = \left[\frac{2m}{\hbar^2} \cdot E_{kin} \cdot \cos\theta \right]^{1/2} \qquad (XIV.2)$$

In order to connect the previous quantities to the energy and wave vector of electrons inside the crystal we make the following assumptions:

-       about the energy, we suppose that the electrons do not suffer any inelastic scattering inside the crystal as well as when they cross the potential barrier. The energy conservation during the two-fold process-optical excitation-escape in the vacuum involves the relationship:

$$E_f = E_{kin} + W \qquad (XIV.3)$$

$$E_i = \hbar\omega - W - E_{kin} \qquad (XIV.4)$$

where $\hbar\omega$ = photon energy > 0

                W = work function > 0

                E_i = binding energy of the electron in the initial state with respect to the Fermi level (< 0)

                E_f = energy of the electron in the final state

-       about the wave vector; the presence of a surface introduces a distinction between the perpendicular and parallel component of the wave vector. The two-fold periodicity remaining the same in the plane of the surface, the parallel component of the electron wave vector will be conserved during the crossing of the potential barrier.

$$K_{II}^{in} = K_{II}^{out} \qquad (XIV.5)$$

In addition, the two-fold periodicity of the surface introduces the so-called "umklapp" processes

$$\vec{K}_{II}^{in} = \vec{K}_{II}^{out} \pm \vec{G}_s \qquad (XIV.6)$$

where $\vec{G}_s$ is a fundamental surface reciprocal lattice vector. When there is no surface reconstruction (i.e. when the atomic arrangement at the surface is the same as in the bulk), $\vec{G}_s$ is always the projection on the surface of a bulk reciprocal lattice vector.

$$\vec{K}_{II}^{in} = \vec{K}_{II}^{out} \pm \vec{G}_b \qquad (XIV.7)$$

Because the periodicity along the normal to the surface is broken

$$K_{\perp}^{out} \neq K_{\perp}^{in} \qquad (XIV.8)$$

-      For two-dimensional states (surface states, layered crystals) the determination of $E_i(\vec{K})$ is trivial since only E_i and K_{II} have to be determined; working at fixed photon energy $\hbar\omega$ equations (XIV.1) and (XIV.4) allow to map the bands $E_i(\vec{K}_\perp)$.

-      For three-dimensional cases, the interpretation of an angle resolved photoemission experiment necessarily proceeds by an additional hypothesis on the final state inside the crystal: the simplest method is to take a free electron final state as a first approximation. Collecting electrons emitted at the normal ($\theta = 0$) and changing the photon energy $\hbar\omega$ it is possible to obtain. $E_i(\vec{K}_\perp)$. Today the valence band of most metals and semiconductors has been experimentally determined using such a technique[1].

XIV.1.2. Superconductivity

The superconductivity in low T_c materials is generally due to the coupling of the electrons with the phonons. If this coupling is too weak there is no superconductivity (e.g. Cu, Ag, Al). The interaction between the electrons and the lattice is characterised by

$$\lambda = 2\int \alpha^2 F(\omega)\frac{d\omega}{\omega} \qquad (XIV.9)$$

where $F(\omega)$ = phonons density of states

$\alpha^2(\omega)$ describes the interaction between the electrons and the phonons.

XIV.1.2.1.   The gas of repulsive electrons of the normal state transforms itself into a quantum fluid of highly correlated pairs of electrons (Cooper pairs). In the case of $\lambda < 1$ (weak coupling), a full theory exists. This is the Bardeen-Cooper-Schrieffer[2] model (B.C.S.). The transition to the superconductivity state is a second order transition. The superconductivity state is characterised by the opening of an energy gap 2Δ centered about the Fermi energy. Figure XIV.2 shows a constant density of the states $N_0(E)$ above T_c and $N_s(E)$ in the superconductivity state

$$N_s(E) = N_0(E).\frac{E}{\left[E^2 - \Delta^2\right]^{1/2}} \text{ for } |E| > \Delta \qquad (XIV.10)$$

-      The number of states removed (A) from the gap is equivalent to the number of states appearing at the border (B).

-      The energy gap 2Δ increases as the temperature drops, reaching a maximum value of $\Delta(O)$ at T=O

-      There is a fundamental relationship between T_c and $\Delta(O)$

$$\frac{2\Delta}{KT_c} = 3.53 \qquad (XIV.11)$$

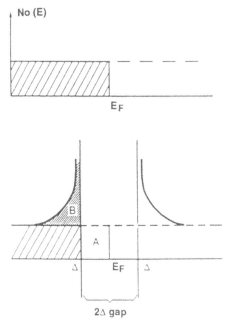

Figure XIV.2. — Density of states of a superconductor, above and below the critical temperature T_c. Notice the opening of an energy gap 2Δ.

XIV.1.2.2. $\lambda > 1$ (strong coupling); deviations from B.C.S. are observed. For example for PbBi alloys $\dfrac{2\Delta}{KT_c} \sim 4.9$. These results can be explained using Nambu-Eliasberg equations[3], which are only solvable in some specific cases.

XIV.1.3. The high T_c superconductors

The most extensively studied high T_c oxide superconductors are $La_{2-x} Sr_x Cu O_4$ ($T_c = 38K$), $Y Ba_2 Cu_3 O_7$ ($T_c = 92K$), $Bi_2 Ca_2 Sr Cu_2 O_8$ ($T_c = 85K$) and $Tl_2 Ca_2 Ba_2 Cu_3 O_{10}$ ($T_c = 125K$). They are cuprate compounds, have a layered structure and are highly anisotropic.

The unit cell is developed from that of a tetragonal perovskite tripled along the c-axis and it consists of a sequence of copper-oxygen layers. $Y Ba_2 Cu_3 O_6$ is an insulator; it becomes a metallic conductor by doping with oxygen. The coherence length (size of the Cooper pair) is very short for the high T_c: 10-15 Å in the CuO_2 layers and few Å along the perpendicular in contrast to 16000 Å in pure Al.

Due to the large number of atoms in the unit cell, these compounds have a complicated band structure, with so many bands that it is impossible to separate them by ARPES and to obtain $E_i(K_\perp)$.

Subsequently, the goals of photoemission will be:
-      to measure the band structure near E_f

-       to check the existence or not of a Fermi surface
-       to measure the gap 2Δ (if existing)
-       to study the anisotropy of 2Δ
-       to study the temperature dependence of 2Δ
-       to determine the symmetry of the superconductivity order parameter (s wave or d wave).

XIV.1.4.   Experimental Results

XIV.1.4.1.   <u>Measurement of the gap 2Δ by angle integrated photoemission</u>

Because the gap is expected to be small, a very high energy resolution (~20meV) is necessary. The first clear evidence of the observation of the gap by photoemission was reported by Imer et al[4]. The measurement was done at $\hbar\omega = 21.2$eV in the angle integrated mode with an energy resolution of 20meV. Data were taken at T=15K and 105K (T$_c$ = 88K). The most striking feature was the appearance of a peak at about 30meV below E$_f$ in the superconducting phase.

Using equation (XIV.10), the B.C.S. density of states of the quasi particle excitation in the superconducting state was calculated from the density of normal state. A gap parameter Δ=30meV was extracted, corresponding to a reduced gap parameter 2Δ/KT$_c$ ~ 8. This value is much larger than the B.C.S. prediction of 3.53 in the weak coupling limit and shows that the use of the B.C.S. model is questionable.

XIV.1.4.2.   <u>Measurement of 2Δ by ARPES</u>

-       First observation of the gap

-       Angle resolved experiments concerning the gap have been performed by various groups [5] [6] [7] [8] [9] [10] [11]. Fig. XIV.3 shows the first data reported by Olson et al.[5]: they have been taken on Bi$_2$ Sr$_2$ Ca Cu$_2$ O$_8$ at $\hbar\omega = 22$eV with an energy resolution of 32meV. The data were obtained at T=20K and above T$_c$ at T=90K. When the polar angle is varied, one observes easily a band crossing the Fermi level (at T=100K). At low temperature, for $\theta = 11°$ one does not observe the opening of the gap due to the large binding energy. Using the B.C.S. model, a value for Δ of 24meV is extracted giving 2Δ/KT$_c$ ~ 7. One should also notice that the superconducting states do not seem to originate from the gap. This could be an artifact due to the normalisation procedure and has been questioned[8] [9].

-       Observation of a pronounced minimum (~-80meV) below the gap peak

This structure was first reported by Dessau et al[7] and later on by Hwu et al[8] and Gunther et al[9]. It confirms that the B.C.S. model in the weak coupling limit is not appropriate to explain the data. Arnold, Mueller and Swihart[12] have analysed these data using Nambu-Eliasberg (strong coupling) theory assuming an unknown boson exchange as the primary mechanism for the superconductivity. The spectral function $\alpha^2F(\omega)$ exhibits a large sharp peak at 10meV, not observed experimentally and the deduced value for λ (8.67) is not credible. The "marginal-Fermi-liquid" model proposed by Varma et al[13] has also been considered.

-       Gap anisotropy: s or d-wave order parameter

In the first study by Olson et al[14] it was shown that the superconductivity gap did not vary in $\bar{K}$ space and therefore provided support for an s-wave order parameter. This has been questioned recently by Wells et al[15] and Zhen et al[10]: they observe a large gap anisotropy and conclude that their data are consistent with the $d_{x^2-y^2}$ order parameter.

Using a different approach (exploiting the polarisation of synchrotron radiation) Kelly et al[11] are reaching the same conclusion. It is obvious that these results are very important for the understanding of the physics of the high T_C superconductors and more data are needed.

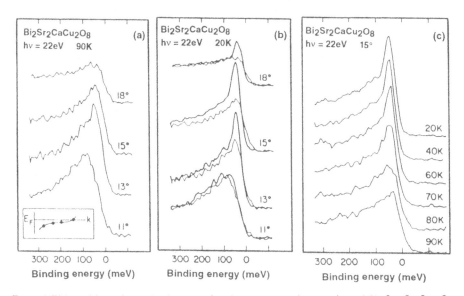

Figure XIV.3. — (a) Angle-resolved energy distribution curves for a surface of $Bi_2 Sr_2 Ca Cu_2 O_8$ cleaved at 20K but measured at 90K. The inset shows the motion of the initial state as the emission angle is changed. (b) Superposition of the 20K (heavier lines) and 90K (lighter lines) spectra. (c) Spectra taken at $\theta=15°$ for various temperatures (from Ref. 5).

XIV.2. High Resolution Energy Loss Spectroscopy

High resolution electron energy loss spectroscopy (HREELS)[16] provides an important complement to photoemission measurements. To study the high T_C superconductors, the incident electrons have energy of few eV, the analyser is operated with a pass energy of 100meV and such that the inelastically scattered electrons within a 1°-1.5° lobe of the elastically scattered beam are analysed.

The energy loss spectra can be analysed by dipole scattering theory[16]; using this formalisation the electron scattering probability is

$$P(K,K') = A(K,K')(n_\omega+1)Img(\omega,q_{//}) \qquad (XIV.12)$$

where A(K,K') is a kinematic term independent of the sample, n_ω is the Bose-Einstein factor and Img is the loss function. The loss function is related to the dielectric properties of the material.

$$\text{Im} g = \text{Im}\left[-\frac{1}{1+\varepsilon(\omega,q_{II})}\right] = \frac{\omega}{4\pi}\text{Re}\rho(\omega) \qquad \text{(XIV.13)}$$

In general, a specific model for $\varepsilon(\omega,q_{II})$ is required to calculate the loss function[17] and it can be written as:

$$\varepsilon(\omega,q_{II}) = 4i\pi\sigma(\omega)/\omega + \delta\varepsilon(\omega,q_{II}) \qquad \text{(XIV.14)}$$

where $\varepsilon(\omega,q_{II})$ and $\sigma(\omega,q_{II})$ are complex numbers and $\delta\varepsilon(\omega,q_{II})$ corresponds to the dielectric contributions from surface phonons and other excitations. However on the limit of high surface conductivity, the dielectric response is reduced to:

$$\varepsilon(\omega,q_{II}) \sim 4i\pi\sigma(\omega)/\omega \qquad \text{(XIV.15)}$$

Thus
$$\text{Im}\left[-\frac{1}{1+\varepsilon(\omega,q_{II})}\right] = \frac{\omega}{4\pi}\text{Re}\rho(\omega) \qquad \text{(16)}$$

and the HREELS directly measure the real part of the complex resistivity.

Temperature dependent electron energy loss spectra for a sample of $Bi_2 Sr_2 Ca Cu_2 O_8$ cleaved at 31K are shown in Fig. XIV.4[18]. The single crystals grown from CuO-rich melts are annealed for several days at 545°C in 12 atm of O_2 (this annealing produces significantly more homogeneous samples). For reference, a spectrum obtained on a TaS_2 crystal at T=31K is also displayed. The resolution, defined as the full width at half maximum of the quasielastic peak is 7.5meV. The same spectrum taken on the high T_c sample shows a broad energy loss peak centered around -60meV. When the temperature is increased the intensity of the loss feature decreases and this feature disappears above T_c. On the basis of the temperature dependence it is assigned to pair-breaking excitations of energy 2Δ.

The bulk excitations (phonons) are not detected because the metallic layers screen the incident electric field to make the measurement sensitive only to the topmost metallic layer.

Fig. XIV.5 shows the difference of the energy loss and energy gain sides of the spectrum to remove background and statistical contribution to the spectra; these curves are proportional to $\omega\rho_s(\omega)$. The magnitude of 2Δ can be assigned to the onset of $\rho_s(\omega)$. At 31K the onset is 46meV, yielding $2\Delta/KT_c \sim 6$, similar to the value obtained by angle resolved photoemission.

The temperature dependence of 2Δ is shown in Fig. XIV.6. One notices then the gap is only weakly dependent on temperature below T_c and that the gap closes rapidly at T_c. It is also apparent that the weak coupling BCS theory (solid curve) is inconsistent with the magnitude of Δ and the sharp opening of Δ (T) for $T \leq T_c$.

314   Y. PETROFF

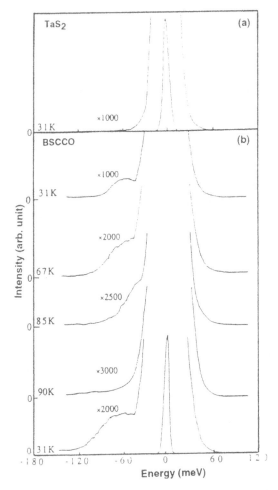

Figure XIV.4. — Temperature dependent energy loss spectra of Bi$_2$ Sr$_2$ Ca Cu$_2$ O$_8$ recorded on a
sample cleaved at 31K (from Ref. 18). The spectra recorded below T$_C$ exhibit an energy loss feature at
-60meV. At the top, the energy loss spectrum obtained at 31K on a non-superconducting material
(TaS$_2$).

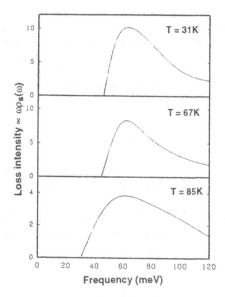

Figure XIV.5. — Plots of the frequency dependent resistivity obtained from the analysis of the data of Fig. XIV.5 (from Ref. 18). The solid lines correspond to the best fit of the energy loss energy gain intensity. The dotted line is an extrapolation to the onset of $\rho_s(\omega)$.

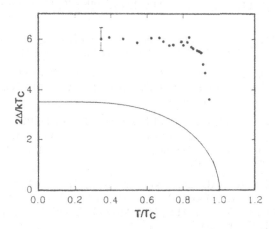

Figure XIV.6. — Temperature dependence of the reduced gap. The experimental data correspond to the solid circles, the solid line is the B.C.S. prediction (from Ref. 18).

XIV.3. Conclusion

The opening of the gap is clearly observed by angle resolved photoemission and energy loss spectroscopy. The value obtained for $2\Delta/KT_c$ (~7) is based on the B.C.S. model which is probably not appropriated. In addition, angle resolved photoemission can give important informations on the symmetry of the superconductivity order parameter.

REFERENCES

1. THIRY P., 1981 - Angle resolved photoemission in "Electronic Structure of Crystal Defects and of disordered Systems". Les Editions de Physique. Les Ulis 435.
    PLUMMER E.W. and EBERHARDT W., 1982 - Adv. Chem. Phys. 49, 533 .

2. BARDEEN J., COOPER L.N., SCHRIEFFER J.R., 1975 - Phys. Rev. 106, 162.

3. ELLIASBERG G.M., 1961 - J.E.T.P. 11, 696.

4. IMER J.M., PATTHEY F., DARDEL B., SCHNEIDER W.D., BAER Y., PETROFF Y., ZETTL A., 1989 - Phys. Rev. Lett. 62, 336.

5. OLSON C.G., LIU R., YANG A.B., LYNCH D.W., ARKO A.J., LIST R.S., VEAL Y.C., CHANG Y.C., JIANG P.Z., PAULIKAS A.P., 1989 - Science 245, 731.

6. MANZKE R., BUSLAPS T., CLAESSEN R., FINK J., 1989 - Europhysics Lett. 9, 447.

7. DESSAU D.S., WELLS B.O., SHEN Z.X., SPICER W.E., ARKO A.J., LIST R.S., MITZI D.B., KAPITULNIK A., 1991 - Phys. Rev. Lett. 66, 2160.

8. HWU Y., LOZZI L., MARSI M., LA ROSA S., VINOKUR M., DAVIS P., ONELLION M., BERGER H., GOZZO F., LEVY F., MARGARITONDO G., 1991 - Phys. Rev. Lett. 67, 2573.

9. GÜNTHER R., TALEB A., INDLEKOFER G., THIRY P., PETROFF Y., - IWEPS Kirchherg, March 1992 (unpublished) and Optical Properties of Semiconductors Ed. G. Martinez, Kluwer Academic Publishers, NATO ASI Series Vol. 228, p 95 (1992).

10. SHEN Z.X., DESSAU D.S., WELLS B.O., KING D.M., SPICER W.E., ARKO A.J., MARSHALL D., LAMBARDO L.W., KAPITULNIK A., DICKINSON P., DONIACH S., DI CARLO J., LOESER T., PARK C.H., 1993 - Phys. Rev. Lett. 70, 1553.

11. KELLY R.J., MA J., MARGARITONDO G., ONELLION M. (unpublished).

12. ARNOLD G.B., MULLER F.M., SWIHART J.C., 1991 - Phys. Rev. Lett. 67, 2569.

13. VARMA J.C., LITTLEWOOD P.B., SCHMITT-RINK S., ABRAHAMS E., RUCKENSTEIN A.E., 1989 - Phys. Rev. lett. 63, 1993.

14. OLSON C.G. et al., 1990 - Solid. State. Commun. 76, 411.

15. WELLS B.O. et al. - Phys. Rev. B (unpublished).

16.   IBACH H. and MILLS D.L. - Electron Energy Loss Spectroscopy and Surface Vibrations (Academic, New York, 1982).

17.   DEMUTH J.E., PERSSON B.N.J., HOLTZBERG F. and

      CHANDRASEKHAR C.V., 1990 - Phys. Rev. Lett. 64, 603.

      PERSSON B.N.J. and DEMUTH J.E., 1990 - Phys. Rev. B 42, 8057.

18.   LI Y., HUANG J.L. and LIEBER C.M., 1992 - Phys. Rev. Lett. 68, 3240.

Index

Dynamical susceptibility		289, 292, 300	
Dynamical theory of diffraction	86, 158, 400		
Dynamics of glassy or amorphous materials	307		
Dynamics of liquids	307		
E.coli ribosomes			131
Effective diffusion			95, 96
Eight-stranded β-barrel			251
Elastic Incoherent Structure Factor			81, 100
Elastic scattering	10, 147		
Elastic constants	199		18
Electrochemical reactions	432		
Electron bunch	41, 43		
Electron deformation density		31, 40, 42	
Electron density		24, 25, 37, 38, 39, 247	49, 50, 51, 52, 53, 211
Electron diffraction			167
Electron microscope	411		
Electron yield detection	355		
Electron-density determination	202		
Electron-electron interactions		259	
Electron-phonon coupling		245, 249, 309	
Electronic structure		**213**, 214, **235**, 242	
Electronic detectors	181		
Embedding method		202	
Emittance	80		
Empirical potential energy functions			285, 286
Energy dispersion	76		
Energy dispersive mode	208		
Energy gap		294, 296, 301, 302, 304, **307**, 309, 310	
Energy loss spectroscopy		307, **312**, 313, 314, 316	
Energy of mixing		262	
Energy resolution	**101, 319**		
Entropic springs			102
Entropy		**123, 124**, 126, 127	
Entropy of mixing		262	23
Enzymes			299
Epithermal neutrons	117		
Equatorial-plane geometry	184		
E.S.C.A. (Electron Spectroscopy for Chemical Analysis)	381		
Ethanols			233
Euler equation			24
Eulerian cradle	181, 182, 186		
Evanescent wave			49, 52
Ewald construction	150, 159, 179		167, 202,